Cancer Sensitizing Agents for Chemotherapy

NOVEL THERAPIES IN HEAD AND NECK CANCER: BEYOND THE HORIZON

VOLUME 9

Cancer Sensitizing Agents for Chemotherapy Series

Series Editor: Benjamin Bonavida, PhD

Volume 1: Nitric Oxide (Donor/Induced) in Chemosensitization
Edited by Benjamin Bonavida, PhD

Volume 2: Role of Nutraceuticals in Cancer Chemosensitization
Edited by Alok Chandra Bharti and Bharat Bhushan Aggarwal

Volume 3: Targeting Cell Survival Pathways to Enhance Response to Chemotherapy
Edited by Daniel M. Johnson, PhD

Volume 4: Protein Kinase Inhibitors as Sensitizing Agents for Chemotherapy
Edited by Zhe-Sheng Chen and Dong-Hua Yang

Volume 5: Breaking Tolerance to Pancreatic Cancer Unresponsiveness to Chemotherapy
Edited by Ganji Purnachandra Nagaraju, PhD, DSc

Volume 6: Improving the Therapeutic Ratio in Head and Neck Cancer
Edited by Randall J. Kimple, MD, PhD

Volume 7: Drug Efflux Pumps in Cancer Resistance Pathways: From Molecular Recognition and Characterization to Possible Inhibition Strategies in Chemotherapy
Edited by Alejandro Sosnik, PhD and Reina Bendayan, PharmD

Volume 8: Drug Resistance in Colorectal Cancer: Molecular Mechanisms and Therapeutic Strategies
Edited by Chi Hin Cho and Tao Hu

Volume 9: Novel Therapies in Head and Neck Cancer: Beyond the Horizon
Edited by Maie A. St. John, MD, PhD, FACS and No-Hee Park, DMD, PhD

Volume 10: pH-Interfering Agents as Chemosensitizers in Cancer Therapy
Edited by Claudiu T. Supuran and Simone Carradori

Upcoming Volumes:

New Targeting in the Reversal of Resistant Glioblastomas
Edited by Ali S. Arbab

Terpenoids As Chemosensitizers Against Drug-resistant Tumors
Edited by Thomas Efferth

Advancing Approaches to Overcoming Cancer Drug Resistance
Edited by Andrew Freywald and Franco Vizeacoumar

Therapeutic Strategies to Overcome ALK Resistance in Cancer
Edited by Luc Friboulet

Reversal of Glioblastoma Resistance to Chemotherapy
Edited by Ramasamy Paulmurugan and Tarik F. Massoud

Overcoming Ovarian Cancer Chemoresistance
Edited by Goli Samimi and Christina Annunziata

Epigenetic Regulation in Cancer Chemotherapy
Edited by Chunfu Wu and Lihui Wang

Cancer Sensitizing Agents for Chemotherapy

NOVEL THERAPIES IN HEAD AND NECK CANCER: BEYOND THE HORIZON

VOLUME 9

Edited by

MAIE A. ST. JOHN, M.D., PH.D., FACS

Thomas C. Calcaterra, M.D., Chair in Head and Neck Surgery; Chair and Executive Medical Director, Department of Head & Neck Surgery; Professor, Head & Neck Surgery and Bioengineering; Co-Director, UCLA Head and Neck Cancer Program; Associate Director Cancer Research and Education, Jonsson Comprehensive Cancer Center; David Geffen School of Medicine at University of California Los Angeles, Los Angeles, CA, United States

NO-HEE PARK, D.M.D., PH.D.

Distinguished Professor of Medicine, David Geffen School of Medicine at University of California Los Angeles; Dean Emeritus and Distinguished Professor of Dentistry, UCLA School of Dentistry, Los Angeles, CA, United States

ELSEVIER

ACADEMIC PRESS

An imprint of Elsevier

Academic Press is an imprint of Elsevier
125 London Wall, London EC2Y 5AS, United Kingdom
525 B Street, Suite 1650, San Diego, CA 92101, United States
50 Hampshire Street, 5th Floor, Cambridge, MA 02139, United States
The Boulevard, Langford Lane, Kidlington, Oxford OX5 1GB, United Kingdom

Library of Congress Cataloging-in-Publication Data
A catalog record for this book is available from the Library of Congress

British Library Cataloguing-in-Publication Data
A catalogue record for this book is available from the British Library

ISSN: 2468-3183

ISBN: 978-0-12-820679-9

For information on all Academic Press publications
visit our website at https://www.elsevier.com/books-and-journals

Publisher: Stacy Masucci
Acquisitions Editor: Rafael Teixeira
Editorial Project Manager: Samantha Allard
Production Project Manager: Punithavathy Govindaradjane
Cover Designer: Greg Harris

Typeset by SPi Global, India

Cover Image Legend

Cancers have been known to mankind since ancient times. The disease was first called cancer by the Greek physician Hippocrates (460–370 BC), who is considered the "Father of Medicine." Hippocrates used the terms carcinos and carcinoma to describe nonulcer forming and ulcer-forming tumors, respectively. In Greek, this means a "crab." The description was named after a crab because the fingerlike spreading projections from a cancer called to mind the shape of a crab. Later the Roman physician Celsus (28–50 BC) translated the Greek term into cancer, the Latin word for crab.

On the cover is the "Crab Nebula," captured in Gilbert, Arizona, by the UCLA engineering student Eric Lazbin. In the year 1054, Chinese astronomers noticed a new "guest star" to appear in the constellation Taurus, which was so bright that it was clearly visible during the day for multiple weeks. It would be 677 years from that point on that the British astronomer John Bevis rediscovered the "guest star" as a now gaseous nebula and decided that it resembled a crab, naming it the "Crab Nebula." The fantastic dust pictured here is visible in the visible-light spectrum, while the pulsar at the core is only visible in the x-ray spectrum. The Crab Nebula is one of the most exciting deep-sky objects and will remain as such for many centuries to come.

Aims and scope for series "Cancer sensitizing agents for chemotherapy"

Current cancer management strategies fail to adequately treat malignancies with chemotherapy, with multivariable dose-restrictive factors such as systemic toxicity and multidrug resistance, hence limiting therapeutic benefits, quality of life, and complete long-term remission rates. The resistance of cancer cells to anticancer drugs is one of the major reasons for the failure of traditional cancer treatments. Cellular components and dysregulation of signaling pathways contribute to drug resistance. If modulated, such perturbations may restore the drug response and its efficacy. The recent understanding of the molecular mechanisms and targets that are implicated for cancer chemoresistance has paved the way to develop a large battery of small molecules (sensitizing agents) that can target resistant factors and reduce the threshold of resistance, thus allowing their combination with chemotherapeutic drugs to be effective and to reverse the chemoresistance. A large variety of chemotherapy-sensitizing agents have been developed, and several have been shown to be effective in experimental models and in cancer patients.

The main objective of the proposed series "Cancer Sensitizing Agents for Chemotherapy" is to publish individualized and focused volumes whereby each volume is edited by an invited expert editor(s). Each volume will dwell on specific chemosensitizing agents with similar targeting activities. The combination treatment of the sensitizing agent with chemotherapy may result in a synergistic/additive activity and the reversal of tumor cells resistance to drugs.

The editor(s) will compile nonoverlapping review chapters on reported findings in various cancers, both experimentally and clinically, with particular emphases on underlying biochemical, genetic, and molecular mechanisms of the sensitizing agent and the combination treatment.

The scope of the series is to provide scientists and clinicians with updated and clinical information that will be valuable in their quest to investigate, develop, and apply novel combination therapies to reverse drug resistance and, thereby, prolong survival and even cure in cancer patients.

Dr. Benjamin Bonavida, PhD (Series Editor)

About the Series Editor

Dr. Benjamin Bonavida, PhD (series editor), is currently a distinguished research professor at the University of California, Los Angeles (UCLA). His research career, thus far, has focused on basic immunochemistry and cancer immunobiology. His research investigations have ranged from the mechanisms of cell-mediated killing and sensitization of resistant tumor cells to chemo-/immunotherapies, characterization of resistant factors in cancer cells, cell-signaling pathways mediated by therapeutic anticancer antibodies, and characterization of a dysregulated NF-κB/Snail/YY1/RKIP/PTEN loop in many cancers that regulate cell survival, proliferation, invasion, metastasis, and resistance. He has also investigated the role of nitric oxide in cancer and its potential antitumor activity. Many of the aforementioned studies are centered on the clinical challenging features of cancer patients' failure to respond to both conventional and targeted therapies. The development and activity of various chemosensitizing agents and their modes of action to reverse chemo- and immunoresistance are highlighted in many refereed publications.

Acknowledgments: The editors wish to acknowledge the excellent editorial assistance of Ms. Inesa Navasardyan and Ms. Brandi Tran who have worked diligently in the completion of this volume, namely, in both the editing and formatting of the various contributions of this volume. The editors also acknowledge the assistance of Rafael Teixeira, acquisitions editor for Elsevier, and Samantha Allard, editorial project manager for Elsevier, for their continuous cooperation during the development of this book.

Aims and Scope of the Volume

The complexity of cancer care in the 21st twenty-first century mandates the input of a spectrum of health care providers to achieve the best possible outcome. Head and neck cancer in particular occurs in an anatomical region unmatched in the number of physiological functions potentially affected with disease and treatment. Further complicating treatment is the wide range of tumor types and tumor locations in head and neck cancer. Each patient also has their unique opinion as to what they would like to prioritize in their treatment plan. In this book, leaders in the field review the multidisciplinary state of the art approaches to head and neck cancer care, emphasizing approaches that can be widely accessible, new in the field and of significant interest and importance. A synthesis of the latest treatments and outcomes is presented here so that all practitioners can be facile with the latest advances in head and neck therapeutics, can build on what is known, and can continue to offer cutting edge validated therapies to all patients.

About the Volume Editors

Maie A. St. John, M.D., Ph.D., FACS

Dr. St. John, a highly renowned surgeon, scientist, and educator, is professor and chair of Department of Head and Neck Surgery

and professor of Bioengineering at UCLA. She also holds the Thomas C. Calcaterra, M.D., chair in Head and Neck Surgery and is codirector of the UCLA Head and Neck Cancer Program. Dr. St. John's laboratory studies the mechanisms of tumor progression and metastasis in head and neck cancer and seeks to identify and study novel genes and pathways for future targeted therapies, while actively incorporating the results of their laboratory-based research into the development of novel therapeutics. Her extramurally funded research portfolio includes multiple NIH grants bridging basic science with clinical research and service. Dr. St. John is an outstanding educator who has provided mentorship to numerous trainees, and serves as a member of several prestigious organizations including the Executive Council of the American Head and Neck Society, Nominating Committee of the Triological Society, American Academy of Otolaryngology-Head & Neck Surgery, American College of Surgeons Board of Governors Advisory Council, and American Laryngological Association. She also is a senior examiner for the American Board of Otolaryngology and a reviewer for the American Academy of Otolaryngology/Head and Neck Surgery CORE grants, and NIH/NIDCR, NCI and NIBIB study sections. She has been the recipient of multiple awards for teaching and research.

Dr. St. John received her undergraduate degree from Stanford University and both M.D. and Ph.D. degrees from Yale University. She completed residency and postdoctoral training at UCLA.

In addition to maintaining a busy surgical practice, Dr. St. John is active in clinical research and is closely involved in training young physician-scientists. Dr. St. John's clinical expertise is in intraoperative tumor margin delineation in the resection and treatment of head and neck cancers to allow for precision surgery. Her laboratory research is translational with a focus on improving therapies and outcomes for patients with head and neck cancer. Her efforts have helped in the development of strong interdepartmental research program in Head & Neck Surgery and Bioengineering. This work has led to clinical trials and patent applications, and has been consistently funded by the NIH, charitable foundations, and industry.

No-Hee Park, D.M.D., Ph.D.

Dr. No-Hee Park is presently the dean emeritus and distinguished professor of the UCLA School of Dentistry. He is also the distinguished professor of Medicine at the UCLA David Geffen School of Medicine. Dr. No-Hee Park is a major scientist in oral and craniofacial research community in the world with over 200 scientific publications in most reputable scientific journals. Dr. Park's contribution to the understanding of the mechanisms of oral cancer development is enormous, which resulted in a new mode of therapy. Dr. Park has received numerous recognitions by his peers and received the Distinguished Scientist Award from the International Association for Dental Research (IADR), AAAS fellowship, the Gies Award as the most achieved dental educator, elected member of the Hall of Fame in Science of Technology from the Republic of Korea, Honorary doctorate from Meikai University, Asahi University and Sejong University, etc. He has served on the editorial board of many research journals, including European J of Cancer, Journal of Dental Research, International Journal of Oncology, Journal of Biotechnology, Odontology, Critical Review of Oncogenesis, Chinese J of Dental Research, and International Journal of Oral Sciences. Dr. Park has also served as a member of a number of NIH grants review committees for both the National Cancer Institute and the National Institute of Dental Research. Dr. Park received his DDS degree from the Seoul National University College of Dentistry, Korea, in 1968 and the clinical certificate in Periodontics in 1971. After serving the Korean military as a captain for over 3 years (1971–74), he migrated to the United States and received a PhD degree in Pharmacology in 1978 from the Georgia Regent University Medical College of Georgia (presently Augusta University in the State of Georgia). Dr. Park completed a postdoctoral fellowship at the Harvard Medical School/Eye Research Institute and also earned a DMD (Doctor of Medicine in Dentistry) degree from Harvard University in 1982.

Preface

It takes a village

The complexity of cancer care in the 21st century mandates the input of a spectrum of health-care providers to achieve the best possible outcome. Head and neck cancer in particular occurs in an anatomical region unmatched in the number of physiological functions potentially affected with disease and treatment including respiration, physical appearance, vocalization, gustation, olfaction, and alimentation. Further complicating treatment is the wide range of tumor types and tumor locations in head and neck cancer, all of which taken separately are relatively rare cancers. Each patient also has their unique opinion as to what they would like to prioritize in their treatment plan (i.e., survival, quality of life, and length of intervention). Patients come to us seeking knowledge, direction, and refuge. They come from every social class, every age group, and every corner of the planet. Thus they come to us "speaking innumerable languages," and we must be able to communicate effectively on every front. The care of the head and neck cancer patient is no longer solely resecting a tumor or offering the latest in immunotherapy. *The care of the head and neck cancer patient is the art of caring for the whole patient*: multimodal surgical and medical management, esthetic considerations, nutrition, pain management, and psychosocial support.

As referenced in the title proverb, a medical "village" functions best when it unites for a common goal of optimal oncologic results while reducing functional deficits and minimizing suffering. This book serves to highlight the multidisciplinary state of the art approaches to head and neck cancer care, emphasizing approaches that can be widely accessible, new in the field, and of significant interest and importance. At its foundation a multidisciplinary approach functions to present a unified voice in the plan of therapy in the heterogeneous and multifaceted disease process that is head and neck cancer. As anyone who cares for cancer patients can attest: "If you take care of cancer, sometimes you win and sometimes you lose; if you take care of the patient, you always win."

Contents

Contributors

Elliot Abemayor Department of Head and Neck Surgery, David Geffen School of Medicine at University of California Los Angeles; UCLA Head and Neck Cancer Program, UCLA Medical Center, Los Angeles, CA, United States

Yazeed M. Alhiyari Department of Head and Neck Surgery, David Geffen School of Medicine at University of California Los Angeles, Los Angeles, CA, United States

Ameya A. Asarkar Department of Otolaryngology/Head Neck Surgery, Ochsner-LSU Health; Surgical Service, Otolaryngology Section, Overton Brooks Veterans Affairs Medical Center, Shreveport, LA, United States

Barbara Burtness Department of Internal Medicine, Yale Cancer Center and Yale School of Medicine, New Haven, CT, United States

Alison G. Chang Division of Hematology and Oncology, David Geffen School of Medicine at University of California Los Angeles, Los Angeles, CA, United States

Dinesh K. Chhetri Department of Head and Neck Surgery, David Geffen School of Medicine at University of California Los Angeles, Los Angeles, CA, United States

Robert K. Chin Department of Radiation Oncology, University of California, Los Angeles, Los Angeles, CA, United States

Ritchell van Dams Department of Radiation Oncology, University of California, Los Angeles, Los Angeles, CA, United States

Andrew Erman Department of Audiology & Speech, University of California, Los Angeles, CA, United States

Donovan Eu Department of Otolaryngology-Head and Neck Surgery/Surgical Oncology, Princess Margaret Cancer Centre/University Health Network, University of Toronto, Toronto, ON, Canada

Robert L. Ferris University of Pittsburgh School of Medicine, UPMC Hillman Cancer Center, Pittsburgh, PA, United States

Jean-Nicolas Gallant Department of Otolaryngology—Head and Neck Surgery, Vanderbilt University Medical Center, Nashville, TN, United States

Shabnam Ghazizadeh Department of Head and Neck Surgery, David Geffen School of Medicine at University of California Los Angeles; UCLA Head and Neck Cancer Program, UCLA Medical Center, Los Angeles, CA, United States

Robert Haddad Dana-Farber Cancer Institute, Boston, MA, United States

Albert Y. Han Department of Head and Neck Surgery, David Geffen School of Medicine at University of California Los Angeles; UCLA Head and Neck Cancer Program, UCLA Medical Center, Los Angeles, CA, United States

John V. Hegde Department of Radiation Oncology, University of California, Los Angeles, Los Angeles, CA, United States

Ali Hosni Department of Radiation Oncology, Princess Margaret Cancer Centre, University of Toronto, Toronto, ON, Canada

Jonathan Irish Department of Otolaryngology-Head and Neck Surgery/Surgical Oncology, Princess Margaret Cancer Centre/University Health Network, University of Toronto, Toronto, ON, Canada

Suchin Khanna Department of Internal Medicine, Yale Cancer Center and Yale School of Medicine, New Haven, CT, United States

Young J. Kim Department of Otolaryngology—Head and Neck Surgery, Vanderbilt University Medical Center, Nashville, TN, United States

Amar U. Kishan Department of Radiation Oncology, University of California, Los Angeles, Los Angeles, CA, United States

Lindsay Lanciault Department of Audiology & Speech, University of California, Los Angeles, CA, United States

Rebecca Levin-Epstein Department of Radiation Oncology, University of California, Los Angeles, Los Angeles, CA, United States

Umair Mahmood Dana-Farber Cancer Institute, Boston, MA, United States

Amanda Miller David Geffen School of Medicine at University of California Los Angeles, Los Angeles, CA, United States

Jeffrey N. Myers Department of Head and Neck Surgery, The University of Texas MD Anderson Cancer Center, Houston, TX, United States

Cherie-Ann O. Nathan Department of Otolaryngology/Head Neck Surgery, Ochsner-LSU Health; Surgical Service, Otolaryngology Section, Overton Brooks Veterans Affairs Medical Center, Shreveport, LA, United States

Neil Parikh Department of Radiation Oncology, University of California, Los Angeles, Los Angeles, CA, United States

David A. Rapkin Department of Head and Neck Surgery, David Geffen School of Medicine at University of California Los Angeles, Los Angeles, CA, United States

Vlad C. Sandulache Bobby R. Alford Department of Otolaryngology Head and Neck Surgery, Baylor College of Medicine, Houston, TX, United States

Nicole C. Schmitt Department of Otolaryngology—Head and Neck Surgery and Winship Cancer Institute, Emory University School of Medicine, Atlanta, GA, United States

Anna Spreafico Division of Medical Oncology and Hematology, Princess Margaret Cancer Centre, University of Toronto, Toronto, ON, Canada

Maie A. St. John Department of Head and Neck Surgery, David Geffen School of Medicine at University of California Los Angeles; UCLA Head and Neck Cancer Program, UCLA Medical Center; Department of Bioengineering, University of California Los Angeles; UCLA Jonsson Comprehensive Cancer Center, Los Angeles, CA, United States

Ryan Stephenson Department of Head and Neck Surgery, David Geffen School of Medicine at University of California Los Angeles, Los Angeles, CA, United States

Kenric Tam Department of Head and Neck Surgery, David Geffen School of Medicine at University of California Los Angeles; UCLA Head and Neck Cancer Program, UCLA Medical Center, Los Angeles, CA, United States

Deborah J. Wong Division of Hematology and Oncology, David Geffen School of Medicine at University of California Los Angeles, Los Angeles, CA, United States

Alisa Zhukhovitskaya Department of Head and Neck Surgery, David Geffen School of Medicine at University of California Los Angeles, Los Angeles, CA, United States

CHAPTER

1

Introduction: Leaps and bounds

Kenric Tam[a,b], *Shabnam Ghazizadeh*[a,b], *Elliot Abemayor*[a,b], *Maie A. St. John*[a,b,c,d]

[a]Department of Head and Neck Surgery, David Geffen School of Medicine at University of California Los Angeles, Los Angeles, CA, United States [b]UCLA Head and Neck Cancer Program, UCLA Medical Center, Los Angeles, CA, United States [c]Department of Bioengineering, University of California Los Angeles, Los Angeles, CA, United States [d]UCLA Jonsson Comprehensive Cancer Center, Los Angeles, CA, United States

Abstract

The landscape of managing head and neck malignancies has changed greatly over the last several decades. With improvements in treatment and adjuvant options, improved survival rates and decreasing morbidity are widely seen. Earlier detection of malignancies and new operative technologies has shifted the focus from needlessly aggressive *open* surgical resections to minimally invasive surgery, while preserving or improving oncologic outcomes. Advancements in targeted radiation therapy delivery have decreased radiation-induced toxicity to adjacent healthy tissue with, for example, the adaptation of intensity-modulated proton therapy (IMPT) for head and neck cancers showing promise in further limiting radiation-related morbidity. Perhaps the most exciting advancement in managing cancer is the advent of a new era of precision medicine in the form of immunotherapy. Targeted checkpoint inhibitors and monoclonal antibodies are currently being used for locoregionally advanced, recurrent, and metastatic head and neck squamous cell carcinoma. The future of head and neck cancer treatment is evolving and appears brighter now and in the future than it did in the past. In this chapter, we review ongoing advances in head and neck cancer treatment with the hope of serving as a roadmap for new frontiers in head and neck cancer care.

Abbreviations

ADCC	antibody-dependent cell-mediated cytotoxicity
chemoRT	chemoradiation therapy
CT	computed tomography
EBRT	external beam radiation therapy
EGFR	epidermal growth factor receptor
FDG-PET	fluorodeoxyglucose positron emission tomography
GSL	glottic and supraglottic larynx
Gy	gray
HNC	head and neck cancer

Novel Therapies in Head and Neck Cancer: Beyond the Horizon
https://doi.org/10.1016/B978-0-12-820679-9.00001-3

HNSCC	head and neck squamous cell carcinoma
HPV+ OPC	HPV-positive oropharyngeal carcinoma
HPV	human papillomavirus
IMPT	intensity-modified proton therapy
LA-HNSCC	locoregionally advanced HNSCC
MRI	magnetic resonance imaging
NK	natural killer
NSCLC	non-small cell lung cancer
OPC	oropharyngeal cancer
PCs	pharyngeal constrictors
PD-L1	programmed death-ligand 1
R/M HNSCC	recurrent/metastatic HNSCC
RT	radiation therapy
SP	da Vinci single-port system
TLM	transoral laser microsurgery
TORS	transoral robotic surgery

Conflict of interest

No potential conflicts of interest were disclosed.

Introduction

Head and neck cancer accounts for more than 650,000 cases each year worldwide, and, in the United States, approximately 53,000 patients develop head and neck cancer (HNC) each year [1,2]. Despite the decline in traditional carcinogen-induced head and neck cancers, the last 30 years has seen a significant increase in HNC in otherwise healthy populations driven by a rise in HPV-associated oropharyngeal cancers (OPC) [3,4]. Notwithstanding the increasing incidence of HNC, we have seen an increased survival for head and neck malignancies, with the overall 5-year relative survival rate increasing from 54.7% in 1992–96 to 65.9% in 2002–06, with the greatest improvement seen in tonsillar and oral tongue carcinomas [3].

Medical advancements have improved our understanding of tumor biology, identifying distinct subsites and etiologies, and have led to the development of targeted treatment options. Identification of the human papillomavirus (HPV) as an acquired cause of OPC, for example, has greatly changed the management of these tumors and introduced new stratification and treatment algorithms. Surgical innovation has allowed for numerous advances in perioperative and intraoperative practices, improving access and minimizing surgical morbidity. Similarly, optimized delivery of radiation to tumor cells using intensity-modified radiation therapy (IMRT), molecular imaging, and intensity-modified proton therapy (IMPT) continue to maximize quality-of-life outcomes for patients while improving oncologic results [5]. Though cytotoxic therapy continues to have a vital role in HNC, perhaps the most exciting advancement in systemic treatments has been the application of immunotherapy. Targeting unique cancer protein antigens is quickly becoming a promising option for HN cancers and has the potential to revolutionize cancer treatment.

Cancer care in the 21st century mandates an understanding of a spectrum of treatment modalities and distinct fields including but not limited to surgery, radiation oncology, systemic therapies, nutrition, pain management, and rehabilitation. HNCs, in particular, represent a histologically and anatomically diverse collection of neoplasms, all of which have unique treatment algorithms. Sequelae of cancer progression and side effects of oncologic treatment in this region can disrupt vital structures and have dramatic effects on quality of life [6,7]. Facial disfigurement, speech and swallowing impairment, tracheostomy and gastrostomy tube dependence, and chronic infections are among the myriad of possible consequences that contribute to the significant morbidity from HNCs. New frontiers in the treatment of HN neoplasms focus on improving oncologic outcomes while minimizing morbidity and long-term toxicities. The synthesis of these therapies with patient centered care is essential to achieve the best possible outcome. This chapter is an overview of the leaps and bounds made in the treatment of HN cancer.

Surgical innovation: *TORS and TLM*

Surgical innovation has allowed for numerous advances in pre-, peri-, and intraoperative practices. Improved access to tumors through surgical techniques such as transoral laser microsurgery (TLM) and, more recently, transoral robotic surgery (TORS) has opened the door for surgical resection of tumors that were managed either with open surgery or with radiation therapy. "Open" surgical approaches requiring mandibulotomy and/or mandibulectomy—splitting or removing the mandible, respectively—were traditionally required to reach tumors of the oral cavity, oropharynx, and lateral pharyngeal wall. A mandibulotomy approach to the oropharynx was first described in 1920 as a primary surgical approach to oropharyngeal cancers offered tumor access, but with a less than ideal aesthetic result [8–10]. Complications from these surgeries were often high, and transection of critical anatomical structures resulted in significant functional morbidity. Postoperative complications including fistula, nonunion, plate exposure, cellulitis, and abscesses are reported to have occurred in up to 31.6% of patients [11]. Furthermore, such open approaches often required the added morbidity of a prophylactic tracheostomy [12]. Despite their efficacy the significant functional and cosmetic morbidity from these radical surgeries led to a search for other equally effective treatment modalities such as radiation and systemic treatments [13].

In the late 1990s the idea of organ preservation was developed to specifically address quality of life and speech and swallowing outcomes for patients with advanced laryngeal carcinoma and compared outcomes using radiation (preservation of the larynx) with primary surgery (laryngectomy) [14]. Several landmark studies evaluating primary radiation with adjuvant chemotherapy for laryngeal cancer showed similar outcomes and survival compared with surgery (95% and 80% 5-year cause specific and overall survival rates) while preserving functional anatomy and limiting the need for gastrostomy tubes and tracheostomies [15]. However, as with any cancer treatment modality, chemotherapy and radiotherapy can have serious adverse effects. In the HN specifically, radiation-induced side effects such as xerostomia (40%), dysphagia (35%), hearing loss (43%), and trismus (25.4%) are seen in patients receiving radiation therapy [16]. In recent years, organ preservation strategies and deintensification of treatment have been applied to malignancies of the oropharynx as well. With HPV-positive oropharyngeal carcinoma (HPV+OPC), a better understanding of tumor phenotype and excellent overall prognosis irrespective of primary treatment modality has led to a shift in treatment paradigm to consider surgery as a means of obtaining more information about the tumor with the aim of personalizing treatment and perhaps deescalating radiation, obviating the need for chemotherapy, and ultimately, diminishing long-term side effects in this population with an already robust survival outcome [17].

While traditional approaches such as mandibulotomy, described in the preceding text, required an understanding of surgical anatomy from lateral to medial or external skin incision toward the aerodigestive tract ("outside in"), transoral approaches necessitate a deeper understanding of aerodigestive tract anatomy starting from the "inside out" [18]. In 1951, Huet first described transoral lateral oropharyngectomy without special instrumentation primarily for small and accessible tonsillar SCC lesions [19]. This early work was further refined with many cadaveric studies that described the complex transoral anatomy of the lateral oropharyngeal wall, base of tongue, and parapharyngeal spaces that have in turn paved the way for transoral endoscopic head and neck surgery either using microscope and laser or robotic-assisted devices [20,21].

Transoral robotic surgery (TORS) and transoral laser microsurgery (TLM) have revolutionized the treatment of upper aerodigestive tract neoplasms, specifically oropharyngeal, hypopharyngeal, and laryngeal cancers through improved tumor visualization and decreased morbidity while maintaining or improving oncologic outcomes. TLM has been used to treat laryngeal and supraglottic pathology since the 1970s and now has an expanding role in oropharyngeal and hypopharyngeal sites. Strong and Jako first connected a CO_2 laser to a microscope to treat benign laryngeal disease and 3 years later in 1975 began treating T1 laryngeal cancer [22,23]. Shortly thereafter, Vaughan began using TLM techniques to resect supraglottic malignancies [24]. Controversially, but strongly and effectively, these approaches were expanded by Steiner and Ambrosch who were among the first to demonstrate that TLM could be safely used to surgically remove tumors of the upper aerodigestive tract with similar oncologic outcomes to commonly used chemoradiation regimens [25,26]. Against all convention, the investigators proposed *cutting through the tumor to assess depth* and thereby ablate disease by microscopically ascertaining malignant from normal tissue [26,27]. Although TLM precludes en bloc tumor resection, the use of CO_2 laser is thought to allow for more precise margins compared with cautery. The technique is not without challenges, however, as there can be difficulty achieving access for adequate tissue manipulation and hemostasis [28].

The indications for TLM include early-stage glottic, supraglottic, and oropharyngeal cancers. Both radiation therapy and TLM achieve similar oncologic and functional outcomes for early glottic cancers, with, perhaps, the exception of cancers involving the anterior commissure. While there are no randomized control trials directly comparing TLM with radiation therapy, numerous case series demonstrate that TLM has similar survival rates and voice outcomes in comparable lesions [15,29,30]. Management of tumors involving the anterior commissure with TLM, however, is controversial. Indeed the earlier staging systems segregated these tumors out due to clinical outcomes that were not comparable with similar-sized lesions occurring elsewhere within the larynx. While some have advocated for radiation therapy for anterior commissure tumors due to superior voice outcomes and difficulty of surgical resection, others have retrospectively demonstrated that TLM had similar voice outcomes and recurrence rates to radiation therapy [31–33].

While TLM was initially developed for the resection of early-stage laryngeal and supraglottic cancers, it has since been adopted for access to the oropharynx. Steiner et al. described the use of TLM for base of tongue malignancies [27]. A recent multicenter, prospective study demonstrated the utility of TLM for advanced stage base of tongue and tonsillar cancers with excellent oncologic and functional outcomes; 87% of patient had normal swallowing or episodic dysphagia and a 3-year overall survival rate of 86%, on par with outcomes of patients treated with chemoradiation or robotic resections [34].

The da Vinci Surgical System is a surgical robotic system (Intuitive Surgical, Sunnyvale, CA) that was initially FDA approved for use in urological, general laparoscopic, and gynecological surgeries but has since been adapted for head and neck surgery. A surgeon sitting at a robotic console controls the robot's operating arms that consists of an endoscope and two surgical instruments (Fig. 1). The endoscope provides a high-definition, binocular view of the anatomy and allows for clear, visual assessment of margins. The robotic console also eliminates hand tremor and allows for precise movements. In contrast to TLM, TORS allows for en bloc removal of tumors with the need for serial sections. However, critics of TORS purport that there is poor tactile feedback, prohibiting the ability to manually palpate the tumor, and high initial capital cost of acquiring a robot, on the order of a two million dollar price tag [35,36].

Weinstein and O'Malley were the first to apply transoral robotic surgery for early-stage T1 and T2 oropharyngeal malignancies and demonstrated favorable oncologic and functional outcomes [37]. In their initial study, O'Malley et al. demonstrated the feasibility of TORS in both human cadavers and ex vivo canines and subsequently performed lingual tonsillectomies in three patients [38]. Shortly thereafter, a phase 1 study demonstrated the efficacy of radical palatine tonsillectomies in 27 patients with negative margins in 93% of patients [39]. Recent data for TORS resection of oropharyngeal malignancies demonstrates 2-year locoregional control rate and overall survival of 91.8% and 91%, respectively [40]. One of the most feared complications of TORS is postoperative hemorrhage from the lingual artery. In an effort to minimize this outcome, cervical lingual arterial ligation has been practiced. Whereas there is no decline in overall number of bleeding events, such ligation does result in a statistically significant reduction in the incidence of major bleeding [41].

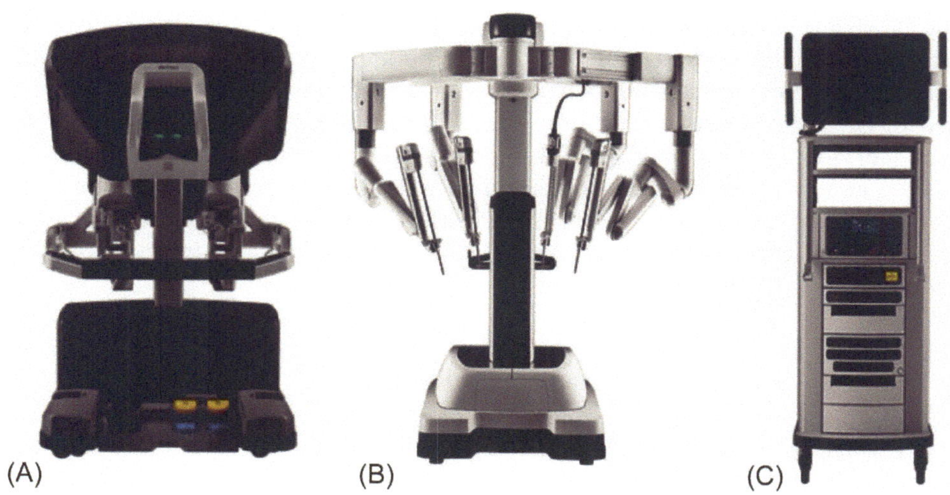

FIG. 1 (A) Surgeon console controls the patient cart. (B) Patient cart holds three endoscopic instruments and cameras. (C) Vision cart communicates between the surgeon console and the patient cart and provides 3-D high-definition vision system.

Further technological advancements in the da Vinci Surgical System, including increased miniaturization of instruments and development of a single-port system, have allowed for increasingly complex transoral oncologic resections. The da Vinci Single-Port System (SP) uses a single 25-mm cannula through which all robotic instruments pass, which allows for greater articulation within small orifices such as the oral cavity, oropharynx, and supraglottis. All instrument tips are triangulated around the target, avoiding external instrument collisions. The da Vinci SP has been used to perform supraglottic laryngectomies in cadavers and the hypopharynx, larynx, oropharynx, and nasopharynx in a recent phase II trial [42,43]. In 2019 the FDA cleared the da Vinci SP system for lateral oropharyngectomy procedures and base of tongue resections. Perhaps this system can increase accessibility for patients with unfavorable anatomy, making previously difficult to resect cancers more amenable to surgery.

With increasing use in many centers, over the past decade, TORS has become an established first-line treatment for T1 and T2 oropharyngeal squamous cell carcinomas. A multiinstitutional study with 410 patients revealed a 91.8% locoregional control rate and 91% overall survival [40]. Despite this promising result, there have been no large-scale phase III controlled trials comparing TORS with radiation therapy for OPC. The oropharynx: radiotherapy versus transoral robotic surgery (ORATOR) phase II trial enrolled 68 patients with T1-T2 N0-2 OPSCC [44]. The MD Anderson Dysphagia Inventory (MDADI) scores were superior in the RT arm compared with the TORS arm ($P = .042$) but did not meet the powered clinically meaningful change. All other quality-of-life metrics including feeding tube rates were similar; there was more neutropenia, constipation, and tinnitus in the RT arm and more trismus in the TORS arm. Overall survival and progression-free survival at 3 years were similar. Compared with patients treated with definitive chemoradiation therapy (chemoRT) for OPC, patients receiving adjuvant chemoRT had fewer dental problems, worse appearance, and worse coughing. Patients undergoing surgery alone compared with adjuvant and definitive chemoRT patients had better salivary, taste, and oral function and less pain and dental disease [45].

While TORS is an established treatment for early-stage OPSCC, its role in the hypopharynx and supraglottis continues to evolve. Studies evaluating the role in hypopharyngeal cancer are limited. Small prospective studies have shown similar oncologic outcomes to the current standard of care (chemoradiation therapy), while functional outcomes of swallowing are improved [46]. In 23 patients with T1–T4 hypopharyngeal cancer, there was an overall survival of 89% at 3 years, and disease-free survival was 84%. There are no studies comparing TORS hypopharyngeal surgical resection with radiation or systemic treatments. As further studies with larger cohorts are published, the role of this technology in resection of tumors of the aerodigestive tract will continue to be elucidated.

Radiation therapy

Since the establishment of radiation oncology as a clinical discipline, external beam therapy has played a crucial role in the treatment of head and neck. Dr. Henri Coutard, one of the early pioneers of radiation oncology, began treating incurable head and neck tumors with radiation in 1919 [47]. In 1922, at the International Congress of Otology in Paris, Coutard and colleagues presented six patients with advanced laryngeal carcinoma who were successfully treated with radiation therapy and subsequently demonstrated no further evidence of disease. Soon after, others began treating tumors of the tonsils and oral cavity [48]. Currently, radiation therapy (RT) is often used for primary treatment of select HN tumors such as advanced stage oropharyngeal and laryngeal cancers and as an adjuvant treatment for high risk and aggressive tumors.

The radiation dose, measured in gray (Gy), corresponds to the absorption of 1 joule of energy per kilogram of matter. The deeper the targeted malignancy, the more Gy that needs to be delivered to penetrate the surrounding healthy tissue. However, local toxicities including desquamation, mucositis, fibrosis, and cellular damage also affect the healthy neighboring tissue in a dose-dependent manner making dosing and treatment challenging [49]. Typically, HN cancers require 2 Gy per day, 5 days a week for 5 weeks, for a total of 66–70 Gy. Radiation to lower risk cervical nodal basins requires lower radiation doses of approximately 50 Gy, but postoperative radiation of microscopic disease requires higher doses of 60–66 Gy to decrease risk of locoregional failure [50]. Oropharyngeal cancers are typically dosed with 60–65 Gy of radiation and are associated with a 4%–7% risk of ORN [51]. Other complications include costochondritis of the larynx, trismus, long-term xerostomia, and acute but debilitating mucositis.

Several techniques in radiation delivery have been developed, namely, external beam radiation therapy (EBRT) and brachytherapy. Head and neck cancers are often treated with EBRT, of which there are several types of radiation particle beams. In head and neck cancers, photons, which are skin sparing and have less desquamation, and electrons, which are better suited for superficial targets such as skin, are the primary radiation particle beams utilized.

Other methods of radiation delivery, such as brachytherapy, in which a radioactive source is placed within or adjacent to the tumor, are primarily used in recurrent head and neck cancer. Brachytherapy allows for a focused dose delivery in a previously irradiated field, thus minimizing toxicity to adjacent, healthy tissue [52]. Brachytherapy can be utilized alone but has better locoregional control rates and overall survival when combined as adjunctive therapy with salvage surgery [52]. Despite the focused dose provided by brachytherapy, patients continue to have risks of fistula formation, dysphagia, and osteoradionecrosis [53].

In the 2000s, intensity-modulated radiotherapy (IMRT), a form of EBRT, emerged as a promising method to better target radiation beams and minimize damage to adjacent healthy tissue. Utilizing imaging modalities including computed tomography (CT), magnetic resonance imaging (MRI), and fluorodeoxyglucose positron emission tomography (FDG- PET), the radiation oncologist is able to precisely map out the tumor and contour radiation beams to deposit primarily on malignant tissue while minimizing delivery to adjacent healthy organs [54]. Current developments in this technology are now focusing on quality-of-life outcomes, such as improving swallowing and xerostomia as well as deintensification [55]. Radiation to structures such as the pharyngeal constrictors (PCs) and the glottic and supraglottic larynx (GSL) can result in long-term dysphagia and aspiration [56]. New IMRT planning conformations allow for dosimetric sparing of the PCs and GSL, resulting in similar oncologic outcome but reduced long-term dysphagia at 1 year [57]. Similarly, new parotid- and submandibular-sparing IMRT has shown significant reduction in self-reported xerostomia with no changes in local-regional control or overall survival [58].

Increased diagnosis of HPV-positive OPC (HPV + OPC) has prompted changes to radiation regimens with or without added systemic therapy. Traditionally, OPC was treated with 70 Gy of radiation with concurrent cisplatin, but studies showing that HPV+ OPC has more favorable outcomes compared with HPV− OPC prompted a move toward deintensification of chemoradiation protocols to minimize toxicity [59,60]. A variety of strategies toward deintensification are being investigated including (1) staged decrease in chemoradiation intensity if favorable response as determined by decrease in imaged tumor size or (2) using F-18 PET to monitor resolution of tumor hypoxia after radiation as a marker of good response [55,61,62].

Despite the advances afforded by IMRT, photon-based radiation therapy still results in irradiation of healthy tissue both superficial and deep to the target tissue. New technologies, including proton beam radiation therapy, are being developed and refined to limit toxicity to healthy tissue. Compared with photon-based radiation therapy, protons have low-energy effects on superficial tissue. In addition, proton radiation is deposited in a more specific and confined area. This tonotopic pattern, selectively affecting the target tissue while minimizing damage to healthy tissue beyond the indicated depth, is called the "Bragg peak" [63]. Despite the advantages of proton beam radiation therapy, its use has been limited by availability at a handful of medical centers due to high cost.

Research into the use of proton beam therapy for oropharyngeal carcinoma is ongoing. Retrospective cohort analyses have shown decreased need for gastrostomy tubes and lower rates of xerostomia compared with IMRT, but no statistically significant difference in mucositis [64]. Dosimetry studies suggest that there is significant sparing of the oral cavity and major salivary glands [65]. Currently, there is an ongoing clinical trial comparing IMRT with intensity-modulated proton beam (IMPT) therapy in oropharyngeal cancer with the primary outcome measure of late onset toxicity up to 2 years after completion of treatment [66]. Similar results have been demonstrated for tumors of the nasopharynx, sinuses, and skull base, although there is a lack of randomized control studies. A retrospective case-control study of nasopharyngeal carcinomas found that proton beam therapy resulted in only 20% of patients needing gastrostomy tube placement compared with 65% in IMRT patients [67]. A review of 40 patients with sinonasal carcinoma found decreased dosing to the oral cavity, esophagus, and parotid glands resulting in lower g-tube dependence and decreased pain requirements. Tumors of the skull base in particular have high morbidity due to proximity to vital neurologic structures. Proton beam therapy has been used to treat skull base tumors since the late 1990s. In a large cohort, 519 patients with skull base chordomas and chondrosarcomas were successfully treated with proton beam therapy with a 5-year survival of 73% for chordomas and 98% for chondrosarcoma. However, fatal brainstem toxicity and temporal lobe injury were reported in eight of these patients [68]. In comparison, treatment of chordoma and chondrosarcoma with image-guided IMRT resulted in 5-year overall survival of 85.6% and 87.7%, respectively [69]. Further prospective studies will help to understand this tool and its role in HN cancer treatment.

Immunotherapy

In addition to advances made in surgery, radiation, and cytotoxic chemotherapy, immunotherapy has emerged as a promising new, less toxic more targeted method of treatment. While immunotherapy is often viewed as the next revolution in cancer care, the concept of immunotherapy for head and neck cancer stems from the late 19th century. In the 1890s Dr. William B. Coley, a surgeon, observed that several patients with inoperable tumors had spontaneous regression of their cancer after suffering from severe bacterial infections. As a careful clinician will do, he began to further investigate this finding. His first subject was a patient with a large, inoperable laryngeal sarcoma who, after being deliberately infected with a virulent strain of streptococcus, had a complete response and resolution of the tumor [70]. While the claims made in Coley's case series of 10 patients are difficult to validate, the idea that the immune system can be harnessed to attack cancer cells has fueled current breakthroughs in immunotherapy with the 2018 Nobel Prize in medicine being awarded to two researchers whose work centers on immune checkpoint inhibitors [71].

Traditional chemotherapy is based upon incorporation of cytotoxic chemicals that bind to the DNA of rapidly expanding cells, thereby interfering with DNA repair and replication. While effective, this mechanism of action also results in disruption of all cells with high turnover rates, for example, healthy epithelia of the gastrointestinal tract. Consequences of damage to healthy cells can result in many systemic cytotoxic effects such as diarrhea, hair loss, and mucositis. In comparison, immunotherapy targets precise molecular antigens located on tumor cells. As part of the human immune system, the body identifies proteins on the surface of tumor cells as "nonself" antigens and creates an antitumor response. In response, cancer cells express cell-surface proteins (i.e., immune checkpoints) that inhibit the immune system's antitumor response (Fig. 2). The two fundamental mechanisms of immunotherapy involve the use of monoclonal antibodies that either target specific cancer cell tumor markers or block immune checkpoint signals. Immunotherapy is a rapidly evolving field, and the identification of new biomarkers and immune checkpoint inhibitors is paramount to developing novel treatments and cures for head and neck cancer.

Immunotherapy research has primarily concentrated on head and neck squamous cell carcinoma (HNSCC), which is the most common type of head and neck cancer. Like many other malignancies, HNSCC overexpresses a number of proteins and signaling pathways that result in unrestrained growth. Most notably, HNSCC overexpresses the epidermal growth factor receptor (EGFR) compared with healthy, physiologic tissue [72,73]. Cetuximab, a human-mouse chimeric IgG1 antibody against EGFR, was the first-FDA approved immunotherapy for HNSCC. Cetuximab likely has multiple mechanisms of action. Studies have shown that it acts as an inhibitor of the EGFR signaling pathway, resulting in decreased tumor growth, invasion, metastasis, and angiogenesis [74,75]. Cetuximab also activates NK cells and T cells and induces antibody-dependent cell-mediated cytotoxicity (ADCC) [76]. In ADCC the constant region of the monoclonal antibody binds to the immune cell—typically a T cell or natural killer (NK) cell—allowing for targeting and direct cytotoxicity against tumor cells [77].

Currently cetuximab is used as an adjunct to chemotherapy regimens for recurrent/metastatic (R/M HNSCC) and locoregionally advanced HNSCC (LA-HNSCC). The landmark trial that led to FDA approval for LA-HNSCC demonstrated a significant increase in overall survival when comparing cetuximab and radiation therapy (RT) with RT alone (24.4 vs 14.9 months) for the treatment of locally advanced oropharyngeal, hypopharyngeal, and laryngeal HNSCC [78]. The addition of cetuximab to cisplatin-based chemoradiation, however, does not appear to change overall outcomes. In a phase III trial, patients randomized to cetuximab + chemoRT had similar overall survival and progression-free survival compared with those treated with chemoRT alone [59].

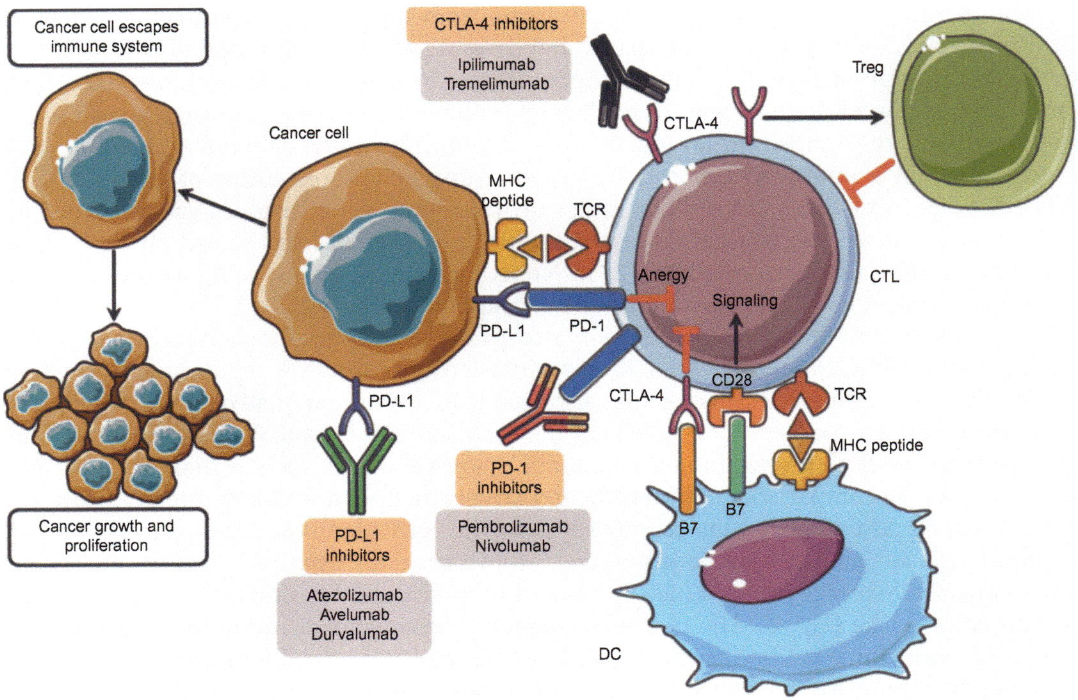

FIG. 2 Cancer cell immune escape mechanisms and immune checkpoint inhibitors. *License CC BY-NC 3.0, Ayoub NM, Al-Shami KM, Yaghan RJ. Immunotherapy for HER2-positive breast cancer: recent advances and combination therapeutic approaches. Breast Cancer Targets Ther 2019;11:53–69.*

The role of cetuximab in R/M HNSCC was established in the Erbitux in First-Line Treatment of Recurrent or Metastatic Head and Neck Cancer (EXTREME) trial. In this trial, 442 patients were randomized to receive cisplatin or carboplatin followed by 5-FU or the same chemotherapy regimen with the addition of cetuximab. The chemoRT + cetuximab group had increased overall survival (10.7 vs 7.4 months) and progression-free survival (5.6 vs 3.3 months) [79].

Until recently, there have been no phase III trials that compared cetuximab plus RT with cisplatin plus RT. Cisplatin-based chemoRT has remained the standard of care for LA-HNSCC based on metaanalysis of over 17,000 patients and shows an absolute survival benefit [80]. For the treatment of HPV-positive oropharyngeal cancer, the radiation therapy oncology group (RTOG) 1016 trial found that RT plus cetuximab demonstrated inferior overall survival and progression-free survival compared with the current standard of care treatment of RT plus cisplatin [17]. The role of cetuximab in HNSCC treatment, particularly with the advent of PD-1/PDL1 immunotherapies, which are also approved for LA-HNSCC and R/M HNSCC, continues to evolve.

HNSCC also employs a number of mechanisms to escape the antitumor immune response, including impairment of tumor-infiltrating lymphocytes and inhibition of NK cell cytotoxicity [81,82]. In particular, programmed death-ligand 1 (PD-L1) inhibits PD-1 positive that

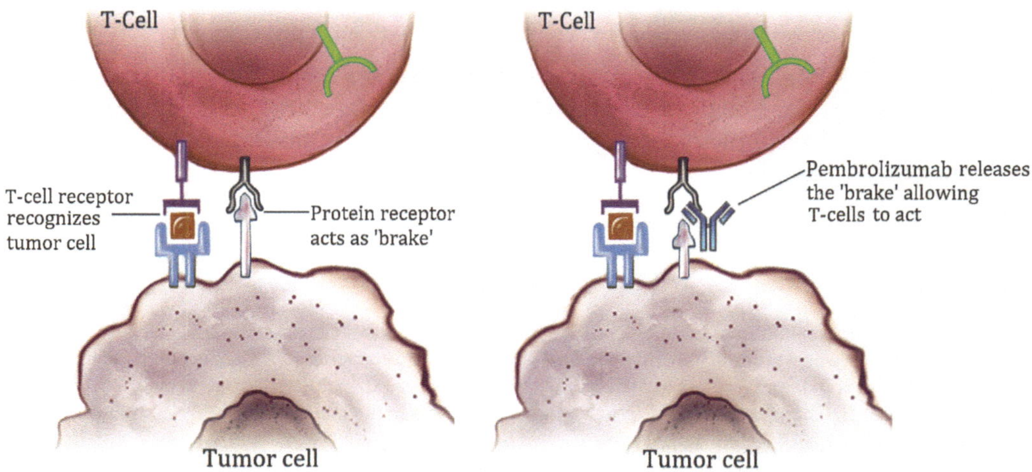

FIG. 3 PD-1 and PD-L1 immune checkpoint. *From Wahid M, Akhter N, Jawed A, Dar SA, Mandal RK, Lohani M, et al. Pembrolizumab's non-cross resistance mechanism of action successfully overthrown ipilimumab. Crit Rev Oncol Hematol 2017;11:1–6.*

activated immune cells such as myeloid cells, T cells, and B cells (Fig. 3) [83]. Fifty to sixty percent of HNSCC express the PD-1/PD-L1 immune checkpoint [84]. Other HNSCC immune checkpoint inhibitors include CEACAM1 inhibition of NK cells and CTLA-4/CD80 blockade of cytotoxic T cells [85,86].

Both pembrolizumab and nivolumab are human monoclonal antibodies directed against the PD-1 receptors and inhibit PD-1/PDL1 binding and activity. Both have a number of indications from non-small cell lung cancer (NSCLC) and melanoma to, more recently, head and neck squamous cell carcinoma. In vitro studies have demonstrated that blockade of this immune checkpoint results in increased activity of B cells and T cells [83,84].

In 2016 the KEYNOTE-012 and CheckMate 141 trials led to FDA approval of pembrolizumab and Nivolumab for the treatment of recurrent or metastatic (R/M) HNSCC. In a phase III trial, patients who had R/M HNSCC within 6 months of receiving platinum-based chemotherapy were randomized to receive either nivolumab or a single-agent systemic therapy—methotrexate, docetaxel, or cetuximab [87]. Patients randomized to the nivolumab cohort had a significantly longer overall survival, approximately 19 percentage points higher at 1 year. The NCCN guidelines, however, listed PD-1 inhibitors as second line systemic therapies for R/M HNSCC without options of salvage or radiation therapy [88,89].

In 2019 KEYNOTE-048 demonstrated the efficacy of PD-1 inhibitors as frontline therapy for R/M HNSCC. Eight hundred eighty-two patients who had not received prior systemic chemotherapy were randomized to either pembrolizumab, pembrolizumab + cisplatin/carboplatin, or cetuximab, and the P + C group demonstrated significantly increased overall survival (14.7 vs 11.0 months). KEYNOTE-048 led to the FDA approval of pembrolizumab as a frontline therapy for R/M disease.

The role of PD-1 inhibitors for head and neck cancer continues to evolve. Several studies are underway to investigate the effect of p16 positivity on the use of immune modulators and even the use of checkpoint inhibitors as frontline therapy for primary disease [90].

Conclusion

The treatment of head and neck cancer is complex, requiring multimodality treatment to achieve both excellent oncologic outcomes while preserving essential native functions. Breakthroughs in surgical technology, radiation therapy, and immunotherapy have made these goals ever more attainable.

References

[1] Bray F, Ferlay J, Soerjomataram I, Siegel RL, Torre LA, Jemal A. Global cancer statistics 2018: GLOBOCAN estimates of incidence and mortality worldwide for 36 cancers in 185 countries. CA Cancer J Clin 2018;68:394–424.

[2] Siegel RL, Miller KD, Jemal A. Cancer statistics, 2019. CA Cancer J Clin 2019;69:7–34.

[3] Pulte D, Brenner H. Changes in survival in head and neck cancers in the late 20th and early 21st century: a period analysis. Oncologist 2010;15:994–1001.

[4] Jakobsen KK, Grønhøj C, Jensen DH, Karnov KKS, Agander TK, Specht L, et al. Increasing incidence and survival of head and neck cancers in Denmark: a nation-wide study from 1980 to 2014. Acta Oncol (Madr) 2018;58:1570–6.

[5] Grégoire V, Langendijk JA, Nuyts S. Advances in radiotherapy for head and neck cancer. J Clin Oncol 2015; 33(29):3277–84.

[6] Terrell JE, Ronis DL, Fowler KE, Bradford CR, Chepeha DB, Prince ME, et al. Clinical predictors of quality of life in patients with head and neck cancer. Arch Otolaryngol Head Neck Surg 2004;130:401–8.

[7] Verdonck-De Leeuw IM, Buffart LM, Heymans MW, Rietveld DH, Doornaert P, De Bree R, et al. The course of health-related quality of life in head and neck cancer patients treated with chemoradiation: a prospective cohort study. Radiother Oncol 2014;110:422–8.

[8] Spiro RH, Gerold FP, Shah JP, Sessions RB, Strong EW. Mandibulotomy approach to oropharyngeal tumors. Am J Surg 1985;150:466–9.

[9] Trotter W. A method of lateral pharyngotomy for the exposure of large growths in the epilaryngeal region. J Laryngol Rhinol Otol 1920;13:196–8.

[10] Gooris PJJ, Worthington P, Evans JR. Mandibulotomy: a surgical approach to oral and pharyngeal lesions. Int J Oral Maxillofac Surg 1989;18:359–64.

[11] Dai TS, Hao SP, Chang KP, Pan WL, Yeh HC, Tsang NM. Complications of mandibulotomy: midline versus paramidline. Otolaryngol Head Neck Surg 2003;128:137–41.

[12] Anehosur VS, Karadiguddi P, Joshi VK, Lakkundi BC, Ghosh R, Gopalkrishnan K. Elective tracheostomy in head and neck surgery: our experience. J Clin Diagn Res 2017;11:ZC36–9.

[13] Rubinstein M, Armstrong WB. Transoral laser microsurgery for laryngeal cancer: a primer and review of laser dosimetry. Lasers Med Sci 2011;26:113–24.

[14] Lefebvre JL. Larynx preservation: the discussion is not closed. Otolaryngol Head Neck Surg 1998;118:389–93.

[15] Mendenhall WM, Werning JW, Hinerman RW, Amdur RJ, Villaret DB. Management of T1-T2 glottic carcinomas. Cancer 2004;100(9):1786–92.

[16] Strojan P, Hutcheson KA, Eisbruch A, Beitler JJ, Langendijk JA, AWM L, et al. Treatment of late sequelae after radiotherapy for head and neck cancer. Cancer Treat Rev 2017;59:79–92.

[17] Gillison ML, Trotti AM, Harris J, Eisbruch A, Harari PM, Adelstein DJ, et al. Radiotherapy plus cetuximab or cisplatin in human papillomavirus-positive oropharyngeal cancer (NRG oncology RTOG 1016): a randomised, multicentre, non-inferiority trial. Lancet 2019;393:40–50.

[18] Asimakopoulos P, Ah-See K. CONTEMPORARY TRANSORAL SURGERY FOR PRIMARY HEAD AND NECK CANCER M.L. Hinni, D.G. Lott Plural Publishing, 2014 ISBN 978 1 59756 555 4 pp 450 Price US $225.00. J Laryngol Otol 2015;129(5):517.

[19] Huet P. L'electro-coagulation dans le epitheliomas de l'amygadale-palatine. Ann Otolaryngol Chir Cervicofac 1951;68:433–42.

[20] Rubin F, Laccourreye O, Weinstein GS, Holsinger FC. Transoral lateral oropharyngectomy. Eur Ann Otorhinolaryngol Head Neck Dis 2017;134:419–22.

[21] Gun R, Durmus K, Kucur C, Carrau RL, Ozer E. Transoral surgical anatomy and clinical considerations of lateral oropharyngeal wall, parapharyngeal space, and tongue base. Otolaryngol Head Neck Surg 2016;154(3):480–5.
[22] Strong MS. Laser excision of carcinoma of the larynx. Laryngoscope 1975;85:1286–9.
[23] Strong MS, Jako GJ. Laser surgery in the larynx. Early clinical experience with continuous CO_2 laser. Ann Otol Rhinol Laryngol 1972.
[24] Vaughan CW, Strong MS, Jako GJ. Laryngeal carcinoma: transoral treatment utilizing the CO_2 laser. Am J Surg 1978;136:490–3.
[25] Steiner W. Results of curative laser microsurgery of laryngeal carcinomas. Am J Otolaryngol Neck Med Surg 1993;14:116–21.
[26] Steiner W, Ambrosch P. Endoscopic laser surgery of the upper aerodigestive tract: with special emphasis on cancer surgery [Internet]. Thieme; 2000. Available from: https://books.google.com/books?id=nLZsAAAAMAAJ.
[27] Steiner W, Fierek O, Ambrosch P, Hommerich CP, Kron M. Transoral laser microsurgery for squamous cell carcinoma of the base of the tongue. Arch Otolaryngol Head Neck Surg 2003;129:36–43.
[28] Goldenberg D, Koch WM. Surgical treatment of early oral cavity and oropharyngeal cancers. Early diagnosis treat. Cancer ser. Head neck cancers; 2010. p. 115–27.
[29] Csandy M, Czigner J, Svay L. Endolaryngeal CO_2 laser microsurgery of early vocal cord cancer; 2015. p. 219–21.
[30] Eckel HE. Local recurrences following transoral laser surgery for early glottic carcinoma: frequency, management, and outcome. Ann Otol Rhinol Laryngol 2001;110:7–15.
[31] Chone CT, Yonehara E, Martins JEF, Altemani A, Crespo AN. Importance of anterior commissure in recurrence of early glottic cancer after laser endoscopic resection. Arch Otolaryngol Head Neck Surg 2007;133:882–7.
[32] Hakeem AH, Tubachi J, Pradhan SA. Significance of anterior commissure involvement in early glottic squamous cell carcinoma treated with trans-oral CO_2 laser microsurgery. Laryngoscope 2013;123:1912–7.
[33] Sachse F, Stoll W, Rudack C. Evaluation of treatment results with regard to initial anterior commissure involvement in early glottic carcinoma treated by external partial surgery or transoral laser microresection. Head Neck 2009;31:531–7.
[34] Haughey BH, Hinni ML, Salassa JR, Hayden RE, Grant DG, Rich JT, et al. Transoral laser microsurgery as primary treatment for advanced-stage oropharyngeal cancer: a United States multicenter study. Head Neck 2011;33:1683–94.
[35] Tam K, Orosco RK, Dimitrios Colevas A, Bedi N, Starmer HM, Beadle BM, et al. Cost comparison of treatment for oropharyngeal carcinoma. Laryngoscope 2019;129:1604–9.
[36] Hans S, Badoual C, Gorphe P, Brasnu D. Transoral robotic surgery for head and neck carcinomas. Eur Arch Otorhinolaryngol 2012;269:1979–84.
[37] O'Malley BW, Weinstein GS, Snyder W, Hockstein NG. Transoral robotic surgery (TORS) for base of tongue neoplasms. Laryngoscope 2006;116:1465–72.
[38] O'Malley BW, Weinstein GS, Snyder W, Hockstein NG. Transoral robotic surgery (TORS) for base of tongue neoplasms. Laryngoscope 2006;116:1465–72.
[39] Weinstein GS, O'Malley BW, Snyder W, Sherman E, Quon H. Transoral robotic surgery: radical tonsillectomy. Arch Otolaryngol Head Neck Surg 2007;133:1220–6.
[40] De Almeida JR, Li R, Magnuson JS, Smith RV, Moore E, Lawson G, et al. Oncologic outcomes after transoral robotic surgery a multi-institutional study. JAMA Otolaryngol Head Neck Surg 2015;141:1043–51. Available from: http://www.ncbi.nlm.nih.gov/pubmed/26402479.
[41] Kubik M, Mandal R, Albergotti W, Duvvuri U, Ferris RL, Kim S. Effect of transcervical arterial ligation on the severity of postoperative hemorrhage after transoral robotic surgery. Head Neck 2017;39:1510–5.
[42] Orosco RK, Tam K, Nakayama M, Holsinger FC, Spriano G. Transoral supraglottic laryngectomy using a next-generation single-port robotic surgical system. Head Neck 2019;41:2143–7.
[43] Chan JYK, Wong EWY, Tsang RK, Holsinger FC, Tong MCF, Chiu PWY, et al. Early results of a safety and feasibility clinical trial of a novel single-port flexible robot for transoral robotic surgery. Eur Arch Otorhinolaryngol 2017;274:3993–6.
[44] Nichols AC, Theurer J, Prisman E, Read NE, Berthelet E, Tran E, et al. A phase II randomized trial for early-stage squamous cell carcinoma of the oropharynx: radiotherapy versus trans-oral robotic surgery (ORATOR). J Clin Oncol 2019;37:6006.
[45] Xu MJ, Plonowska KA, Gurman ZR, Humphrey AK, Ha PK, Wang SJ, et al. Treatment modality impact on quality of life for human papillomavirus–associated oropharynx cancer. Laryngoscope 2020;130:E48–56.

[46] Park YM, Kim WS, De Virgilio A, Lee SY, Seol JH, Kim SH. Transoral robotic surgery for hypopharyngeal squamous cell carcinoma: 3-year oncologic and functional analysis. Oral Oncol 2012;48:560–6.
[47] Holsti LR. Development of clinical radiotherapy since 1896. Acta Oncol (Madr) 1995;34:995–1003.
[48] Thames HD. Early fractionation methods and the origins of the NSD concept. Acta Oncol (Madr) 1988;27:89–103.
[49] Coutard H. Roentgen therapy of epitheliomas of the tonsillar region hypopharynx and larynx from 1920 to 1926. Am J Roentegenol 1932;28:313–31.
[50] Yeh S-A. Radiotherapy for head and neck cancer. Semin Plast Surg 2010;24:127–36. Available from, http://www.ncbi.nlm.nih.gov/pubmed/22550433.
[51] Moon DH, Moon SH, Wang K, Weissler MC, Hackman TG, Zanation AM, et al. Incidence of, and risk factors for, mandibular osteoradionecrosis in patients with oral cavity and oropharynx cancers. Oral Oncol 2017;72:98–103.
[52] Rodin J, Bar-Ad V, Cognetti D, Curry J, Johnson J, Zender C, et al. A systematic review of treating recurrent head and neck cancer: a reintroduction of brachytherapy with or without surgery. J Contemp Brachytherapy 2018;10:454–62.
[53] Narayana A, Cohen GN, Zaider M, Chan K, Lee N, Wong RJ, et al. High-dose-rate interstitial brachytherapy in recurrent and previously irradiated head and neck cancers-preliminary results. Brachytherapy 2007;6:157–63.
[54] Bortfeld T. IMRT: a review and preview. Phys Med Biol 2006;.
[55] Hegde JV, Shaverdian N, Felix C, Wang PC, Veruttipong D, Hsu S, et al. Functional outcomes after de-escalated chemoradiation therapy for human papillomavirus–positive oropharyngeal cancer: secondary analysis of a phase 2 trial. Int J Radiat Oncol Biol Phys 2018;100:647–51.
[56] Eisbruch A, Schwartz M, Rasch C, Vineberg K, Damen E, Van As CJ, et al. Dysphagia and aspiration after chemoradiotherapy for head-and-neck cancer: which anatomic structures are affected and can they be spared by IMRT? Int J Radiat Oncol Biol Phys 2004;60:1425–39.
[57] Feng FY, Kim HM, Lyden TH, Haxer MJ, Worden FP, Feng M, et al. Intensity-modulated chemoradiotherapy aiming to reduce dysphagia in patients with oropharyngeal cancer: clinical and functional results. J Clin Oncol 2010;28:2732–8.
[58] Tam M, Riaz N, Kannarunimit D, Peña AP, Schupak KD, Gelblum DY, et al. Sparing bilateral neck level IB in oropharyngeal carcinoma and xerostomia outcomes. Am J Clin Oncol Cancer Clin Trials 2015;38:343–7.
[59] Ang KK, Harris J, Wheeler R, Weber R, Rosenthal DI, Nguyen-Tân PF, et al. Human papillomavirus and survival of patients with oropharyngeal cancer—NEJM. N Engl J Med 2010;363:24–35. Available from: http://www.nejm.org/doi/full/10.1056/NEJMoa0912217.
[60] Rischin D, Young RJ, Fisher R, Fox SB, Le QT, Peters LJ, et al. Prognostic significance of p16INK4Aand human papillomavirus in patients with oropharyngeal cancer treated on TROG 02.02 phase III trial. J Clin Oncol 2010;28:4142–8.
[61] Masterson L, Moualed D, Masood A, Dwivedi RC, Benson R, Sterling JC, et al. De-escalation treatment protocols for human papillomavirus-associated oropharyngeal squamous cell carcinoma. Cochrane Database Syst Rev 2014;2:CD010271.
[62] Yom SS, Gillison ML, Trotti AM. Dose de-escalation in human papillomavirus-associated oropharyngeal cancer: first tracks on powder. Int J Radiat Oncol Biol Phys 2015;93:986–8.
[63] Leeman JE, Romesser PB, Zhou Y, McBride S, Riaz N, Sherman E, et al. Proton therapy for head and neck cancer: expanding the therapeutic window. Lancet Oncol 2017;18:e254–65.
[64] Sio TT, Lin HK, Shi Q, Gunn GB, Cleeland CS, Lee JJ, et al. Intensity modulated proton therapy versus intensity modulated photon radiation therapy for oropharyngeal cancer: first comparative results of patient-reported outcomes. Int J Radiat Oncol Biol Phys 2016;95:1107–14.
[65] Van De Water TA, Lomax AJ, Bijl HP, De Jong ME, Schilstra C, Hug EB, et al. Potential benefits of scanned intensity-modulated proton therapy versus advanced photon therapy with regard to sparing of the salivary glands in oropharyngeal cancer. Int J Radiat Oncol Biol Phys 2011;79:1216–24.
[66] Identifier NCT01893307. Randomized trial of intensity-modulated proton beam therapy (IMPT) versus intensity-modulated photon therapy (IMRT) for the treatment of oropharyngeal cancer of the head and neck [Internet]. Available from: https://clinicaltrials.gov/ct2/show/NCT01893307; 2013.
[67] Holliday EB, Garden AS, Rosenthal DI, Fuller CD, Morrison WH, Gunn GB, et al. Proton therapy reduces treatment-related toxicities for patients with nasopharyngeal cancer: a case-match control study of intensity-modulated proton therapy and intensity-modulated photon therapy. Int J Part Ther 2015;2:19–28.
[68] Noel G, Gondi V. Proton therapy for tumors of the base of the skull. Chin Clin Oncol 2016;5:51.

[69] Sahgal A, Chan MW, Atenafu EG, Masson-Cote L, Bahl G, Yu E, et al. Image-guided, intensity-modulated radiation therapy (IG-IMRT) for skull base chordoma and chondrosarcoma: preliminary outcomes. Neuro-Oncology 2015;17:889–94.
[70] Coley WB. The treatment of malignant tumors by repeated inoculations of erysipelas: with a report of ten original cases. Clin Orthop Relat Res 1991;3–11.
[71] NobelPrize.org. The Nobel Prize in physiology or medicine. Nobel Media AB; 2018. p. 2019.
[72] Grandis JR, Chakraborty A, Melhem MF, Zeng Q, Tweardy DJ. Inhibition of epidermal growth factor receptor gene expression and function decreases proliferation of head and neck squamous carcinoma but not normal mucosal epithelial cells. Oncogene 1997;15:409–16.
[73] Kalyankrishna S, Grandis JR. Epidermal growth factor receptor biology in head and neck cancer. J Clin Oncol 2006;24:2666–72.
[74] Huang SM, Li J, Harari PM. Molecular inhibition of angiogenesis and metastatic potential in human squamous cell carcinomas after epidermal growth factor receptor blockade. Mol Cancer Ther 2002;1:507–14.
[75] Baselga J. The EGFR as a target for anticancer therapy – focus on cetuximab. Eur J Cancer 2001;37:16–22.
[76] Ferris RL, Jaffee EM, Ferrone S. Tumor antigen-targeted, monoclonal antibody-based immunotherapy: clinical response, cellular immunity, and immunoescape. J Clin Oncol 2010;28:4390–9.
[77] Zahavi D, AlDeghaither D, O'Connell A, Weiner LM. Enhancing antibody-dependent cell-mediated cytotoxicity: a strategy for improving antibody-based immunotherapy. Antib Ther 2018;1:7–12.
[78] Bonner JA, Harari PM, Giralt J, Cohen RB, Jones CU, Sur RK, et al. Radiotherapy plus cetuximab for locoregionally advanced head and neck cancer: 5-year survival data from a phase 3 randomised trial, and relation between cetuximab-induced rash and survival. Lancet Oncol 2010;11:21–8.
[79] Vermorken JB, Mesia R, Rivera F, Remenar E, Kawecki A, Rottey S, et al. Platinum-based chemotherapy plus cetuximab in head and neck cancer. N Engl J Med 2008;359:1116–27.
[80] Pignon JP, le Maître A, Maillard E, Bourhis J. Meta-analysis of chemotherapy in head and neck cancer (MACH-NC): an update on 93 randomised trials and 17,346 patients. Radiother Oncol 2009;92:4–14.
[81] Dasgupta S, Bhattacharya-Chatterjee M, O'Malley BW, Chatterjee SK. Inhibition of NK cell activity through TGF-β1 by down-regulation of NKG2D in a murine model of head and neck cancer. J Immunol 2005;175:5541–50.
[82] Ferris RL. Progress in head and neck cancer immunotherapy: can tolerance and immune suppression be reversed? Orl 2004;66:332–40.
[83] Topalian SL, Drake CG, Pardoll DM. Immune checkpoint blockade: a common denominator approach to cancer therapy. Cancer Cell 2015;27:450–61.
[84] Lyford-Pike S, Peng S, Young GD, Taube JM, Westra WH, Akpeng B, et al. Evidence for a role of the PD-1:PD-L1 pathway in immune resistance of HPV-associated head and neck squamous cell carcinoma. Cancer Res 2013;73:1733–41.
[85] Tam K, Schoppy DW, Shin JH, Tay JK, Moreno-Nieves U, Mundy DC, et al. Assessing the impact of targeting CEACAM1 in head and neck squamous cell carcinoma. Otolaryngol Head Neck Surg 2018;159:76–84. https://doi.org/10.1177/0194599818756627.
[86] Forster MD, Devlin MJ. Immune checkpoint inhibition in head and neck cancer. Front Oncol 2018;8:310.
[87] Ferris RL, Blumenschein G, Fayette J, Guigay J, Colevas AD, Licitra L, et al. Nivolumab for recurrent squamous-cell carcinoma of the head and neck. N Engl J Med 2016;375:1856–67.
[88] Seiwert TY, Burtness B, Mehra R, Weiss J, Berger R, Eder JP, et al. Safety and clinical activity of pembrolizumab for treatment of recurrent or metastatic squamous cell carcinoma of the head and neck (KEYNOTE-012): an open-label, multicentre, phase 1b trial. Lancet Oncol 2016;17:956–65.
[89] Harrington KJ, Ferris RL, Blumenschein G, Colevas AD, Fayette J, Licitra L, et al. Nivolumab versus standard, single-agent therapy of investigator's choice in recurrent or metastatic squamous cell carcinoma of the head and neck (CheckMate 141): health-related quality-of-life results from a randomised, phase 3 trial. Lancet Oncol 2017;18:1104–15.
[90] Cohen EEW, Soulières D, Le Tourneau C, Dinis J, Licitra L, Ahn MJ, et al. Pembrolizumab versus methotrexate, docetaxel, or cetuximab for recurrent or metastatic head-and-neck squamous cell carcinoma (KEYNOTE-040): a randomised, open-label, phase 3 study. Lancet 2019;393:156–67.

CHAPTER

2

Chemotherapy: Where we have been and where we are going

Alison G. Chang, Deborah J. Wong

Division of Hematology and Oncology, David Geffen School of Medicine at University of California Los Angeles, Los Angeles, CA, United States

Abstract

The treatment of squamous cell carcinomas of the head and neck should include a multidisciplinary approach with surgeons, radiation oncologists, medical oncologists, and allied health specialists. For early stage disease the treatment entails surgical resection or radiotherapy (RT) as the primary therapy, without the addition of chemotherapy. Historically, for locally advanced disease, cytotoxic chemotherapy with cisplatin is often incorporated into the multimodal treatment regime in the postoperative setting in combination with RT, or concurrently with RT as definitive therapy. In select cases, chemotherapy with a platinum-based regimen may also be given as induction therapy prior to definitive treatment.

For patients with recurrent or metastatic disease that are not amenable to surgery or radiation, treatment with single- or multiagent cytotoxic chemotherapy had long been the only option for palliative therapy. The treatment resulted in only modest response rates of 10%–20% for single-agent chemotherapy with no definite improvement in overall survival. The addition of cetuximab to cytotoxic chemotherapy resulted in a modest improvement in outcomes for patients with metastatic disease but with significant toxicity. More recently, first-line treatment with the combination of platinum, 5-FU, and the anti-programmed death-1 (PD-1) antibody pembrolizumab has further improved survival rates regardless of the programmed death ligand 1 (PD-L1) status and has been a recent practice-changing development for patients with this disease.

Abbreviations

5-FU	5-fluorouracil
AE	adverse event
AUC	area under the curve
CCRT	concurrent/concomitant chemoradiotherapy
CI	confidence interval
CIV	continuous IV
CPS	combined positive score
DFS	disease free survival
ENE	extranodal extension

https://doi.org/10.1016/B978-0-12-820679-9.00002-5

HPV	human papillomavirus
HR	hazard ratio
LVI	lymphovascular invasion
NCCN	National Comprehensive Cancer Network
OS	overall survival
PD-1	programmed death 1
PD-L1	programmed death ligand 1
PFS	progression free survival
PNI	perineural invasion
RT	radiation therapy
SCCHN	squamous cell carcinoma of the head and neck

Conflict of interest

No potential conflicts of interest were disclosed.

Introduction

Squamous cell carcinomas of the head and neck (SCCHN) most commonly arise from the oral cavity, oropharynx, hypopharynx, and larynx, and much of the data on the role of chemotherapy in the optimal treatment of head and neck cancers focus on SCCHN arising from these primary sites. Local therapy with surgery or radiation therapy (RT) is typically used with early stage I and II diseases. However, given the anatomic complexity of cancers arising in the head and neck, with stage III and IV locoregional diseases, the management often requires a multidisciplinary input and collaboration with surgeons, radiation oncologists, and medical oncologists, and a combined modality treatment approach is typically needed given the greater risk of locoregional recurrence and distant metastases. Recurrent and metastatic disease without surgical or radiation options are treated with palliative systemic therapy or best supportive care.

Chemotherapy agents with activity in squamous cell carcinoma of the head and neck (SCCHN) include platinums such as cisplatin and carboplatin, taxanes like docetaxel and paclitaxel, and antimetabolites such as fluorouracil and methotrexate. These can be used as single agents or in combination, with targeted therapy or immunotherapy. Toxicities are greater with the multidrug regimens (Table 1).

TABLE 1 Systemic therapy for nonnasopharyngeal SCCHN.

Primary systemic therapy and concurrent RT	
Category 1	Cisplatin 100 mg/m^2 every 3 weeks for 3 cycles [1]
	Carboplatin/5-FU every 3 weeks for three cycles [2] Carboplatin 70 mg/m^2/day on days 1–4 5-FU 600 mg/m^2/day on days 1–4
Category 2B	Cisplatin 40 mg/m^2 weekly [3,4]
	Cetuximab 400 mg/m^2 initial dose then 250 mg/m^2 weekly for 7 weeks [5]
	Carboplatin/paclitaxel weekly [6] Carboplatin 100 mg/m^2 on day 1 Paclitaxel 45 mg/m^2 on day 1

TABLE 1 Systemic therapy for nonnasopharyngeal SCCHN—cont'd

Primary systemic therapy and concurrent RT	
	Cisplatin/paclitaxel weekly [7] Cisplatin 20 mg/m^2 on day 1 Paclitaxel 30 mg/m^2 on day 1
	Cisplatin/5-FU every 2 weeks for seven cycles [8] Cisplatin 60 mg/m^2 on day 1 5-FU 800 mg/m^2/day CIV on days 1–5
	5-FU/hydroxyurea [7] 5-FU 800 mg/m^2/day with each daily fraction of radiation Hydroxyurea 1 g every 12 h for 11 doses per radiation cycle
Postoperative chemoradiation	
Category 1 for high-risk[a] nonoropharyngeal cancers	Cisplatin 100 mg/m^2 every 3 weeks for three cycles [9–11]
Induction/sequential chemotherapy	
Category 1 if induction is chosen	Docetaxel/cisplatin/5-FU (TPF) [12] Docetaxel 75 mg/m^2 every 3 weeks for three cycles Cisplatin 100 mg/m^2 every 3 weeks for three cycles 5-FU 1000 mg/m^2/day for 4 days every 3 weeks for three cycles
Category 2B	Following induction, CCRT with weekly carboplatin, cisplatin, or cetuximab
Recurrent, unresectable, or metastatic without surgery or RT option, first line	
Category 1	Platinum (cisplatin or carboplatin)/5-FU/cetuximab (EXTREME) every 3 weeks for six cycles [13] Cisplatin 100 mg/m^2 or carboplatin AUC 5 on day 1 5-FU 1000 mg/m^2/day CIV on days 1–4 Cetuximab 400 mg/m^2 initial dose then 250 mg/m^2 weekly
Suggested OS advantage over EXTREME	Pembrolizumab/platinum (cisplatin or carboplatin)/5-FU every 3 weeks [14] Pembrolizumab 200 mg on day 1 for 24 months Cisplatin 100 mg/m^2 or carboplatin AUC 5 on day 1 for six cycles 5-FU 1000 mg/m^2/day for 4 days on day 1 for six cycles
Recurrent, unresectable, or metastatic without surgery or RT option, second line	
	Nivolumab 240 mg every 2 weeks or 480 mg every 4 weeks [15]
	Pembrolizumab 200 mg every 3 weeks [16–18]
Recurrent, unresectable, or metastatic without surgery or RT option, subsequent lines	
	Combination therapy: cisplatin/cetuximab, platinum (cisplatin or carboplatin)/taxane (docetaxel or paclitaxel), cisplatin/5-FU, platinum (cisplatin or carboplatin)/taxane (docetaxel or paclitaxel)/cetuximab
	Single agents: cisplatin, carboplatin, paclitaxel, docetaxel, 5-FU, methotrexate, cetuximab, capecitabine
Category 2B	Afatinib

[a] *High-risk includes extranodal extension and/or positive margins.*
Adapted from the National Comprehensive Cancer Network (NCCN) Guidelines, version 2.2019.

In locally advanced cancers, chemotherapy has been used concurrently with RT in the postoperative adjuvant setting, as primary definitive treatment with RT, or in combination as induction chemotherapy prior to surgery and/or RT. The decisions regarding the modalities and sequence of therapy are influenced by tumor location and burden, patient's performance status, preexisting comorbidities, and potential treatment-related morbidities.

Chemotherapy can also be used a part of a palliative systemic regimen for metastatic or recurrent disease without options for definitive therapy. Agents have been used as monotherapy, in combination with targeted therapy, or more recently, paired with immunotherapy.

Although HPV-related oropharyngeal cancers have been recognized as a biologically distinct disease with a better prognosis than non-HPV related SCCHN, the current standard treatment approach for HPV-related oropharyngeal cancer remains the same as non-HPV related SCCHN. Additional details on the HPV-related oropharyngeal cancers will be presented in Chapter 5.

Nasopharyngeal carcinoma (NPC) is a subtype of SCCHN that is pathophysiologically distinct from SCCHN arising from the oral cavity, oropharynx, hypopharynx, and larynx. For local or locoregional disease, due to its deep anatomic location, surgery is usually not utilized. RT is the main therapeutic modality. The addition of chemotherapy to RT is utilized for patients with stage II and greater disease. Multiagent chemotherapy is the mainstay of therapy for metastatic NPC in patients with good performance status.

Chemotherapy in locally advanced SCCHN

Postoperative concurrent chemoradiotherapy

In patients who undergo surgical resection but have positive surgical margins or extranodal extension (ENE), the addition of adjuvant chemotherapy delivered concurrently with RT, also known as concurrent chemoradiotherapy (CCRT), improves the outcomes compared with those who receive postoperative RT alone. Historically the role of adjuvant CCRT was explored in those patients with features of high-risk disease including positive margins, pT3 or pT4 primary, pN2 or pN3 nodal disease, extranodal extension (ENE), perineural invasion (PNI), and lymphovascular invasion (LVI). The two largest randomized phase III trials comparing postoperative CRRT with RT in high-risk patients are the European Organization for Research and Treatment of Cancer (EORTC) 22,931 and the Radiation Therapy Oncology Group (RTOG) 9501 trials. In the EORTC 22931, 334 patients with locally advanced SCCHN with the aforementioned high-risk features were randomized to receive CCRT with cisplatin 100 mg/m^2 every 3 weeks or RT alone. Patients in the CCRT group had improved 5-year progression free survival (PFS, 47% vs. 36%), locoregional control (82% vs. 69%), and overall survival (OS, 53% vs. 40%) compared with patients in the postoperative RT alone group [19]. Similarly the RTOG 9501 randomized 459 patients to CCRT with cisplatin 100 mg/m^2 every 3 weeks or RT alone. The initial RTOG 9501 results at 46 months showed that CCRT improved locoregional control (82% vs. 72%) and DFS (54% vs. 45%) but not OS [20]. However, the benefit seen in DFS and locoregional control disappeared at 10 years [21]. Unplanned subgroup analyses of both trials showed that in patients with resected SCCHN with adverse features of extranodal extension (ENE) and/or positive margins, CCRT with high-dose cisplatin 100 mg/m^2

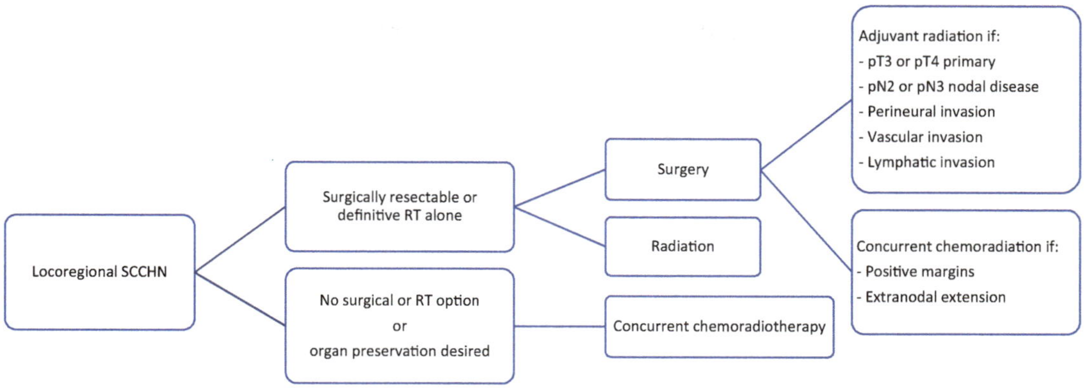

FIG. 1 General management of locoregional SCCHN.

every 3 weeks for 3 cycles improved locoregional control, DFS, and OS over RT alone [22]. Therefore postoperative CCRT is the standard for patients with positive surgical margins or ENE (Fig. 1).

High-dose cisplatin is highly emetogenic and can cause other acute and long-term toxicities, including ototoxicity, nephrotoxicity, and peripheral neuropathy. Due to concerns for toxicity, lower but more frequent dosing of cisplatin 30 or 40 mg/m^2 administered weekly is also commonly utilized in practice. A comparison of cisplatin 30 mg/m^2 weekly versus standard every 3-week dosing with cisplatin 100 mg/m^2 in 300 patients demonstrated fewer grade 3 or higher adverse events (AEs) (71.6% vs. 84.6%) but an inferior 2-year locoregional control rate of 58.5% versus 73.1% [23]. The patient population included those receiving adjuvant postoperative CCRT and definitive CCRT, with over 84% of the 300 patients with oral cavity cancers and over 90% receiving CCRT in the postoperative adjuvant setting. The results from JCOG1008, an ongoing phase II/III trial in Japan comparing higher dose cisplatin 40 mg/m^2 weekly with cisplatin 100 mg/m^2 every 3 weeks in the postoperative setting, are pending [24]. Until then the standard-of-care adjuvant therapy in postoperative patients with high-risk pathologic features remains concurrent high-dose cisplatin and RT for patients who are fit enough to tolerate it.

Definitive combined chemoradiotherapy

For patients who have locally advanced but unresectable disease or who wish to pursue a nonsurgical, organ-preserving approach, the combined modality treatment with chemotherapy and RT can be curative. While the standard treatment entails definitive CCRT with cisplatin, historically, induction or neoadjuvant chemotherapy prior to surgery and/or RT, definitive CCRT, and sequential chemoradiotherapy with induction chemotherapy followed by CCRT have all been evaluated as curative intent treatment regimens. The following section summarizes the data evaluating chemotherapy and radiotherapy as definitive treatments for locally advanced SCCHN.

Definitive concurrent chemoradiotherapy (CCRT)

The addition of concurrent chemotherapy to radiotherapy was associated with an improvement in 5-year OS of 6.5%, a 12%–13% improvement in locoregional control, and about a 3% decrease in distant metastases compared with RT alone, but the magnitude of benefit also appears to decline with advancing age [25]. A variety of regimens including chemotherapy, monotherapy, and combination therapy with RT have been studied. High-dose cisplatin 100 mg/m^2 every 3 weeks concurrently with RT is the current standard of care for those patients with good performance status. Alternatives include weekly cisplatin, cisplatin and paclitaxel, cisplatin and 5-FU, carboplatin and 5-FU, carboplatin and paclitaxel, and 5-FU and hydroxyurea. For patients with renal insufficiency, hearing loss, peripheral neuropathy, or other contraindications to platinum therapy, concurrent cetuximab can be used.

A 2003 phase III Intergroup trial randomized patients with unresectable SCCHN to daily fractionated RT alone, daily fractionated RT with 3 cycles of concurrent cisplatin 100 mg/m^2 every 3 weeks, or a split course of RT with 3 cycles of 5-FU 1000 mg/m^2/day for 4 days and cisplatin 75 mg/m^2 every 4 weeks [26]. Surgical resection was allowed and encouraged in the split course RT group. Complete response (CR) rates were higher in the two CCRT arms: 27.4%, 40.2%, and 49.4% for RT alone, CCRT with cisplatin, and CCRT with cisplatin and 5-FU, respectively. Furthermore, at a median follow-up of 41 months, the 3-year OS was significantly better for CCRT with high-dose cisplatin (37%, $P = .014$) versus daily fractionated RT alone (23%). Despite a higher CR rate, there was no overall survival advantage seen in the group who received split course with 5-FU and cisplatin (3-year OS rate 27%). Unsurprisingly, toxicity was greater with CCRT with higher rates of myelosuppression, mucositis, nausea, and vomiting compared with RT alone.

The RTOG 91–11 trial also showed superior outcomes with CCRT for laryngeal preservation [27] compared with induction chemotherapy followed by RT, or RT alone. Five hundred forty-seven patients with locally advanced laryngeal cancer received up to three cycles of PF induction chemotherapy followed by RT, CCRT with three cycles of cisplatin 100 mg/m^2 every 3 weeks, or RT alone. At 2 years, the laryngeal preservation rates were 75% for the induction chemotherapy group, 88% ($P = .005$) with cisplatin CCRT, and 70% ($P < .001$) with RT alone. The locoregional disease control was also better with CCRT at 78% as opposed to 61% in the induction group and 56% in the RT alone group. Similar to the Intergroup study, 82% of patients in the CCRT group had grade 3 or greater toxicities, with mucosal toxicity occurring at almost twice the rate compared with the other groups. At 10-year follow-up, laryngectomy-free survival rates were better in the arms with chemotherapy with HR 0.75 (95% CI, 59%–95%; $P = .02$) in the induction chemotherapy group and HR 0.78 (95% CI, 78%–98%; $P = .03$) in the CCRT group versus RT alone [28]. OS did not differ significantly among the arms. Based on the data from the 2003 phase III Intergroup study and the RTOG 91-11, CCRT with cisplatin is the standard definitive treatment for locally advanced, nonresectable SCCHN.

As in the postoperative adjuvant CCRT setting, alternate dosing schedules for cisplatin have been evaluated; however, there are a limited prospective, randomized studies directly comparing outcomes with high dose with every 3 week dose of cisplatin 100 mg/m^2 to weekly cisplatin 40 mg/m^2. A phase I/II trial using the hyperfractionated accelerated RT approach with weekly cisplatin in patients with locally advanced SCCHN of the oropharynx and hypopharynx determined a 2-year OS rate of 67% [29]. However, the majority of patients were only able to complete four cycles of cisplatin chemotherapy due to mucosal toxicity. Another phase II study conducted in Spain also used a boost RT regimen with four cycles of concurrent weekly cisplatin 40 mg/m^2 [30]. Of the 94 participants, 95% completed the entire protocol. Major toxicities included mucositis, but there was minimal myelosuppression.

Though there are limited prospective data, a metaanalysis of 52 prospective trials with over 4200 patients examining the high dose and weekly dose schedules in both the adjuvant and definitive settings did not show a significant survival difference in outcomes between high-dose and weekly cisplatin [31]. Specifically, for definitive CCRT, weekly cisplatin was associated with less myelosuppression, nausea, vomiting, and nephrotoxicity than three weekly cisplatin but did not reach significance for mucositis. There were also no significant differences in the overall response rate, complete response rates, or overall survival. Similar conclusions were made in a retrospective cohort study of 2901 Veterans Affairs patients and a metaanalysis of 39 definitive CCRT trials with a total of 3668 patients [32,33].

Beyond CCRT with cisplatin alone, the combination therapy with cisplatin has also been utilized concurrently with RT. The RTOG 97–03 randomized patients with locally advanced SCC of the oral cavity, oropharynx, or hypopharynx to definitive CCRT with cisplatin 10 mg/m^2 daily and 5-FU 400 mg/m^2/day for the final 10 days of treatment; hydroxyurea 1 g every 12 h and 5-FU 800 mg/m^2/day with each fraction of radiation every other week; or weekly paclitaxel 30 mg/m^2 and cisplatin 20 mg/m^2 [34]. The estimated 2-year overall survival rates were 57.4% for cisplatin and 5-FU, 69.4% for hydroxyurea and 5-FU, and 66.6% for cisplatin and paclitaxel.

An alternate cisplatin and 5-FU dosing regimen is listed as a treatment option in the NCCN guidelines. This was derived from a study evaluating concurrent chemoradiotherapy with cisplatin 60 mg/m^2 on day 1 and 5-FU 800 mg/m^2 CIV on days 1–5 [35]. The 5-year OS was 43% with lower failure rates in patients with tongue and glottic malignancies.

For patients who are not appropriate candidates for cisplatin, CCRT with carboplatin-based regimens are an option. In the 94–01 French Head and Neck Oncology and Radiotherapy Group trial, patients with locally advanced oropharyngeal carcinoma were randomized to RT alone or CCRT with three cycles of carboplatin 70 mg/m^2/day for 4 days and 5-FU 600 mg/m^2/day for 4 days every 3 weeks [36]. Five-year OS, DFS, and locoregional control rates were 22% and 16% ($P = .05$), 27% and 15% ($P = .01$), and 48% and 25% ($P = .002$) for CCRT and RT alone, respectively, with each outcome favoring CCRT. Fifty-six percent of patients receiving CCRT had grade 3 or greater AEs compared with 30% for the group who received RT alone.

In addition, the benefit of CCRT was not overcome by an accelerated RT schedule. In the GORTEC 99-02 study, 840 participants were randomized to one of three arms: CCRT with conventionally fractionated RT (70 Gy over 7 weeks) with carboplatin and 5-FU, CCRT with accelerated RT (70 Gy in 6 weeks) with carboplatin and 5-FU, or very accelerated RT alone (64.8 Gy in 3.5 weeks) [37]. For the primary end point, 3-year PFS was 37.6% (95% CI, 32.1%–43.4%) with conventional CCRT, 34.1% (95% CI, 28.7%–39.8%) with accelerated CCRT, and 32.2% (95% CI, 27.0%–37.9%) with very accelerated RT alone. Compared with patients in the very accelerated RT alone group, the patients who received conventional CCRT had a statistically significant improvement in PFS (HR 0.82, 0.67–0.99, $P = .041$).

Weekly paclitaxel 40–45 mg/m^2 and carboplatin 100 mg/m^2 with concurrent daily RT is another consideration. In this trial, 98% of patients were able to complete the protocol with a CR rate of 75% and a 3-year OS of 48% [38]. The most frequent grade 3 toxicities were mucositis (70%), dysphagia (33%), leukopenia (30%), and skin desquamation (25%).

Finally, for patients who are not platinum eligible, the epidermal growth factor receptor (EGFR) inhibitor cetuximab can be used concurrently with RT. However, recent data in patients with low-risk HPV-related oropharyngeal SCCHN demonstrated that concurrent cetuximab-RT is inferior to standard high-dose cisplatin-RT. In the European phase III De-ESCALaTE HPV trial, 334 nonsmokers and smokers with less than a 10 pack-year history with locally advanced HPV-positive SCCHN were randomized to receive CCRT with cisplatin 100 mg/m^2 weekly for three cycles or cetuximab 400 mg/m^2 loading dose followed by 250 mg/m^2 weekly [39]. The primary end point was rate of grade 3 or greater AEs. At 24 months the frequency of toxicities was not significantly different between the two groups, with a mean of 4.8 events per patient in both groups. There was a greater incidence of myelosuppression and gastrointestinal toxicity in the cisplatin group and more skin toxicity in the cetuximab group. However, the 2-year OS rate was 97.5% with cisplatin versus 89.4% with cetuximab (HR 5.0 [95% CI, 1.7 to 14.7; P = .001]). Disease recurrence was also more frequent in the cetuximab than in the cisplatin group, 16.1% versus 6.0%, (HR 3.4 [95% CI, 1.6–7.2; P = .0007]).

The RTOG 1016 also demonstrated an inferior OS with concurrent cetuximab-RT compared with concurrent cisplatin-RT in patients with HPV-related oropharyngeal SCC [40]. In this study, 849 participants were randomized to receive either cetuximab at a loading dose of 400 mg/m^2 5–7 days prior to the start of RT then cetuximab 250 mg/m^2 weekly for 7 weeks or cisplatin 100 mg/m^2 every 3 weeks for two cycles with concurrent RT. The estimated 5-year OS was 77.9% (95% CI, from 73.4% to 72.2%) for cetuximab and 84.6% (95% CI, 80.6%–88.6%) for cisplatin. Thus the cetuximab arm did not meet noninferiority criteria. In addition, locoregional failure was higher for cetuximab than cisplatin, 17.3% versus 9.9%, respectively. As in the De-ESCALaTE HPV trial, toxicity rates were similar between the two groups at approximately 77.5%. Given the data from both the De-ESCALaTE HPV and RTOG 1016, cisplatin-RT remains the standard of care for platinum-eligible patients with locally advanced SCCHN, even in patients with low-risk HPV-related oropharyngeal carcinoma.

Induction chemotherapy

In a metaanalysis of 14 randomized controlled trials dating back to the 1960s, cisplatin-based induction chemotherapy when added to a definitive locoregional therapy (surgery, RT, CCRT, or surgery followed by adjuvant RT or CCRT) decreased the rate of distant metastases by 7% (95% CI, 0%–13%; P = .05) [41]. However, despite excellent response rates, in a metaanalysis of 93 trials, the addition of induction chemotherapy to locoregional treatment did not improve the locoregional control or the OS over platinum-based CCRT [25]. Indeed a greater OS benefit was seen with the addition of chemotherapy to the locoregional treatment in the CCRT group (HR 0.81 [0.78–0.86]) than in the group who received induction chemotherapy followed by locoregional treatment (HR 0.96 [0.90–1.02]; P < .0001). Thus induction chemotherapy has not routinely been utilized. However, multidrug combinations of chemotherapy have been utilized as induction therapy in patients with good performance status who have higher risk of distant metastases, bulky N3 disease, a delay in surgery or RT, and/or desire for an organ-preserving treatment plan.

The induction chemotherapy regimens commonly incorporate a combination of platinum, taxane, and fluorouracil chemotherapies based on several studies comparing the treatment with a platinum and 5FU doublet to triple therapy with platinum, 5FU, and taxane. In a 2005 phase III study from Spain, patients were randomized to paclitaxel 175 mg/m^2 on day 1, cisplatin 100 mg/m^2 on day 2, and 5-FU 500 mg/m^2 continuous IV (CIV) over 24 h on day 2–6 (PCF), or cisplatin 100 mg/m^2 and 5-FU 1000 mg/m^2 CIV over 24 h for 5 days (CF) every 21 days for three cycles [42]. Those with a complete response (CR) then received CCRT with cisplatin 100 mg/m^2 every 3 weeks for three cycles. At a median follow-up of 23 months, the OS rates were 51% with PCF and 43% with CF ($P = .063$). The CR rate was 14% in the CF arm versus 33% in the PCF arm ($P < .001$). Rates of grade 2–4 mucositis were greater in the CF group than PCF group: 53% versus 15% ($P < .001$).

TAX 323 was a European trial that randomized patients with stage III or IV locally advanced SCCHN to docetaxel 75 mg/m^2, cisplatin 75 mg/m^2, 5-FU 750 mg/m^2/day CIV over 24 h for 5 days (TPF), or cisplatin 100 mg/m^2 and 5-FU 1000 mg/m^2/day every 3 weeks for four cycles followed by RT alone [43]. At a median follow-up of 32.5 months, patients treated with TPF had a relative reduction of 28% in the risk of disease progression or death, as compared with PF. The median PFS was 11.0 months in the TPF group and 8.2 months in the PF group (HR 0.72; $P = .007$). The most common AEs were myelotoxicity, alopecia, stomatitis, infections, nausea, vomiting, anorexia, diarrhea, and hearing loss. Grade 3 or greater neutropenia occurred in 76.9% versus 52.5% in patients receiving TPF and PF, respectively. However, mortality rates from toxicity were 2.3% in the TPF group and 5.5% in the PF group.

In the US TAX 324 trial, patients were randomized to receive induction chemotherapy with either docetaxel 75 mg/m^2, cisplatin 100 mg/m^2, and 5-FU 1000 mg/m^2/day CIV over 24 h for 4 days (TPF) or cisplatin 100 mg/m^2 and 5-FU 1000 mg/m^2/day CIV over 24 h for 5 days (PF) every 3 weeks for three cycles followed by weekly carboplatin AUC 1.5 and concurrent RT [44]. The overall response rates with the induction chemotherapy was 72% in the TPF group and 64% in the PF group ($P = .07$). The 3-year OS was 62% and 48% for TPF versus PF, respectively (HR 0.70; $P = .006$). However, 99% of the patients had at least one AE, and rates of grade 3 or greater neutropenia was much higher with TPF than PF (83% vs. 56%).

The GORTEC trial compared induction chemotherapy with PF versus TAX 323 TPF in patients with laryngeal and hypopharyngeal cancers [45]. The primary end point was laryngeal preservation. Patients received three cycles of either PF or TPF. If patients had a response to induction therapy, they received RT with or without concurrent chemotherapy, and if they did not have a response, they received a total laryngectomy followed by adjuvant radiation with or without concurrent chemotherapy. Overall the response rate was higher in the TPF group versus the PF group (80.0% vs. 59.2%), respectively. The 3-year larynx preservation rate was 70.3% with TPF versus 57.5% with PF. These effects were maintained in the 10-year follow-up [46]. Larynx preservation rates were 70.3% (95% CI, 58%–80%) with TPF versus 46.5% (95% CI, 31%–63%) with PF. However, there was no statistically significant difference in locoregional control rates, DFS, or OS.

Sequential chemoradiotherapy

While the aforementioned induction chemotherapy studies compared two different induction regimens with locoregional RT or CCRT, a few studies have attempted to determine prospectively whether adding induction chemotherapy to CCRT, termed sequential chemoradiotherapy, improved outcomes compared with definitive CCRT alone. Of note, if CCRT is used after induction chemotherapy, high-dose cisplatin is not recommended due to significant toxicity per NCCN guidelines. In the phase II Southwestern Oncology Group (SWOG) 0216 trial, 5% of patients died from treatment-related complications and 91% of patients experienced at least one grade 3 or greater toxicity during concurrent chemoradiotherapy with cisplatin 100 mg/m^2 every 3 weeks for two cycles after completing two cycles of induction TPF [47]. The subsequent phase II TREMPLIN study randomized patients with locally advanced laryngeal or hypopharyngeal squamous cell carcinoma to either CCRT with cisplatin 100 mg/m^2 every 3 weeks for three cycles or cetuximab 400 mg/m^2 loading dose then 250 mg/m^2 weekly after three cycles of induction TPF [48]. At a median follow-up of 36 months, the OS was not significantly different between the two arms: 75% (95% CI, 62%–85%) for cisplatin and 73% (95% CI, 60%–84%) for cetuximab. The larynx preservation rate at 3 months was also not significantly different, 95% (95% CI, 86%–98%) for cisplatin versus 93% (95% CI, 83%–97%) for cetuximab. However, there was increased toxicity in the group treated with concurrent cisplatin, with 57% of patients requiring protocol modifications due to toxicity, compared with 29% in the group who received cetuximab.

The PARADIGM trial sought to determine whether sequential chemoradiotherapy with TPF improved OS compared with CCRT alone [49].Treatment-naïve patients with unresectable locally advanced SCCHN were randomized to CCRT with two cycles of high-dose cisplatin every 3 weeks or to three cycles of induction TPF followed by CCRT with weekly chemotherapy. Participants in the sequential chemoradiotherapy arm received concurrent weekly docetaxel 20 mg/m^2 for 4 weeks if they had a poor response, partial response, or stable disease with TPF. Those with a complete response received weekly carboplatin AUC 1.5 for 7 weeks as in the TAX 324 trial. The study was terminated early due to slow accrual, but after a median follow-up of 49 months, 3-year OS was 78% (95% CI, 66%–86%) in the CCRT group and 73% (95% CI, 60%–82%) in the sequential chemoradiotherapy group (HR 1.09; $P = .77$). Although there was no significant difference between the arms, the authors concluded that induction chemotherapy was still an appropriate approach in patients at high risk for local or distant failure.

In the DeCIDE trial, treatment-naïve patients with N2 or N3 disease were randomized to receive either twice daily RT every other week with docetaxel 25 mg/m^2, fluorouracil 600 mg/m^2/day for 5 days, and hydroxyurea 500 mg every 12 h for 6 days or two cycles of induction TPF followed by CCRT with the same three chemotherapy drugs [50]. At 30 months follow-up, there were no statistically significant differences in rates of distant failure-free survival, recurrence-free survival, or overall survival.

In summary, when chemotherapy is given prior to definitive locoregional therapy for locally advanced SCCHN, docetaxel, cisplatin, and 5-FU is the standard induction regimen. However, despite a modest improvement in the rate of distant metastases of 7% and overall response rates up to 80%, the addition of induction chemotherapy to CCRT has not shown an OS or locoregional control benefit over CCRT alone. Nevertheless, induction chemotherapy has been utilized in patients with high risk for distant metastases, bulky disease, significant delays in RT, and/or desire for laryngeal preservation. If CRRT is pursued as the local therapy after completion of induction chemotherapy, concurrent weekly carboplatin, cisplatin, or cetuximab is recommended over high-dose cisplatin every 3 weeks due to significant toxicity of high-dose cisplatin.

Systemic therapy for recurrent, unresectable, or metastatic disease

Unfortunately, up to 50% of patients with locally advanced SCCHN will develop recurrent/metastatic disease. The choice of regimen for patients with recurrent or metastatic disease without surgical or RT options should be based on performance status, comorbidities, symptoms, and prior lines of therapy. For patients without prior systemic treatment, options include combinations of cytotoxic chemotherapy, chemotherapy with targeted therapy, and immunotherapy with or without chemotherapy. In these cases, tumor tissue programmed death ligand 1 (PD-L1) combined positive score (CPS), defined as the proportion of PD-L1-positive cells (tumor cells, lymphocytes, and macrophages) compared with viable tumor cells, can also help determine eligibility for immunotherapy monotherapy or combination with cytotoxic chemotherapy as frontline treatment. The optimal regimen for patients who have previously received systemic therapy is not clear and can be determined by performance status, comorbidities, prior regimens, response to prior therapies, and time to relapse.

Palliative single agent cytotoxic chemotherapy has a response rate around 10%–20% in recurrent or metastatic SCCHN. Monotherapy with cisplatin 100 mg/m^2 has an ORR 17% and that with 5-FU 1000 mg/m^2/day CIV over 4 days is 13% [51]. Capecitabine 1250 mg/m^2 twice daily for 2 weeks on the 1 week off had an ORR of 24.2% with most common AEs being myelosuppression, fatigue, palmar-plantar erythrodysesthesia, mucositis, dysphagia, and diarrhea [52]. Methotrexate 40 g/m^2 weekly had an ORR of 10% but did not have a significantly improved OS over cisplatin or carboplatin with 5-FU [53]. Docetaxel 40 mg/m^2 weekly had greater ORR 27% versus 15% for methotrexate, but again, OS was not significantly different [54]. Response rates to single agent chemotherapy are similar to responses to cetuximab monotherapy, which has about a 13% response rate [55].

Combination therapy is often considered in patients with good performance status given improved response rates over single agent therapy. Doublet chemotherapy regimens include platinum and taxane, cisplatin and 5-FU, and cisplatin and cetuximab. Carboplatin AUC 6 and docetaxel 65 mg/m^2 every 3 weeks had an ORR of 25% and median OS of 7.4 months [56]. Cisplatin 100 mg/m^2 with 5-FU 1000 mg/m^2/day CIV for 4 days every 3 weeks and cisplatin 75 mg/m^2 with paclitaxel 175 mg/m^2 every 3 weeks had similar median survival of 8.7 and 8.1 months, respectively [57]. A phase III study comparing cisplatin 100 mg/m^2 every 4 weeks to cisplatin at same dosing with cetuximab loading dose of 400 mg/m^2 followed by 250 mg/m^2 weekly showed an improved ORR of 26% versus 10% and improved median OS of 9.2 versus 8.0 months with the addition of cetuximab [58].

The phase III EXTREME trial was the first triplet therapy to demonstrate a statistically significant improvement in overall survival compared with doublet therapy [59]. Four hundred forty-two patients with untreated recurrent or metastatic SCCHN were randomized to cisplatin 100 mg/m^2 or carboplatin AUC 5 on day 1, fluorouracil 1000 mg/m^2/day for 4 days every 3 weeks for up to six cycles or the same chemotherapy with cetuximab 400 mg/m^2 initially then 250 mg/m^2 weekly for up to six cycles. At a median follow-up of 18 months, median OS was 7.4 months in the chemotherapy-alone group as compared with 10.1 months in the chemotherapy-cetuximab group (HR 0.80 [95% CI, 0.64–0.99; P = .04]). There were similar rates of grade 3 or 4 AEs between the two arms, but there was a trend toward increased frequency of sepsis, hypomagnesemia, skin rash, and infusion reactions with cetuximab. Given the improved OS, combination therapy with platinum, 5FU and cetuximab has been a standard first-line chemotherapy-targeted therapy combination regimen for many years.

In 2019 the US Food and Drug Administration approved pembrolizumab, either as monotherapy or, in combination with platinum and 5FU, as frontline treatment for recurrent or metastatic SCCHN. This approval was based on data from the practice-changing KEYNOTE-048 study, wherein 882 patients with unresectable, recurrent, or metastatic SCCHN who had not previously received systemic therapy were randomized to three groups: pembrolizumab 200 mg/m^2 every 3 weeks, pembrolizumab 200 mg/m^2 every 3 weeks plus platinum (cisplatin 100 mg/m^2 or carboplatin AUC 5 every 3 weeks), and 5-FU 1000 mg/m^2/day for 4 days every 3 weeks, or treatment as per the EXTREME regimen with cetuximab (400 mg/m^2 loading dose then 250 mg/m^2 weekly) plus platinum and 5-FU for up to 6 cycles [60]. Patients were stratified by PD-L1 CPS $\geq$20 and $\geq$1. For the pembrolizumab-chemotherapy group, compared with cetuximab-chemotherapy, with a median follow-up of 25 months, the median OS for patients with CPS $\geq$20 was 14.7 months for the pembrolizumab-chemotherapy group versus 11.0 months for cetuximab-chemotherapy (HR 0.60 [95% CI, 0.45 to 0.82; $P = .0004$]). In CPS $\geq$1, median OS was 13.6 months versus 10.4 months for pembrolizumab-chemotherapy and cetuximab-chemotherapy, respectively (HR 0.65 [95% CI, 0.53 to 0.80; $P < .0001$]). In the total population irrespective of PD-L1 status, a statistically significant improvement in median OS was also seen for pembrolizumab-chemotherapy with 13.0 months versus 10.7 months for cetuximab-chemotherapy (HR 0.77 [95% CI, 0.63–0.93, $P = .0034$]). The magnitude of improvement in overall response rate was more pronounced with CPS $\geq$20, 42.9% versus 38.2%, than with CPS $\geq$1, 36.4% versus 35.7%. For pembrolizumab monotherapy, compared with cetuximab-chemotherapy, median OS for CPS $\geq$20 was 14.8 months versus 10.7 months (HR 0.58 [95% CI, 0.44–0.78]), and in those with CPS $\geq$1, OS was 22% with 12.3 months versus 8% with 10.3 months (HR 0.74, [95% CI 0.61–0.90]). The median OS for the total population was not statistically significant between the two arms. ORR was lower with pembrolizumab at 16.9% versus 36.0%, but duration of response was longer, 22.6 versus 4.5 months.

As there was no data on patients with CPS between 1 and 20, it is not clear whether the benefit with pembrolizumab monotherapy was largely driven by the CPS $\geq$20 population. However, given the results from KEYNOTE-048, patients with PD-L1 CPS $\geq$1 can still be considered for pembrolizumab monotherapy. For those without PD-L1 expression or with rapidly progressive disease, pembrolizumab plus platinum and 5-FU is an established option for frontline treatment of recurrent or metastatic SCCHN (Fig. 2).

The addition of taxane chemotherapy has also been evaluated as cytotoxic therapy for recurrent and metastatic SCCHN. A phase II trial in 2005 examined the relative benefit of adding cetuximab 400 mg/m^2 loading then 250 mg/m^2 weekly to cisplatin 75 or 100 mg/m^2 every 3 weeks if there was stable (SD) or progressive disease (PD) after two cycles of either cisplatin 75 mg/m^2 and paclitaxel 175 mg/m^2 or cisplatin 100 mg/m^2 and 5-FU 1000 mg/m^2 [61]. If patients developed neurotoxicity or nephrotoxicity, they were changed to carboplatin AUC 6. About 18% with SD, 20% with PD after two cycles of cisplatin-based doublet, and 6% with PD within 3 months of initial therapy had a response. The median OS for the groups were 11.7, 6.1, and 4.3 months.

The phase II GORTEC 2008-03 used a different taxane, docetaxel, in combination with cisplatin and cetuximab [62]. Patients ineligible for curative locoregional treatment received

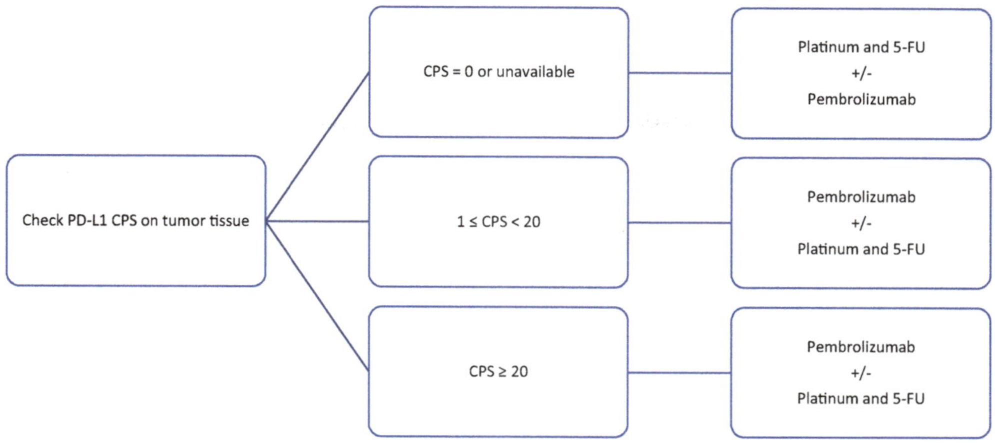

FIG. 2 Recurrent, unresectable, or metastatic disease in patients with no prior systemic therapy.

cisplatin 75 mg/m^2, docetaxel 75 mg/m^2, and cetuximab with a 400 mg/m^2 loading dose then 250 mg/m^2 weekly every 3 weeks for four cycles. Cetuximab 500 mg/m^2 every 2 weeks was continued as maintenance therapy. Of the 54 patients enrolled, 44.4% had an objective response with TPEx. The most common AEs were rash (16.6%) and neutropenia (20.4%).

Nasopharyngeal cancer

In contrast to SCCHN arising from other primary sites, due to its radiosensitivity and anatomic location, nasopharyngeal carcinoma (NPC) is usually treated with radiation rather than surgery. Keratinizing SCC (formerly WHO type I) is more common in North America and follows a pattern of locoregional recurrence. Of the nonkeratinizing subtypes, differentiated (WHO type II) and undifferentiated (WHO type III), the undifferentiated subtype is the most common in endemic areas of southern China and northern Africa. These also have strong associations with Epstein-Barr virus (EBV) and tend to have later, more distant metastases involving bone, lung, liver, and lymph nodes. Irrespective of the subtype, for NPC stage II disease and higher, the combined modality treatment with platinum-based chemotherapy is recommended (Table 2).

The role of CCRT followed by adjuvant chemotherapy was first established by the Intergroup 0099 trial [63]. Patients with treatment-naïve stage III and IV nasopharyngeal cancer were randomized to receive RT alone or CCRT with cisplatin 100 mg/m^2 every 3 weeks for three cycles followed by adjuvant cisplatin 80 mg/m^2 and 5-FU 1000 mg/m^2/day CIV for 4 days every 4 weeks for three cycles. The majority of patients had the WHO type I or keratinizing squamous cell cancer subtype. The 3-year PFS rate was 24% for the RT alone group and 69% for the CCRT and adjuvant chemotherapy group ($P < .001$). There was also improvement with CCRT in the 3-year OS rate, 46% versus 76% ($P < .001$). The median survival was 34 months with RT and was not reached with CCRT.

TABLE 2 Systemic therapy for nasopharyngeal SCCHN.

Chemotherapy followed by adjuvant chemotherapy	
Cisplatin 100 mg/m^2 every 3 weeks or 40 mg/m^2 every week with RT then cisplatin 80 mg/m^2 and 5-FU 1000 mg/m^2/day for 4 days every 4 weeks for three cycles [71,72]	
Category 2B	Cisplatin/RT then carboplatin AUC 5 and 5-FU 1000 mg/m^2/day for 4 days every 4 weeks for three cycles [73]
	Cisplatin/RT without adjuvant chemotherapy [74]
Induction/sequential chemotherapy [a]	
Category 1 if EBV+ disease, category 2A for EBV-	Docetaxel/cisplatin/5-FU every 3 weeks for three cycles [75] Docetaxel 60 mg/m^2 on day 1 Cisplatin 60 mg/m^2 on day 1 5-FU 600 mg/m^2/day CIV days 1–5
	Cisplatin/gemcitabine every 3 weeks for three cycles [76] Cisplatin 80 mg/m^2 on day 1 Gemcitabine 1000 mg/m^2 on days 1 and 8
Category 2B	Docetaxel/cisplatin every 3 weeks for two cycles [77] Docetaxel 75 mg/m^2 on day 1 Cisplatin 75 mg/m^2 on day 1
Recurrent, unresectable or metastatic without surgery or RT option, first line	
Category 1	Cisplatin/gemcitabine every 3 weeks for up to six cycles [78] Cisplatin 80 mg/m^2 on day 1 Gemcitabine 1000 mg/m^2 on days 1 and 8
Recurrent, unresectable or metastatic without surgery or RT option, second line	
	Combination therapy: cisplatin/5-FU, platinum (cisplatin or carboplatin)/taxane (docetaxel or paclitaxel), carboplatin/cetuximab
	Single agents: cisplatin, carboplatin, paclitaxel, docetaxel, 5-FU, methotrexate, gemcitabine, capecitabine
Category 2B	Nivolumab 240 mg every 2 weeks or 480 mg every 4 weeks
	Pembrolizumab 200 mg every 3 weeks

[a] *Follow induction/sequential chemotherapy with concurrent chemotherapy with RT.*
Adapted from the National Comprehensive Cancer Network (NCCN) Guidelines, version 2.2019.

Another phase III trial using concurrent weekly cisplatin 40 mg/m2 also showed a survival benefit over RT alone [64]. With a total of 350 patients and a median follow-up of 5.5 years, the 5-year OS was 58.6% with RT and 70.3% with CCRT. However, in the subgroup analysis, the improvement was only seen in patients with T3/T4 tumors (HR 0.51 [95% CI, 0.3–0.88]).

A potential alternative to concurrent cisplatin is a carboplatin-based regimen. A phase III trial compared the Intergroup 0099 regimen with CCRT with weekly carboplatin 100 mg/m^2 for 6 weeks followed by adjuvant carboplatin AUC 5 and 5-FU 1000 mg/m^2/day for 4 days every 4 weeks for three cycles [65]. With a median follow-up of 26.3 months, 59% of the 101 patients who received cisplatin and 73% of the 105 patients who received carboplatin were able to complete CCRT. Three-year OS was 77.7% (95% CI, 67.6%–87.8%) for cisplatin and 79.2% (95% CI, 69.8%–88.6%) for carboplatin ($P = .9884$). Overall there were lower incidences of nephrotoxicity, anemia, leukopenia, nausea, and vomiting but more thrombocytopenia with carboplatin. However, as follow-up was shorter than that published for other NPC trials, cisplatin is still preferred over carboplatin in appropriate patients.

The Meta-Analysis of Chemotherapy in Nasopharynx Carcinoma (MAC-NPC) collaborative group analyzed 19 trials and 4806 patients and found a survival benefit of 6.3% at 5 years with the addition of chemotherapy to RT [66]. The greatest benefit was seen with CCRT followed by adjuvant chemotherapy (HR 0.65 [95% CI, 0.56–0.76]). CCRT without adjuvant chemotherapy had a HR of 0.80 (95% CI, 0.70–0.93) compared with RT alone. There was no advantage seen with adjuvant chemotherapy or induction chemotherapy without CCRT.

An ongoing trial, NRG-HN001 (NCT02135042), is attempting to determine whether EBV DNA PCR is a predictive biomarker to select patients who would benefit from adjuvant chemotherapy. After completing CCRT, patients with stage II to IVB EBV-related NPC will be stratified by EBV DNA PCR status. If EBV DNA is undetectable, patients will be randomized to observation or adjuvant cisplatin and 5-FU. If EBV DNA is present, then patients will receive either cisplatin and 5-FU or cisplatin and gemcitabine.

Beyond CCRT followed by adjuvant chemotherapy, another option is induction chemotherapy followed by CCRT. A phase III trial published by Sun et al. randomized 480 treatment-naïve patients with stage III–IVB NPC to CCRT with cisplatin 100 mg/m^2 every 3 weeks or induction chemotherapy with docetaxel 60 mg/m2, cisplatin 60 mg/m2, and 5-FU 600 mg/m2/day CIV for 5 days every 3 weeks followed by CCRT [67]. There was an improvement 3-year failure-free survival of 80% versus 72% (HR 0.68 [95% CI, 0.48–0.97]) and OS of 92% versus 86% (HR 0.59 [95% CI, 0.36–0.95]) with the addition of induction chemotherapy. This regimen has a category 1 recommendation by the NCCN for EBV-associated NPC and a category 2A recommendation for non-EBV associated disease.

A more recent 2019 phase III trial also from China has established cisplatin and gemcitabine as an induction chemotherapy regimen prior to CCRT. In this study, 480 patients with locally advanced NPC received CCRT with cisplatin 100 mg/m^2 every 3 weeks or induction gemcitabine 1000 mg/m^2 on days 1 and 8 plus cisplatin 80 mg/m^2 on day 1 every 3 weeks for three cycles followed by CCRT with high-dose cisplatin [68]. The 3-year recurrence-free survival was 85.3% versus 76.5% (HR 0.51 [95% CI, 0.34–0.77]) for induction and CCRT alone, respectively. The OS was 94.6% with induction and 90.3% with CCRT alone (HR 0.43 [95% CI, 0.24–0.77]). Unsurprisingly, there were more grade 3 and higher myelosuppression, nausea, and vomiting events in the induction group.

Although these trials used high-dose cisplatin with RT, the general recommendations are to use concurrent weekly platinum dosing due to cumulative toxicity. At least in Asia, induction chemotherapy appears to improve survival by decreasing the rate of distant metastases. Whether these results can be generalized to other populations is unclear. In addition, long-term survival and toxicity data have not matured enough to determine whether induction or adjuvant chemotherapy improves outcomes over CCRT alone.

For recurrent, unresectable, or metastatic NPC without options for local therapy, cisplatin and gemcitabine is the preferred first-line treatment regimen. A phase III trial with 362 patients compared cisplatin 80 mg/m^2 on day 1 and gemcitabine 1000 mg/m^2 on days 1 and 8 every 3 weeks with cisplatin 80 mg/m^2 on day 1 and 5-FU 1000 mg/m^2/day for 4 days every 3 weeks for up to six cycles [69]. At a median follow-up of 19.4 months, median PFS was 7.0 months with cisplatin-gemcitabine compared with 5.6 months with cisplatin-5-FU. There was a greater incidence of leukopenia, neutropenia, and thrombocytopenia in the gemcitabine group but increased mucositis in the group who received 5-FU.

Other cytotoxic chemotherapy combinations for NPC include cisplatin and 5-FU, platinum and taxanes, and carboplatin and cetuximab. In a comparison of 5 cisplatin-based regimens including cisplatin and 5-fluororacil (PF); paclitaxel and cisplatin (TP); gemcitabine and cisplatin (GP); paclitaxel, cisplatin, and 5-fluororacil (TPF); and bleomycin, cisplatin, and 5-fluororacil (BPF), there was no significant difference in PFS or OS [70]. There were higher rates of response with GP and TPF compared with PF and TP.

Options for second-line palliative treatment are similar to that for other SCCHN. These include paclitaxel, docetaxel, methotrexate, and capecitabine.

Conclusion

The treatment of head and neck cancers has evolved over the last few decades but primarily still involves local therapy with surgery and RT when feasible. Chemotherapy does not play a role in the early stage disease. In patients with locally advanced SCCHN, chemotherapy is often incorporated into the multimodal treatment plan and administered concurrently with RT in the postoperative setting or as definitive therapy for SCCHN. In patients with recurrent and metastatic diseases that are not amenable to local therapy, multidrug or single agent chemotherapy and/or targeted therapy has long been the standard, but combination cytotoxic chemotherapy with pembrolizumab, the anti-PD-1 immunotherapy, has further improved survival rates in the frontline setting. The addition of immunotherapy to standard definitive treatment for earlier stage SCCHN is a subject of many ongoing studies.

References

[1] Adelstein DJ, Li Y, Adams GL, Wagner Jr H, Kish JA, Ensley JF, Schuller DE, Forastiere AA. An intergroup phase III comparison of standard radiation therapy and two schedules of concurrent chemoradiotherapy in patients with unresectable squamous cell head and neck cancer. J Clin Oncol 2003;21(1):92–8.

[2] Bourhis J, Sire C, Graff P, Grégoire V, Maingon P, Calais G, Gery B, Martin L, Alfonsi M, Desprez P, Pignon T, Bardet E, Rives M, Geoffrois L, Daly-Schveitzer N, Sen S, Tuchais C, Dupuis O, Guerif S, Lapeyre M, Favrel V, Hamoir M, Lusinchi A, Temam S, Pinna A, Tao YG, Blanchard P, Aupérin A. Concomitant chemoradiotherapy versus acceleration of radiotherapy with or without concomitant chemotherapy in locally advanced head and neck carcinoma (GORTEC 99-02): an open-label phase 3 randomised trial. Lancet Oncol 2012;13(2):145–53.

[3] Beckmann GK, Hoppe F, Pfreundner L, Flentje MP. Hyperfractionated accelerated radiotherapy in combination with weekly cisplatin for locally advanced head and neck cancer. Head Neck 2005;27(1):36–43.

[4] Medina JA, Rueda A, de Pasos AS, Contreras J, Cobo M, Moreno P, Benavides M, Villanueva A, Alba E. A phase II study of concomitant boost radiation plus concurrent weekly cisplatin for locally advanced unresectable head and neck carcinomas. Radiother Oncol 2006;79(1):34–8.

[5] Bonner JA, Harari PM, Giralt J, Cohen RB, Jones CU, Sur RK, Raben D, Baselga J, Spencer SA, Zhu J, Youssoufian H, Rowinsky EK, Ang KK. Radiotherapy plus cetuximab for locoregionally advanced head and neck cancer: 5-year survival data from a phase 3 randomised trial, and relation between cetuximab-induced rash and survival. Lancet Oncol 2010;11(1):21–8.

[6] Suntharalingam M, Haas ML, Conley BA, Egorin MJ, Levy S, Sivasailam S, Herman JM, Jacobs MC, Gray WC, Ord RA, Aisner JA, Van Echo DA. The use of carboplatin and paclitaxel with daily radiotherapy in patients with locally advanced squamous cell carcinomas of the head and neck. Int J Radiat Oncol Biol Phys 2000;47(1):49–56.

[7] Garden AS, Harris J, Vokes EE, Forastiere AA, Ridge JA, Jones C, Horwitz EM, Glisson BS, Nabell L, Cooper JS, Demas W, Gore E. Preliminary results of Radiation Therapy Oncology Group 97-03: a randomized phase ii trial of concurrent radiation and chemotherapy for advanced squamous cell carcinomas of the head and neck. J Clin Oncol 2004;22(14):2856–64.

[8] Taylor 4th SG, Murthy AK, Vannetzel JM, Colin P, Dray M, Caldarelli DD, Shott S, Vokes E, Showel JL, Hutchinson JC. Randomized comparison of neoadjuvant cisplatin and fluorouracil infusion followed by radiation versus concomitant treatment in advanced head and neck cancer. J Clin Oncol 1994;12(2):385–95.

[9] Bernier J, Domenge C, Ozsahin M, Matuszewska K, Lefèbvre JL, Greiner RH, Giralt J, Maingon P, Rolland F, Bolla M, Cognetti F, Bourhis J, Kirkpatrick A, van Glabbeke M, European Organization for Research and Treatment of Cancer Trial 22931. Postoperative irradiation with or without concomitant chemotherapy for locally advanced head and neck cancer. N Engl J Med 2004;350(19):1945–52.

[10] Cooper JS, Pajak TF, Forastiere AA, Jacobs J, Campbell BH, Saxman SB, Kish JA, Kim HE, Cmelak AJ, Rotman M, Machtay M, Ensley JF, Chao KS, Schultz CJ, Lee N, Fu KK, Radiation Therapy Oncology Group 9501/Intergroup. Postoperative concurrent radiotherapy and chemotherapy for high-risk squamous-cell carcinoma of the head and neck. N Engl J Med 2004;350(19):1937–44.

[11] Noronha V, Joshi A, Patil VM, Agarwal J, Ghosh-Laskar S, Budrukkar A, Murthy V, Gupta T, D'Cruz AK, Banavali S, Pai PS, Chaturvedi P, Chaukar D, Pande N, Chandrasekharan A, Talreja V, Vallathol DH, Mathrudev V, Manjrekar A, Maske K, Bhelekar AS, Nawale K, Kannan S, Gota V, Bhattacharjee A, Kane S, Juvekar SL, Prabhash K. Once-a-week versus once-every-3-weeks cisplatin chemoradiation for locally advanced head and neck cancer: a phase III randomized noninferiority trial. J Clin Oncol 2018;36(11):1064–72.

[12] Posner MR, Hershock DM, Blajman CR, Mickiewicz E, Winquist E, Gorbounova V, Tjulandin S, Shin DM, Cullen K, Ervin TJ, Murphy BA, Raez LE, Cohen RB, Spaulding M, Tishler RB, Roth B, Viroglio Rdel C, Venkatesan V, Romanov I, Agarwala S, Harter KW, Dugan M, Cmelak A, Markoe AM, Read PW, Steinbrenner L, Colevas AD, Norris Jr CM, Haddad RI, TAX 324 Study Group. Cisplatin and fluorouracil alone or with docetaxel in head and neck cancer. N Engl J Med 2007;357(17):1705–15.

[13] Vermorken JB, Mesia R, Rivera F, Remenar E, Kawecki A, Rottey S, Erfan J, Zabolotnyy D, Kienzer HR, Cupissol D, Peyrade F, Benasso M, Vynnychenko I, De Raucourt D, Bokemeyer C, Schueler A, Amellal N, Hitt R. Platinum-based chemotherapy plus cetuximab in head and neck cancer. N Engl J Med 2008; 359(11):1116–27.

[14] Rischin D, Harrington KJ, Greil R, Soulieres D, Tahara M, de Castro G, Psyrri A, Baste N, Neupane PC, Bratland A, Fuereder T, Hughes BGM, Mesia R, Ngamphaiboon N, Rordorf T, Ishak WZW, Zhang Y, Jin F, Gumuscu B, Burtness B. Protocol-specified final analysis of the phase 3 KEYNOTE-048 trial of pembrolizumab (pembro) as first-line therapy for recurrent/metastatic head and neck squamous cell carcinoma (R/M HNSCC). J Clin Oncol 2019;37 [suppl; abstr 6000].

[15] Ferris RL, Blumenschein Jr G, Fayette J, Guigay J, Colevas AD, Licitra L, Harrington K, Kasper S, Vokes EE, Even C, Worden F, Saba NF, Iglesias Docampo LC, Haddad R, Rordorf T, Kiyota N, Tahara M, Monga M, Lynch M, Geese WJ, Kopit J, Shaw JW, Gillison ML. Nivolumab for recurrent squamous-cell carcinoma of the head and neck. N Engl J Med 2016;375(19):1856–67.

[16] Chow LQM, Haddad R, Gupta S, Mahipal A, Mehra R, Tahara M, Berger R, Eder JP, Burtness B, Lee SH, Keam B, Kang H, Muro K, Weiss J, Geva R, Lin CC, Chung HC, Meister A, Dolled-Filhart M, Pathiraja K, Cheng JD, Seiwert TY. Antitumor activity of pembrolizumab in biomarker-unselected patients with recurrent and/or metastatic head and neck squamous cell carcinoma: results from the phase Ib KEYNOTE-012 expansion cohort. J Clin Oncol 2016;34(32):3838–45.

[17] Cohen EEW, Soulières D, Le Tourneau C, Dinis J, Licitra L, Ahn MJ, Soria A, Machiels JP, Mach N, Mehra R, Burtness B, Zhang P, Cheng J, Swaby RF, Harrington KJ, KEYNOTE-040 Investigators. Pembrolizumab versus methotrexate, docetaxel, or cetuximab for recurrent or metastatic head-and-neck squamous cell carcinoma (KEYNOTE-040): a randomised, open-label, phase 3 study. Lancet 2019;393(10167):156–67.

[18] Seiwert TY, Burtness B, Mehra R, Weiss J, Berger R, Eder JP, Heath K, McClanahan T, Lunceford J, Gause C, Cheng JD, Chow LQ. Safety and clinical activity of pembrolizumab for treatment of recurrent or metastatic squamous cell carcinoma of the head and neck (KEYNOTE-012): an open-label, multicentre, phase 1b trial. Lancet Oncol 2016;17:956–65.

[19] Bernier J, Domenge C, Ozsahin M, Matuszewska K, Lefèbvre JL, Greiner RH, Giralt J, Maingon P, Rolland F, Bolla M, Cognetti F, Bourhis J, Kirkpatrick A, van Glabbeke M, European Organization for Research and Treatment of Cancer Trial 22931. Postoperative irradiation with or without concomitant chemotherapy for locally advanced head and neck cancer. N Engl J Med 2004;350(19):1945–52.

[20] Cooper JS, Pajak TF, Forastiere AA, Jacobs J, Campbell BH, Saxman SB, Kish JA, Kim HE, Cmelak AJ, Rotman M, Machtay M, Ensley JF, Chao KS, Schultz CJ, Lee N, Fu KK, Radiation Therapy Oncology Group 9501/Intergroup. Postoperative concurrent radiotherapy and chemotherapy for high-risk squamous-cell carcinoma of the head and neck. N Engl J Med 2004;350(19):1937–44.

[21] Cooper JS, Zhang Q, Pajak TF, Forastiere AA, Jacobs J, Saxman SB, Kish JA, Kim HE, Cmelak AJ, Rotman M, Lustig R, Ensley JF, Thorstad W, Schultz CJ, Yom SS, Ang KK. Long-term follow-up of the RTOG 9501/intergroup phase III trial: postoperative concurrent radiation therapy and chemotherapy in high-risk squamous cell carcinoma of the head and neck. Int J Radiat Oncol Biol Phys 2012;84(5):1198–205.

[22] Bernier J, Cooper JS, Pajak TF, van Glabbeke M, Bourhis J, Forastiere A, Ozsahin EM, Jacobs JR, Jassem J, Ang KK, Lefèbvre JL. Defining risk levels in locally advanced head and neck cancers: a comparative analysis of concurrent postoperative radiation plus chemotherapy trials of the EORTC (#22931) and RTOG (# 9501). Head Neck 2005; 27(10):843–50.

[23] Noronha V, Joshi A, Patil VM, Agarwal J, Ghosh-Laskar S, Budrukkar A, Murthy V, Gupta T, D'Cruz AK, Banavali S, Pai PS, Chaturvedi P, Chaukar D, Pande N, Chandrasekharan A, Talreja V, Vallathol DH, Mathrudev V, Manjrekar A, Maske K, Bhelekar AS, Nawale K, Kannan S, Gota V, Bhattacharjee A, Kane S, Juvekar SL, Prabhash K. Once-a-week versus once-every-3-weeks cisplatin chemoradiation for locally advanced head and neck cancer: a phase III randomized noninferiority trial. J Clin Oncol 2018;36(11):1064–72.

[24] Kunieda F, Kiyota N, Tahara M, Kodaira T, Hayashi R, Ishikura S, Mizusawa J, Nakamura K, Fukuda H, Fujii M, Head and Neck Cancer Study Group of the Japan Clinical Oncology Group. Randomized phase II/III trial of post-operative chemoradiotherapy comparing 3-weekly cisplatin with weekly cisplatin in high-risk patients with squamous cell carcinoma of head and neck: Japan Clinical Oncology Group Study (JCOG1008). Jpn J Clin Oncol 2014;44(8):770–4.

[25] Pignon JP, le Maître A, Maillard E, Bourhis J, MACH-NC Collaborative Group. Meta-analysis of chemotherapy in head and neck cancer (MACH-NC): an update on 93 randomised trials and 17,346 patients. Radiother Oncol 2009;92(1):4–14. https://doi.org/10.1016/j.radonc.2009.04.014.

[26] Adelstein DJ, Li Y, Adams GL, Wagner Jr H, Kish JA, Ensley JF, Schuller DE, Forastiere AA. An intergroup phase III comparison of standard radiation therapy and two schedules of concurrent chemoradiotherapy in patients with unresectable squamous cell head and neck cancer. J Clin Oncol 2003;21(1):92–8.

[27] Forastiere AA, Goepfert H, Maor M, Pajak TF, Weber R, Morrison W, Glisson B, Trotti A, Ridge JA, Chao C, Peters G, Lee DJ, Leaf A, Ensley J, Cooper J. Concurrent chemotherapy and radiotherapy for organ preservation in advanced laryngeal cancer. N Engl J Med 2003;349(22):2091–8.

[28] Forastiere AA, Zhang Q, Weber RS, Maor MH, Goepfert H, Pajak TF, Morrison W, Glisson B, Trotti A, Ridge JA, Thorstad W, Wagner H, Ensley JF, Cooper JS. Long-term results of RTOG 91-11: a comparison of three nonsurgical treatment strategies to preserve the larynx in patients with locally advanced larynx cancer. J Clin Oncol 2013;31(7):845–52.

[29] Beckmann GK, Hoppe F, Pfreundner L, Flentje MP. Hyperfractionated accelerated radiotherapy in combination with weekly cisplatin for locally advanced head and neck cancer. Head Neck 2005;27(1):36–43.

[30] Medina JA, Rueda A, de Pasos AS, Contreras J, Cobo M, Moreno P, Benavides M, Villanueva A, Alba E. A phase II study of concomitant boost radiation plus concurrent weekly cisplatin for locally advanced unresectable head and neck carcinomas. Radiother Oncol 2006;79(1):34–8.

[31] Szturz P, Wouters K, Kiyota N, Tahara M, Prabhash K, Noronha V, Castro A, Licitra L, Adelstein D, Vermorken JB. Weekly low-dose versus three-weekly high-dose cisplatin for concurrent chemoradiation in locoregionally advanced non-nasopharyngeal head and neck cancer: a systematic review and meta-analysis of aggregate data. Oncologist 2017;22(9):1056–66.

[32] Bauml JM, Vinnakota R, Anna Park YH, Bates SE, Fojo T, Aggarwal C, Limaye S, Damjanov N, Di Stefano J, Ciunci C, Genden EM, Wisnivesky JP, Ferrandino R, Mamtani R, Langer CJ, Cohen RB, Sigel K. Cisplatin every 3 weeks versus weekly with definitive concurrent radiotherapy for squamous cell carcinoma of the head and neck. J Natl Cancer Inst 2019;111(5):490–7.

[33] Mohamed A, Twardy B, Zordok MA, Ashraf K, Alkhoder A, Schrapp K, Steuer C, Chen Z, Pakkala S, Pillai R, Trad Wadsworth J, Higgins K, Beitler JJ, Ramalingam SS, Owonikoko TK, Khuri FR, Shin DM, Behera M,

Saba NF. Concurrent chemoradiotherapy with weekly versus triweekly cisplatin in locally advanced squamous cell carcinoma of the head and neck: comparative analysis. Head Neck 2019;41(5):1490–8.

[34] Garden AS, Harris J, Vokes EE, Forastiere AA, Ridge JA, Jones C, Horwitz EM, Glisson BS, Nabell L, Cooper JS, Demas W, Gore E. Preliminary results of Radiation Therapy Oncology Group 97-03: a randomized phase ii trial of concurrent radiation and chemotherapy for advanced squamous cell carcinomas of the head and neck. J Clin Oncol 2004;22(14):2856–64.

[35] Taylor 4th SG, Murthy AK, Griem KL, Recine DC, Kiel K, Blendowski C, Hurst PB, Showel JT, Hutchinson Jr JC, Campanella RS, Chen S, Caldarelli DD. Concomitant cisplatin/5-FU infusion and radiotherapy in advanced head and neck cancer: 8-year analysis of results. Head Neck 1997;19(8):684–91.

[36] Denis F, Garaud P, Bardet E, Alfonsi M, Sire C, Germain T, Bergerot P, Rhein B, Tortochaux J, Calais G. Final results of the 94-01 French Head and Neck Oncology and Radiotherapy Group randomized trial comparing radiotherapy alone with concomitant radiochemotherapy in advanced-stage oropharynx carcinoma. J Clin Oncol 2004;22(1):69–76.

[37] Bourhis J, Sire C, Graff P, Grégoire V, Maingon P, Calais G, Gery B, Martin L, Alfonsi M, Desprez P, Pignon T, Bardet E, Rives M, Geoffrois L, Daly-Schveitzer N, Sen S, Tuchais C, Dupuis O, Guerif S, Lapeyre M, Favrel V, Hamoir M, Lusinchi A, Temam S, Pinna A, Tao YG, Blanchard P, Aupérin A. Concomitant chemoradiotherapy versus acceleration of radiotherapy with or without concomitant chemotherapy in locally advanced head and neck carcinoma (GORTEC 99-02): an open-label phase 3 randomised trial. Lancet Oncol 2012;13(2):145–53.

[38] Suntharalingam M, Haas ML, Conley BA, Egorin MJ, Levy S, Sivasailam S, Herman JM, Jacobs MC, Gray WC, Ord RA, Aisner JA, Van Echo DA. The use of carboplatin and paclitaxel with daily radiotherapy in patients with locally advanced squamous cell carcinomas of the head and neck. Int J Radiat Oncol Biol Phys 2000;47(1):49–56.

[39] Mehanna H, Robinson M, Hartley A, Kong A, Foran B, Fulton-Lieuw T, Dalby M, Mistry P, Sen M, O'Toole L, Al Booz H, Dyker K, Moleron R, Whitaker S, Brennan S, Cook A, Griffin M, Aynsley E, Rolles M, De Winton E, Chan A, Srinivasan D, Nixon I, Grumett J, Leemans CR, Buter J, Henderson J, Harrington K, McConkey C, Gray A, Dunn J, De-ESCALaTE HPV Trial Group. Radiotherapy plus cisplatin or cetuximab in low-risk human papillomavirus-positive oropharyngeal cancer (De-ESCALaTE HPV): an open-label randomised controlled phase 3 trial. Lancet 2019;393(10166):51–60.

[40] Gillison ML, Trotti AM, Harris J, Eisbruch A, Harari PM, Adelstein DJ, Sturgis EM, Burtness B, Ridge JA, Ringash J, Galvin J, Yao M, Koyfman SA, Blakaj DM, Razaq MA, Colevas AD, Beitler JJ, Jones CU, Dunlap NE, Seaward SA, Spencer S, Galloway TJ, Phan J, Dignam JJ, Le QT. Radiotherapy plus cetuximab or cisplatin in human papillomavirus-positive oropharyngeal cancer (NRG Oncology RTOG 1016): a randomised, multicentre, non-inferiority trial. Lancet 2019;393(10166):40–50.

[41] Ma J, Liu Y, Huang XL, Zhang ZY, Myers JN, Neskey DM, Zhong LP. Induction chemotherapy decreases the rate of distant metastasis in patients with head and neck squamous cell carcinoma but does not improve survival or locoregional control: a meta-analysis. Oral Oncol 2012;48(11):1076–84.

[42] Hitt R, López-Pousa A, Martínez-Trufero J, Escrig V, Carles J, Rizo A, Isla D, Vega ME, Martí JL, Lobo F, Pastor P, Valentí V, Belón J, Sánchez MA, Chaib C, Pallarés C, Antón A, Cervantes A, Paz-Ares L, Cortés-Funes H. Phase III study comparing cisplatin plus fluorouracil to paclitaxel, cisplatin, and fluorouracil induction chemotherapy followed by chemoradiotherapy in locally advanced head and neck cancer. J Clin Oncol 2005;23(34):8636–45.

[43] Vermorken JB, Remenar E, van Herpen C, Gorlia T, Mesia R, Degardin M, Stewart JS, Jelic S, Betka J, Preiss JH, van den Weyngaert D, Awada A, Cupissol D, Kienzer HR, Rey A, Desaunois I, Bernier J, Lefebvre JL, EORTC 24971/TAX 323 Study Group. Cisplatin, fluorouracil, and docetaxel in unresectable head and neck cancer. N Engl J Med 2007;357(17):1695–704.

[44] Posner MR, Hershock DM, Blajman CR, Mickiewicz E, Winquist E, Gorbounova V, Tjulandin S, Shin DM, Cullen K, Ervin TJ, Murphy BA, Raez LE, Cohen RB, Spaulding M, Tishler RB, Roth B, Viroglio Rdel C, Venkatesan V, Romanov I, Agarwala S, Harter KW, Dugan M, Cmelak A, Markoe AM, Read PW, Steinbrenner L, Colevas AD, Norris Jr CM, Haddad RI, TAX 324 Study Group. Cisplatin and fluorouracil alone or with docetaxel in head and neck cancer. N Engl J Med 2007;357(17):1705–15.

[45] Pointreau Y, Garaud P, Chapet S, Sire C, Tuchais C, Tortochaux J, Faivre S, Guerrif S, Alfonsi M, Calais G. Randomized trial of induction chemotherapy with cisplatin and 5-fluorouracil with or without docetaxel for larynx preservation. J Natl Cancer Inst 2009;101(7):498–506.

[46] Janoray G, Pointreau Y, Garaud P, Chapet S, Alfonsi M, Sire C, Jadaud E, Calais G. Long-term results of a multicenter randomized phase III trial of induction chemotherapy with cisplatin, 5-fluorouracil, ± docetaxel for larynx preservation. J Natl Cancer Inst 2015;108(4). pii: djv368.

[47] Adelstein DJ, Moon J, Hanna E, Giri PG, Mills GM, Wolf GT, Urba SG. Docetaxel, cisplatin, and fluorouracil induction chemotherapy followed by accelerated fractionation/concomitant boost radiation and concurrent cisplatin in patients with advanced squamous cell head and neck cancer: a Southwest Oncology Group phase II trial (S0216). Head Neck 2010;32(2):221–8.

[48] Lefebvre JL, Pointreau Y, Rolland F, Alfonsi M, Baudoux A, Sire C, de Raucourt D, Malard O, Degardin M, Tuchais C, Blot E, Rives M, Reyt E, Tourani JM, Geoffrois L, Peyrade F, Guichard F, Chevalier D, Babin E, Lang P, Janot F, Calais G, Garaud P, Bardet E. Induction chemotherapy followed by either chemoradiotherapy or bioradiotherapy for larynx preservation: the TREMPLIN randomized phase II study. J Clin Oncol 2013; 31(7):853–9.

[49] Haddad R, O'Neill A, Rabinowits G, Tishler R, Khuri F, Adkins D, Clark J, Sarlis N, Lorch J, Beitler JJ, Limaye S, Riley S, Posner M. Induction chemotherapy followed by concurrent chemoradiotherapy (sequential chemoradiotherapy) versus concurrent chemoradiotherapy alone in locally advanced head and neck cancer (PARADIGM): a randomised phase 3 trial. Lancet Oncol 2013;14(3):257–64.

[50] Cohen EE, Karrison TG, Kocherginsky M, Mueller J, Egan R, Huang CH, Brockstein BE, Agulnik MB, Mittal BB, Yunus F, Samant S, Raez LE, Mehra R, Kumar P, Ondrey F, Marchand P, Braegas B, Seiwert TY, Villaflor VM, Haraf DJ, Vokes EE. Phase III randomized trial of induction chemotherapy in patients with N2 or N3 locally advanced head and neck cancer. J Clin Oncol 2014;32(25):2735–43.

[51] Jacobs C, Lyman G, Velez-García E, Sridhar KS, Knight W, Hochster H, Goodnough LT, Mortimer JE, Einhorn LH, Schacter L, et al. A phase III randomized study comparing cisplatin and fluorouracil as single agents and in combination for advanced squamous cell carcinoma of the head and neck. J Clin Oncol 1992; 10(2):257–63.

[52] Martinez-Trufero J, Isla D, Adansa JC, Irigoyen A, Hitt R, Gil-Arnaiz I, Lambea J, Lecumberri MJ, Cruz JJ. Phase II study of capecitabine as palliative treatment for patients with recurrent and metastatic squamous head and neck cancer after previous platinum-based treatment. Br J Cancer 2010;102(12):1687–91.

[53] Forastiere AA, Metch B, Schuller DE, Ensley JF, Hutchins LF, Triozzi P, Kish JA, McClure S, VonFeldt E, Williamson SK, et al. Randomized comparison of cisplatin plus fluorouracil and carboplatin plus fluorouracil versus methotrexate in advanced squamous-cell carcinoma of the head and neck: a Southwest Oncology Group study. J Clin Oncol 1992;10(8):1245–51.

[54] Guardiola E, Peyrade F, Chaigneau L, Cupissol D, Tchiknavorian X, Bompas E, Madroszyk A, Ronchin P, Schneider M, Bleuze JP, Blay JY, Pivot X. Results of a randomised phase II study comparing docetaxel with methotrexate in patients with recurrent head and neck cancer. Eur J Cancer 2004;40(14):2071–6.

[55] Vermorken JB, Trigo J, Hitt R, Koralewski P, Diaz-Rubio E, Rolland F, Knecht R, Amellal N, Schueler A, Baselga J. Open-label, uncontrolled, multicenter phase II study to evaluate the efficacy and toxicity of cetuximab as a single agent in patients with recurrent and/or metastatic squamous cell carcinoma of the head and neck who failed to respond to platinum-based therapy. J Clin Oncol 2007;25(16):2171–7.

[56] Samlowski WE, Moon J, Kuebler JP, Nichols CR, Gandara DR, Ozer H, Williamson SK, Atkins JN, Schuller DE, Ensley JF. Evaluation of the combination of docetaxel/carboplatin in patients with metastatic or recurrent squamous cell carcinoma of the head and neck (SCCHN): a Southwest Oncology Group Phase II study. Cancer Invest 2007;25(3):182–8.

[57] Gibson MK, Li Y, Murphy B, Hussain MH, DeConti RC, Ensley J, Forastiere AA, Eastern Cooperative Oncology Group. Randomized phase III evaluation of cisplatin plus fluorouracil versus cisplatin plus paclitaxel in advanced head and neck cancer (E1395): an intergroup trial of the Eastern Cooperative Oncology Group. J Clin Oncol 2005;23(15):3562–7.

[58] Burtness B, Goldwasser MA, Flood W, Mattar B, Forastiere AA, Eastern Cooperative Oncology Group. Phase III randomized trial of cisplatin plus placebo compared with cisplatin plus cetuximab in metastatic/recurrent head and neck cancer: an Eastern Cooperative Oncology Group study. J Clin Oncol 2005;23(34):8646–54.

[59] Vermorken JB, Mesia R, Rivera F, Remenar E, Kawecki A, Rottey S, Erfan J, Zabolotnyy D, Kienzer HR, Cupissol D, Peyrade F, Benasso M, Vynnychenko I, De Raucourt D, Bokemeyer C, Schueler A, Amellal N, Hitt R. Platinum-based chemotherapy plus cetuximab in head and neck cancer. N Engl J Med 2008; 359(11):1116–27.

[60] Rischin D, Harrington KJ, Greil R, Soulieres D, Tahara M, de Castro G, Psyrri A, Baste N, Neupane PC, Bratland A, Fuereder T, Hughes BGM, Mesia R, Ngamphaiboon N, Rordorf T, Ishak WZW, Zhang Y, Jin F, Gumuscu B, Burtness B. Protocol-specified final analysis of the phase 3 KEYNOTE-048 trial of pembrolizumab (pembro) as first-line therapy for recurrent/metastatic head and neck squamous cell carcinoma (R/M HNSCC). J Clin Oncol 2019;37 [suppl; abstr 6000].

[61] Herbst RS, Arquette M, Shin DM, Dicke K, Vokes EE, Azarnia N, Hong WK, Kies MS. Phase II multicenter study of the epidermal growth factor receptor antibody cetuximab and cisplatin for recurrent and refractory squamous cell carcinoma of the head and neck. J Clin Oncol 2005;23(24):5578–87.

[62] Guigay J, Fayette J, Dillies AF, Sire C, Kerger JN, Tennevet I, Machiels JP, Zanetta S, Pointreau Y, Bozec Le Moal L, Henry S, Schilf A, Bourhis J. Cetuximab, docetaxel, and cisplatin as first-line treatment in patients with recurrent or metastatic head and neck squamous cell carcinoma: a multicenter, phase II GORTEC study. Ann Oncol 2015;26(9):1941–7.

[63] Al-Sarraf M, LeBlanc M, Giri PG, Fu KK, Cooper J, Vuong T, Forastiere AA, Adams G, Sakr WA, Schuller DE, Ensley JF. Chemoradiotherapy versus radiotherapy in patients with advanced nasopharyngeal cancer: phase III randomized Intergroup study 0099. J Clin Oncol 1998;16(4):1310–7.

[64] Chan AT, Leung SF, Ngan RK, Teo PM, Lau WH, Kwan WH, Hui EP, Yiu HY, Yeo W, Cheung FY, Yu KH, Chiu KW, Chan DT, Mok TS, Yau S, Yuen KT, Mo FK, Lai MM, Ma BB, Kam MK, Leung TW, Johnson PJ, Choi PH, Zee BC. Overall survival after concurrent cisplatin-radiotherapy compared with radiotherapy alone in locoregionally advanced nasopharyngeal carcinoma. J Natl Cancer Inst 2005;97(7):536–9.

[65] Chitapanarux I, Lorvidhaya V, Kamnerdsupaphon P, Sumitsawan Y, Tharavichitkul E, Sukthomya V, Ford J. Chemoradiation comparing cisplatin versus carboplatin in locally advanced nasopharyngeal cancer: randomised, non-inferiority, open trial. Eur J Cancer 2007;43(9):1399–406.

[66] Blanchard P, Lee A, Marguet S, Leclercq J, Ng WT, Ma J, Chan AT, Huang PY, Benhamou E, Zhu G, Chua DT, Chen Y, Mai HQ, Kwong DL, Cheah SL, Moon J, Tung Y, Chi KH, Fountzilas G, Zhang L, Hui EP, Lu TX, Bourhis J, Pignon JP, MAC-NPC Collaborative Group. Chemotherapy and radiotherapy in nasopharyngeal carcinoma: an update of the MAC-NPC meta-analysis. Lancet Oncol 2015;16(6):645–55.

[67] Sun Y, Li WF, Chen NY, Zhang N, Hu GQ, Xie FY, Sun Y, Chen XZ, Li JG, Zhu XD, Hu CS, Xu XY, Chen YY, Hu WH, Guo L, Mo HY, Chen L, Mao YP, Sun R, Ai P, Liang SB, Long GX, Zheng BM, Feng XL, Gong XC, Li L, Shen CY, Xu JY, Guo Y, Chen YM, Zhang F, Lin L, Tang LL, Liu MZ, Ma J. Induction chemotherapy plus concurrent chemoradiotherapy versus concurrent chemoradiotherapy alone in locoregionally advanced nasopharyngeal carcinoma: a phase 3, multicentre, randomised controlled trial. Lancet Oncol 2016; 17(11):1509–20.

[68] Zhang Y, Chen L, Hu GQ, Zhang N, Zhu XD, Yang KY, Jin F, Shi M, Chen YP, Hu WH, Cheng ZB, Wang SY, Tian Y, Wang XC, Sun Y, Li JG, Li WF, Li YH, Tang LL, Mao YP, Zhou GQ, Sun R, Liu X, Guo R, Long GX, Liang SQ, Li L, Huang J, Long JH, Zang J, Liu QD, Zou L, Su QF, Zheng BM, Xiao Y, Guo Y, Han F, Mo HY, Lv JW, Du XJ, Xu C, Liu N, Li YQ, MLK C, Xie FY, Sun Y, Ma J. Gemcitabine and cisplatin induction chemotherapy in nasopharyngeal carcinoma. N Engl J Med 2019;381:1124–35.

[69] Zhang L, Huang Y, Hong S, Yang Y, Yu G, Jia J, Peng P, Wu X, Lin Q, Xi X, Peng J, Xu M, Chen D, Lu X, Wang R, Cao X, Chen X, Lin Z, Xiong J, Lin Q, Xie C, Li Z, Pan J, Li J, Wu S, Lian Y, Yang Q, Zhao C. Gemcitabine plus cisplatin versus fluorouracil plus cisplatin in recurrent or metastatic nasopharyngeal carcinoma: a multicentre, randomised, open-label, phase 3 trial. Lancet 2016;388(10054):1883–92.

[70] Jin Y, Shi YX, Cai XY, Xia XY, Cai YC, Cao Y, Zhang WD, Hu WH, Jiang WQ. Comparison of five cisplatin-based regimens frequently used as the first-line protocols in metastatic nasopharyngeal carcinoma. J Cancer Res Clin Oncol 2012;138(10):1717–25.

[71] Al-Sarraf M, LeBlanc M, Giri PG, Fu KK, Cooper J, Vuong T, Forastiere AA, Adams G, Sakr WA, Schuller DE, Ensley JF. Chemoradiotherapy versus radiotherapy in patients with advanced nasopharyngeal cancer: phase III randomized Intergroup study 0099. J Clin Oncol 1998;16(4):1310–7.

[72] Chan AT, Leung SF, Ngan RK, Teo PM, Lau WH, Kwan WH, Hui EP, Yiu HY, Yeo W, Cheung FY, Yu KH, Chiu KW, Chan DT, Mok TS, Yau S, Yuen KT, Mo FK, Lai MM, Ma BB, Kam MK, Leung TW, Johnson PJ, Choi PH, Zee BC. Overall survival after concurrent cisplatin-radiotherapy compared with radiotherapy alone in locoregionally advanced nasopharyngeal carcinoma. J Natl Cancer Inst 2005;97(7):536–9.

[73] Chitapanarux I, Lorvidhaya V, Kamnerdsupaphon P, Sumitsawan Y, Tharavichitkul E, Sukthomya V, Ford J. Chemoradiation comparing cisplatin versus carboplatin in locally advanced nasopharyngeal cancer: randomised, non-inferiority, open trial. Eur J Cancer 2007;43(9):1399–406.

[74] Chen L, Hu CS, Chen XZ, Hu GQ, Cheng ZB, Sun Y, Li WX, Chen YY, Xie FY, Liang SB, Chen Y, Xu TT, Li B, Long GX, Wang SY, Zheng BM, Guo Y, Sun Y, Mao YP, Tang LL, Chen YM, Liu MZ, Ma J. Concurrent chemoradiotherapy plus adjuvant chemotherapy versus concurrent chemoradiotherapy alone in patients with locoregionally advanced nasopharyngeal carcinoma: a phase 3 multicentre randomised controlled trial. Lancet Oncol 2012;13(2):163–71.

[75] Sun Y, Li WF, Chen NY, Zhang N, Hu GQ, Xie FY, Sun Y, Chen XZ, Li JG, Zhu XD, Hu CS, Xu XY, Chen YY, Hu WH, Guo L, Mo HY, Chen L, Mao YP, Sun R, Ai P, Liang SB, Long GX, Zheng BM, Feng XL, Gong XC, Li L, Shen CY, Xu JY, Guo Y, Chen YM, Zhang F, Lin L, Tang LL, Liu MZ, Ma J. Induction chemotherapy plus concurrent chemoradiotherapy versus concurrent chemoradiotherapy alone in locoregionally advanced nasopharyngeal carcinoma: a phase 3, multicentre, randomised controlled trial. Lancet Oncol 2016;17(11):1509–20.
[76] Zhang Y, Chen L, Hu GQ, Zhang N, Zhu XD, Yang KY, Jin F, Shi M, Chen YP, Hu WH, Cheng ZB, Wang SY, Tian Y, Wang XC, Sun Y, Li JG, Li WF, Li YH, Tang LL, Mao YP, Zhou GQ, Sun R, Liu X, Guo R, Long GX, Liang SQ, Li L, Huang J, Long JH, Zang J, Liu QD, Zou L, Su QF, Zheng BM, Xiao Y, Guo Y, Han F, Mo HY, Lv JW, Du XJ, Xu C, Liu N, Li YQ, MLK C, Xie FY, Sun Y, Ma J. Gemcitabine and cisplatin induction chemotherapy in nasopharyngeal carcinoma. N Engl J Med 2019;381:1124–35.
[77] Hui EP, Ma BB, Leung SF, King AD, Mo F, Kam MK, Yu BK, Chiu SK, Kwan WH, Ho R, Chan I, Ahuja AT, Zee BC, Chan AT. Randomized phase II trial of concurrent cisplatin-radiotherapy with or without neoadjuvant docetaxel and cisplatin in advanced nasopharyngeal carcinoma. J Clin Oncol 2009;27(2):242–9.
[78] Zhang L, Huang Y, Hong S, Yang Y, Yu G, Jia J, Peng P, Wu X, Lin Q, Xi X, Peng J, Xu M, Chen D, Lu X, Wang R, Cao X, Chen X, Lin Z, Xiong J, Lin Q, Xie C, Li Z, Pan J, Li J, Wu S, Lian Y, Yang Q, Zhao C. Gemcitabine plus cisplatin versus fluorouracil plus cisplatin in recurrent or metastatic nasopharyngeal carcinoma: a multicentre, randomised, open-label, phase 3 trial. Lancet 2016;388(10054):1883–92.

CHAPTER

3

Targeted therapy: Precision comes to life

Umair Mahmood, Robert Haddad

Dana-Farber Cancer Institute, Boston, MA, United States

Abstract

Traditional treatment approaches for advanced head and neck cancers have been associated with limited clinical outcomes that necessitate the development of novel therapies that can address this unmet clinical need. Precision medicine initiatives employing the use of small molecule inhibitors and monoclonal antibodies have demonstrated promising potential for the management of head and neck squamous cell carcinoma and salivary gland tumors. Such molecular therapies have been used to effectively target aberrant elements composed of tyrosine kinases and phosphoinositide 3-kinase/AKT/mammalian target of rapamycin, cell cycle, and epigenetic pathways. Future studies using innovative trial designs and biomarker enrichment strategies as well as technological advances could pave the way for personalized medicine to become an integral component of standard of care practices to truly optimize patient outcomes.

Abbreviations

5-FU fluorouracil
ACC adenoid cystic carcinoma
AEs adverse events
ASCO American Society of Clinical Oncology
CDK cyclin-dependent kinase
CI confidence interval
CR complete response
ctDNA circulating tumor DNA
DFS disease-free survival
DLTs dose-limiting toxicities
EGFR epidermal growth factor receptor
FDA Food and Drug Administration
FGF2 fibroblast growth factor 2
FLT3 Fms-related tyrosine kinase 3
HDAC histone deacetylase
HER2 human epidermal growth factor receptor 2

Novel Therapies in Head and Neck Cancer: Beyond the Horizon
https://doi.org/10.1016/B978-0-12-820679-9.00003-7

HNSCC	head and neck squamous cell carcinoma
HPV	human papillomavirus
MTD	maximum tolerated dose
mTOR	mammalian target of rapamycin
mTORC1	mammalian target of rapamycin complex 1
OR	odds ratio
ORR	overall response rate
OS	overall survival
PDGFR	platelet-derived growth factor receptor
PFS	progression-free survival
PR	partial response
RMSGCs	recurrent and/or metastatic salivary gland carcinomas
SAE	serious adverse event
SD	stable disease
SGCs	salivary gland carcinomas
T-DM1	ado-trastuzumab emtansine
TILs	tumor-infiltrating lymphocytes
TRAEs	treatment-related adverse events
TRK	tropomyosin receptor kinase
VEGF-A	vascular endothelial growth factor A
VEGFR	vascular endothelial growth factor receptor

Conflict of interest

No potential conflicts of interest were disclosed.

Introduction

In an era where conventional treatment for head and neck cancers using surgery-, radiation-, and chemotherapy-based modalities is associated with limited efficacy, significant toxicity, drug resistance, and high morbidity, there is a great need to devise novel, personalized treatments tailored toward a patient's particular tumor. A selective treatment approach offered by precision medicine is vital in addressing these issues. This is especially true in patients who recur after primary therapy, or those with mutational aberrations are associated with poor prognosis. These efforts have been aided by the development of molecular therapies targeting either defunct cellular proteins or complete pathways with multiple proteins with great specificity. Such an approach often utilizes small molecule inhibitors or monoclonal antibodies that are aimed at exploiting tumor biology by altering the tumor microenvironment or its vasculature, thereby abrogating the growth and metastasis of the tumor cell [1]. Given the wide array of potential targets, this chapter will discuss the most promising therapies that are being developed or have been approved for clinical use (see Table 1).

Employing the precision medicine model for the management of head and neck squamous cell carcinoma

Targeting the epidermal growth factor receptor

The epidermal growth factor receptor (EGFR) belongs to the ErbB family of receptor tyrosine kinases and is an important regulator of cell cycle progression and cellular proliferation

TABLE 1 Molecular targeted agents investigated for the treatment of head and neck cancers.

Drug	Molecular target	Head and neck cancer subtype	Results	Common AEs (≥ grade 3)	References
Cetuximab	EGFR	HNSCC	ORR = 36% (EXTREME regimen)	Neutropenia, anemia, thrombocytopenia, skin reactions, leukopenia	Vermorken et al. [2]
Afatinib	EGFR, pan-HER	HNSCC	ORR = 28% (LUX-Head and Neck 3 trial)	Diarrhea, acne	Guo et al. [3]
Rapamycin	mTORC1	HNSCC	4 RECIST responses, 1 CR and 3 PRs	Neutropenia, thrombocytopenia	Day et al. [4]
Temsirolimus	mTORC1	Nonnasopharynx HNSCC	ORR = 41.7%	Hematologic toxicities	Dunn et al. [5]
Buparlisib	Pan-PIK3	HNSCC	ORR = 39% (BERIL-1 trial)	Hyperglycemia, anemia, fatigue, neutropenia	Soulieres et al. [6]
Palbociclib	CDK4/6	HNSCC	ORR = 39% (platinum-resistant arm) and 19% (cetuximab-resistant arm)	Neutropenia	Adkins et al. [7]
Ribociclib	CDK4/6	HNSCC	RECIST response rates; PR = 10.5%, SD = 58%	Neutropenia	Seront et al. [8]
Vorinostat	HDAC	ACC	ORR = 7% (vorinostat alone)	Lymphopenia	Goncalves et al. [9]
			ORR = 16% (vorinostat + pembrolizumab)	Renal insufficiency, fatigue, nausea	Rodriguez et al. [10]
Sorafenib	VEGFR1–3, BRAF, PDGFR, RET, c-KIT, FLT3	SGCs	ORR = 16%; 11% in ACC, 22% in non-ACC SGCs	Lung infection	Locati et al. [11]
Lenvatinib	VEGFRs 1–3, FGFRs 1–3, KIT, PDGFR α and β, RET	SGCs	ORR = 15.6%	Hypertension, oral pain	Tchekmedyian et al. [12]
Larotrectinib	TRK	Secretory sarcoma of salivary gland	RECIST responses, 16/18	Elevated ALT	Hong et al. [13]
Ado-trastuzumab emtansine	HER2	Salivary duct carcinoma	ORR = 90%	Thrombocytopenia	Li et al. [14]

via signal transduction cascades [15, 16]. Pathogenic alterations in the EGFR protein, namely, high expression levels on the cell surface, have been implicated in 80% of head and neck squamous cell carcinoma (HNSCC) cases and are associated with low overall survival (OS) and disease-free survival (DFS) [17]. Many therapeutic strategies have been aimed at this actionable molecular target including both extracellular inhibition via monoclonal antibodies and intracellular inhibition using small molecule inhibitors.

Cetuximab is a monoclonal antibody that targets the extracellular ligand-binding region of the EGFR protein, thereby effectively shutting down the intracellular signaling mechanism [15]. This drug has been extensively studied in large-scale trials examining its synergistic effects in tandem with radiation or chemotherapy. Cetuximab's radiosensitizing effects were previously reported in a multicenter, randomized Phase 3 trial comparing the use of cetuximab with definitive radiotherapy with radiotherapy alone in the locoregionally advanced HNSCC population [15]. The investigators reported a greater survival advantage (median OS = 49.0 months vs 29.3 months; $P = 0.03$) and a prolonged duration of locoregional disease control (median duration = 24.4 months vs 14.9 months; $P = 0.005$) in the combined treatment arm compared with the radiotherapy alone arm. Another randomized Phase 3 trial studied the efficacy of standard platinum-based chemotherapy (cisplatin/carboplatin) and fluorouracil (5-FU) versus cetuximab added to the aforementioned chemotherapeutic regimen as a first-line treatment for patients with recurrent and/or metastatic HNSCC [2]. This clinical trial also demonstrated the overall survival advantage (median OS = 10.1 months vs 7.4 months; $P = 0.04$) and an increased overall response rate (ORR = 36% vs 20%; $P < 0.001$) obtained by the addition of cetuximab to platinum-fluorouracil chemotherapy. The authors also reported a statistically significant progression-free survival (PFS) time in the experimental cetuximab arm as compared with the control treatment arm (median PFS = 5.6 months vs 3.3 months; $P < 0.001$). These data then led to approval of the only targeted therapy (at that time) by the US Food and Drug Administration (FDA) [18]. This approval permits the use of cetuximab as a standard treatment of locoregionally advanced HNSCC in combination with radiation therapy or in the recurrent locoregional/metastatic setting with platinum-based therapy and 5-FU or in the same setting after failing platinum-based therapy [18]. Commonly observed adverse events (AEs) that were grade 3 or higher and associated with cetuximab treatment include neutropenia, anemia, thrombocytopenia, skin reactions, and leukopenia [2].

Afatinib, a small molecule inhibitor of EGFR, has also been evaluated for its anticancer efficacy via multiple clinical studies in HNSCC patients. Afatinib is known to strongly bind to intracellular kinase domains of EGFR (and other proteins in the ErbB family) to prevent downstream signaling of EGFR [19]. For instance, the LUX-Head and Neck 1 study, a global, randomized Phase 3 trial evaluated the efficacy of afatinib versus methotrexate as a second-line treatment for patients with recurrent or metastatic HNSCC [20]. Although an improvement in OS and ORR was not observed, it did provide evidence suggesting that afatinib use was associated with an increase in PFS (median PFS = 2.6 months vs 1.7 months; $P = 0.03$), especially among human papillomavirus (HPV)–negative tumors, which comprised the largest cohort of subjects in this trial. This finding is certainly of clinical importance since HPV-negative HNSCCs show poor outcomes. Further investigation might be necessary to validate the role of EGFR inhibition for the treatment of HPV-negative HNSCCs. To this effect a Phase 2 trial (NCT02979977) with a biomarker focus is underway, which is evaluating the role of dual inhibition of EGFR via afatinib and cetuximab treatment. More recently, results from the follow-up LUX-Head and Neck 3 study directed toward Asian patients further confirmed findings from the LUX-Head and Neck 1 trial where patients treated with afatinib demonstrated an improvement in PFS (median PFS = 2.9 months vs 2.6 months; $P = 0.0005$) and ORR (ORR = 28% vs 13%, odds ratio (OR) = 2.76; $P = 0.0016$) when compared with subjects receiving methotrexate [3]. These findings were observed among a large cohort of patients with HPV-negative cancers, but no additional improvement in OS was reported from this investigation. The LUX-Head and Neck 1 study also stated that elderly patients (≥65 years) derived a similar clinical benefit (ORR and PFS) with afatinib treatment as compared with their younger counterparts in the study without developing an increased risk of toxicity [21]. The most commonly occurring grade 3 or 4 AEs attributed to afatinib were diarrhea and acne. Afatinib also demonstrated a higher safety profile than methotrexate with respect to fewer treatment-related dose reductions, discontinuations, and deaths [20].

Targeting the phosphoinositide 3-kinase—AKT—Mammalian target of rapamycin axis

The phosphoinositide 3-kinase/AKT/mammalian target of rapamycin (mTOR) pathway is pivotal in regulating cellular apoptosis and chromatin remodeling [16]. Mutant proteins in this pathway have been implicated in 40% of HNSCC cases and thus represent potential actionable therapeutic targets [22].

Rapamycin is a small molecule inhibitor of intracellular mTOR Complex 1 (mTORC1) with published clinical activity in a window of opportunity trial within a neoadjuvant setting [4]. Day and associates prescribed rapamycin therapy to a group of 16 subjects daily for 21 days before the patients received definitive surgical or chemoradiation treatment. They reported a total of four RECIST responses (one complete response (CR) and three partial responses (PRs)) and 12 subjects with stable disease (SD) in this small patient population. These findings, in addition to tumor shrinkage seen during clinical exams, is intriguing considering the brief duration of treatment among advanced HNSCC patients (especially since the three PRs were noted among subjects who were tested to have p16-negative disease). Molecular analysis of pre- and posttreatment biopsied tumor specimens revealed a decrease in phosphorylated S6 and Akt^{S473} proteins among the vast majority of subjects and diminished expression in all subjects of Ki67, an important marker of cellular proliferation. mTORC1 helps promote cell growth by activating these proteins through downstream signaling. Hence, these results support successful inhibition of the target protein and provide a possible explanation of the clinical activity attributed to the study drug. Rapamycin was also well tolerated, with only a single reported case of hypokalemia that was considered to be a grade 3 AE.

The mTORC1 has also been evaluated as a potential target within a definitive treatment setting among individuals with recurrent and/or metastatic HNSCC [5]. Dunn and collaborators performed a Phase 2 trial prescribing temsirolimus, an mTORC1 inhibitor, along with low-dose carboplatin and paclitaxel to study subjects with nonnasopharynx HNSCC on days one and eight of a 21-day cycle. The ORR was noted as 41.7% (95% confidence interval (CI), 27.9–60.3), which was attributed to PRs (no CRs were observed). Temsirolimus also demonstrated a median OS of 12.8 months (95% CI, 9.8–15.8) and a median PFS of 5.9 months (95% CI, 4.8–7.1). It is interesting to note that these findings illustrate favorable results when compared with the EXTREME regimen in a first-line setting, given the fact that 31% of patients had received at least one line of prior systemic therapy. These results underscore that pronounced efficacy of the experimental medications cannot be solely attributed to individual agents but may be due to a synergistic effect caused by the interaction of mTORC1 inhibitors with chemotherapy agents. With respect to AEs (nonattributable to a specific agent), the most frequently reported grade 3 or 4 events were hematologic toxicities and 8% of subjects developed neutropenic fever while participating on the study. Moreover the rate of treatment discontinuation due to AEs in the EXTREME trial was reported to be 20%, whereas patients in the present study were able to better tolerate the treatment with an AE-related discontinuation rate of 13.9%. Although it can be challenging to draw formal conclusions via cross-study comparisons, the aforementioned favorable results do warrant further research to investigate the potential of combining temsirolimus with low-dose cytotoxic chemotherapy.

Alterations in the PIK3 signaling pathway have often been associated with HNSCC, with aberrations ranging from gain-of-function mutations and amplifications in *PIK3CA* to overexpression or activation of downstream or upstream signaling molecules, including the EGFR [23]. Consequently, buparlisib has been utilized as a pan-PIK3 blocker in clinical studies such as the BERIL-1 trial along with paclitaxel in contrast to paclitaxel and placebo treatment in a second-line setting for patients with recurrent and/or metastatic HNSCC [6]. The ORR (median ORR = 39% vs 14%; $P = 0.00031$) from this randomized and double-blind study was similar to that of the temsirolimus trial. Additionally, the OS rates (median OS = 10.4 months vs 6.5 months, hazard ratio (HR), 0.72 (95% CI, 0.49–1.04; $P = 0.041$)) from the BERIL-1 study were also comparable with those from the EXTREME trial. Both outcomes suggest that treatment refractory patients can still derive and maintain similar clinical benefits as observed in a first-line setting. From a biomarker standpoint, correlative studies were successful in identifying biomarkers of response where patients with TP53 alterations, HPV-negative status, low mutational load, or high infiltration of tumor-infiltrating lymphocytes (TILs) or CD8-positive cells derived survival benefit in the combined buparlisib and paclitaxel arm [23]. The TP53 protein acts as a negative modulator of PIK3 pathway, and mutations in this protein have been associated with a poor prognosis. Dysregulation of TP53 leads to increased activity of the PIK3 signaling pathway and makes it more sensitive to buparlisib inhibition, explaining the survival benefit in subjects with TP53 mutations. The majority of subjects (67%) who were treated on the experimental arm were HPV negative but had an OS of 14.95 months versus 5.85 months in the control arm, indicating the potential use of combined PIK3 inhibition and chemotherapy in a traditionally difficult to treat patient population. The exploratory analysis also showed that this combination therapy can serve patients with a low mutational load in circulating tumor DNA (ctDNA) in lieu of checkpoint inhibitors, since the latter treatment option is more efficacious among patients having a high mutational burden. Finally the relationship between immune infiltration status and the corresponding response to combined treatment is noteworthy as it highlights the ability of the current treatment regimen to prime the immune system. It would be interesting to assess OS rates among patients who develop resistance to anti-PIK3 and chemotherapy treatment and opt for immunotherapy in future clinical studies. The BERIL-1 trial also demonstrated a manageable safety profile of the combined treatment, with grade 3 and 4 (nonattributable) AEs consisting of hyperglycemia, anemia, fatigue, and neutropenia. On treatment, death rates were similar across both arms (20% in experimental arm vs 22% in placebo arm), most of which were attributed to progression of existing disease and none were related to protocol treatment [6]. Given these encouraging results from the BERIL-1 trial, a Phase 3 trial called the BURAN study is being planned to further corroborate these findings [24].

Single agent and combinatorial approaches to targeting the cell cycle pathway

One of the mechanisms for the modest therapeutic efficacy of cetuximab in HPV-negative and recurrent/metastatic HNSCCs is the development of EGFR resistance due to a dysregulated cyclin D1 and downstream hyperactivation of CDK4/6 proteins [7]. These proteins together form the cyclin D1–CDK4/6 complex that plays a vital role in regulating cell cycle progression [7]. More specifically, cyclin D1 (encoded by *CCND1*) is responsible for the G1-to-S phase transition, and the entire complex is negatively regulated by p16^{INK4A} (encoded by *CDKN2A*) [8]. Among patients with HPV-negative HNSCC, the Tumor Cancer Genome Atlas reports 31% of cases containing *CCND1* amplifications and 58% of cases that harbored an inactivating mutation in *CDKN2A* [25]. Therefore a novel treatment strategy aimed toward overexpressed cyclin D1 or hyperactive CDK4/6 proteins is needed to effectively treat this patient population.

Palbociclib is a selective CDK4/6 inhibitor that has been evaluated for the first time in a Phase 2 trial among the aforementioned patient population [7]. Subjects were enrolled into either one of two arms (platinum refractory or cetuximab refractory) but were prescribed the same protocol therapy consisting of palbociclib and cetuximab. The ORR reported for single-agent treatment with PD-1 inhibitors or cetuximab in the current diagnostic setting are 13%–15% and 13%, respectively. Conversely the palbociclib trial reported an ORR of 39% (95% CI, 22–59) in the platinum-resistant arm (Group 1) and 19% (95% CI, 6–38) in the cetuximab-resistant arm (Group 2). This finding not only indicates a significant improvement in ORRs compared with previously approved therapies but also suggests that increases in ORR in the cetuximab-resistant arm are due to palbociclib's remedial effect of overcoming EGFR resistance. An OS benefit was also evidenced among subjects in Group 1 (median OS=9.5 months (95% CI, 5.3–16.50)) over cetuximab monotherapy (median OS=6 months) but was comparable with PD-1 immunotherapy (median OS=7.5–8.4 months). Furthermore, data from PD-1 inhibitor studies suggest that PFS among platinum-resistant HNSCC is approximately 2 months whereas the PFS for patients with cetuximab-resistant disease is 2.1 months. The current trial revealed superior PFS results among both groups, with a median PFS value of 5.4 months (95% CI, 3.4–7.0) in the platinum-resistant group and a median PFS of 3.7 (95% CI, 2.9–4.3) months in the cetuximab-resistant cohort. The study did not report any deaths caused by protocol therapy but noted neutropenia as the most common grade 3–4 event that was attributed to palbociclib.

Ribociclib is another CDK4/6 inhibitor that has been studied in combination with cetuximab among p16-negative patients with recurrent and/or metastatic HNSCC, 70% of whom had previously failed cetuximab treatment [8]. This experiment was performed in two phases, a dose-escalation phase followed by an expansion phase. No dose-limiting toxicities (DLTs) were observed at the first dose level (400 mg/day), and one DLT was recorded at the second dose level to conclude that the maximum tolerated dose (MTD) in the current setting was 600 mg/day. Neutropenia was acknowledged as the most frequently occurring grade 3–4 treatment-related AE (TRAE). Median OS and PFS on this treatment plan were 8.3 and 3.5 months, respectively. With regard to response rates, 10.5% of study subjects attained a PR, whereas 58% of participants achieved SD. These findings further lend credence to the toxicity and OS results from the palbociclib trial and highlight the need to perform Phase 3 trials combining CDK4/6 blockers with chemotherapy among individuals with HPV negative, advanced HNSCC that have failed cetuximab treatment.

Precision medicine initiatives for the management of salivary gland carcinomas

Leveraging the epigenetic and immunogenic role of histone deacetylase inhibitors

Salivary gland carcinomas (SGCs) account for 2%–6% of head and neck cancers where the adenoid cystic carcinoma (ACC) subtype represents 50% of metastatic SGC cases [26]. Due to the high rate of malignant transformation in ACC, limited efficacy of chemotherapy treatment, and lack of FDA-approved therapies, there is great need to develop new treatment modalities [26]. The epigenetic regulatory pathway plays a profound role in maintaining chromatin architecture and its functional integrity and can be impaired in SGCs such as ACC [26]. Previous studies have noted an inverse relationship between acetyl-H3 histone expression and Ki67 and elevated chromatin hypoacetylation levels in malignant SG tumors compared with benign tissues [27]. A systemic treatment approach could involve interfering with the acetylation pattern of tumor histones via the development of histone deacetylase (HDAC) inhibitors.

Vorinostat is a small molecule inhibitor of the HDAC that has been studied in a Phase 2 trial among patients with locally advanced, recurrent, or metastatic ACC [9]. The ORR was determined as 7% (95% CI, 2–21), which resulted from two PRs and a clinical benefit rate (PR+SD) of 97% (95% CI, 83–99). HDAC inhibitors have been shown to upregulate major histocompatibility complex class I and II molecules and enhance the antiinflammatory effects of macrophages and dendritic cells [28, 29]. These physiological changes can possibly explain the observance of late responses since the two PRs were reported much later in the course of therapy during the 8th and 10th cycles of protocol treatment [9]. This highlights the potential dual role of vorinostat acting as a modulator of epigenetic and immunostimulatory mechanisms and its potential of being used in combination with immunotherapeutic agents. The two PRs were also durable that lasted for 53 and 7.2 months among the two subjects. The median stable disease duration and PFS were both reported to be 11.4 months (90% CI, 7.1–20.5), whereas the median OS endpoint was not reached in the present study. The study drug was well tolerated, with lymphopenia being the most regularly noted grade 3 event among study participants. Dose reductions occurred among 33% of the patients, with 10% of subjects discontinuing treatment due to toxic side effects.

Given the aforementioned modulatory effects of HDAC inhibitors on the immune system, vorinostat was also studied in combination with pembrolizumab among individuals with progressive, recurrent, and/or metastatic SGCs and HNSCCs [10]. The results from the SGC cohort were presented at the American Society of Clinical Oncology (ASCO) conference in 2018 where the investigators reported an ORR of 16% (only PRs) and an SD rate of 56% [10]. The study team also stated that patients with SGCs had a median PFS of 7 months and a median OS of

13 months. With regard to side effects inclusive of both cohorts, approximately 36% of all study subjects had grade 3 or higher AEs. Side effects from the study drugs mainly included renal insufficiency, fatigue, and nausea. Dose reductions for vorinostat were performed among 18% of all study subjects, and three patients died on study due to myocardial infarction, aspiration pneumonia, and presumed pneumonitis acquired from pembrolizumab treatment.

Employing a multitargeted approach to tyrosine kinase inhibition

A commonly occurring chromosomal defect in ACC is the t(6;9)(q22–23;p23–24) translocation, which is present in almost half of all ACC cases and subsequently causes the fusion of *MYB* and *NFIB* genes [12]. This fusion and the translocation-independent overexpression of the MYB transcription factor can promote upregulation of their target proteins [12]. These include the vascular endothelial growth factor A (VEGF-A), fibroblast growth factor 2 (FGF2), and KIT that play an angiogenic role in tumorigenesis [12]. In SGCs the receptor for VEGF (VEGFR) and its expression in particular has been noted to correlate with lymph node metastasis, poor local control of disease, and worse prognosis [11]. Other key players involved in SGC formation include tropomyosin receptor kinase (TRK) and human epidermal growth factor receptor 2 (HER2). An antiangiogenic approach targeting multiple tyrosine kinase proteins offers another avenue of treating SGCs to maximize the chances of eliciting a therapeutic benefit.

Sorafenib is a multikinase inhibitor that acts as an antagonist for VEGFR1–3, BRAF, platelet-derived growth factor receptor (PDGFR), RET, c-KIT, and Fms-related tyrosine kinase 3 (FLT3) [11]. This drug was the first antiangiogenic agent to demonstrate clinical activity in the context of recurrent and/or metastatic SGCs (RMSGCs) in a Phase 2 study [11]. The combined ORR was 16% (all PRs), which were observed among six subjects; two of them had a histologically confirmed diagnosis of ACC (ORR=11%), whereas the remaining four patients were confirmed to have non-ACC-based SGCs (ORR=22%). Approximately 59% of patients had SD, whose duration exceeded the established metric period of 6 months in 50% of the patients and was more common in ACC subjects than their non-ACC counterparts. The median OS was reported as 26.4 months for ACC and 12.3 months for non-ACC SGC patients ($P = 0.1963$). Median PFS was 8.9 months for ACC and 4.2 months for non-ACC SGC subjects ($P = 0.0386$). During molecular analysis of tumor specimens, it was interesting to note that there was no significant correlation between *MYB-NFIB* fusion or *MYB* expression levels and response rates. However, all six responders depicted an enriched PDGFRβ component that was restricted to the stroma but absent from the tumor cells. PDGFRβ is involved in angiogenesis, and its blockade renders the immature vasculature of tumor cells more vulnerable to the anti-VEGFR effect of sorafenib. PDGFRβ could serve as a clinically meaningful biomarker predicting response and could be vital in guiding treatment decisions for SGC patients using antiangiogenic therapies. Toxicities attributed to sorafenib treatment that were grade 3 or higher were seen in 29.7% patients and the most frequently occurring TRAE was lung infection. There were 14 subjects who suffered from a serious adverse event (SAE), although only four cases were ascribed directly to the study drug. These four SAE cases consisted of tumor necrosis, aspergillus infection, artery thrombosis, and fatal meningitis. With regard to these four SAEs, three of them were considered to be due to disease relapse within the field of previous treatment (surgery, radiotherapy, etc.), which emphasizes the need for proactive monitoring of patients with these clinical characteristics.

Another antiangiogenic drug that has been evaluated among RMSGC patients is lenvatinib that is directed toward VEGFRs 1–3, FGFRs 1–3, KIT, PDGFR α and β, and RET [12]. With an ORR of 15.6% (all PRs), the drug in this Phase 2 trial appears to fare better than other TKIs profiled in a similar clinical setting whose ORRs have typically ranged from 6% to 10% [12, 26]. The clinical benefit rate lasting ≥6 months was observed among 40.6% of study patients. Moreover, lenvatinib also demonstrated superior PFS results (median PFS = 17.5 months; 95% CI, 7.2 months to not reached) when compared with sorafenib. The study also emphasized the importance of cautious management of grade 3–4 TRAEs, which were noted among 62.5% of study subjects, with hypertension and oral pain being the most commonly occurring TRAEs.

With the advent of innovative clinical trial designs, the concept of a tumor agnostic approach to confer greater sensitivity to targeted therapies has enhanced the speed of biomarker discovery and deployment of effective new treatments [16, 30]. Larotrectinib (LOXO-101) was first evaluated using a basket trial design to enroll patients with various unresectable or metastatic tumor types harboring a *NTRK* fusion [30]. The drug is a small molecule inhibitor with a high selectivity for TRK proteins whose aberrant constitutive activation leads to tumor cell proliferation and survival [30, 31]. Updated results from the expanded dataset presented by Hong and associates during the ASCO Annual Meeting in 2019 indicated an objective response among 16 out of 18 subjects with the rare secretory sarcoma [13]. The ORR and PFS reported together for patients with all tumor types were 68% (95% CI, 55–79) and 25.8 months (range = 0.03+ to 39.7+ months), respectively. Most frequently noted, grade 3 or higher TRAEs were found in 16% of all study subjects, which is only composed of elevated levels of alanine aminotransferase. As a result of larotrectinib's promising efficacy findings, the drug received a breakthrough designation by the FDA and received priority review before eventually becoming the second tissue-agnostic drug to receive FDA approval for cancer treatment [32].

Application of small molecule TKIs can also be complemented via the use of novel antibody-drug conjugates such as ado-trastuzumab emtansine (T-DM1) [33]. T-DM1 targets the angiogenesis promoting HER2 that is overexpressed in 30% of salivary duct carcinomas, an aggressive histological subtype of SGCs [14]. Receptor-mediated endocytosis leads to lysosomal degradation of the conjugate and releases the cytotoxic DM1-containing catabolites that disrupt microtubule polymerization and induce apoptosis of the tumor cell [33]. Efficacy results of this agent among non-ACC-based SGC patients with *HER2* amplification were delivered by Li and colleagues during an oral presentation at the 2019 ASCO conference that demonstrated an ORR of 90% (95% CI, 56–100) [14]. Based on the RECIST and PERCIST criteria, CRs were observed among 50% of study subjects, all of whom were heavily pretreated with trastuzumab, pertuzumab, and antiandrogen therapy prior to this trial. At the time of this presentation, patients had been followed for a median period of 12 months (range = 4–20 months), and the median PFS (95% CI, 4–22+ months) and DOR (range = 2–19+ months) had not been reached. T-DM1 was also well tolerated since the most commonly occurring grade 3 or higher TRAE was thrombocytopenia, and no treatment-related deaths occurred on this trial. These encouraging findings demonstrate a dramatic improvement in durable response rates and an enhanced safety profile in certain SGCs compared with previously evaluated TKIs. The trial continues to enroll more patients in an expansion cohort to further validate these findings.

Concluding remarks

The adoption of a precision medicine approach has revolutionized treatments for certain head and neck cancers, such as HNSCC and advanced salivary gland cancers. However, a complex molecular landscape in the recurrent/metastatic setting has limited the efficacy of such agents due to activation of multiple molecular pathways that promote drug resistance [16]. Additionally, there is the challenge of patient compliance, since individuals with advanced cancers might not be able to tolerate an oral drug therapy due to their limited swallowing ability [16].

With respect to trial design, conventional clinical studies in the past utilized an unselected subject accrual strategy that could confound outcomes, since patients with specific molecular signatures could be expected to derive greater clinical benefits than the rest of the population in a given disease. A biomarker enriched strategy is vital in assessing clinical outcomes for difficult to treat patient populations, ranging from HPV-negative HNSCC patients that are refractory to standard treatments to orphan diseases such as ACC for which there is a lack of approved effective therapies. Such strategies appear to be gaining momentum and have benefited from novel trial designs such as basket trials that not only have streamlined the drug development pipeline but also have expedited the regulatory approval process. Moreover, they have allowed validation of predictive biomarkers and have extended the benefits of using a single drug for multiple tumor types harboring the same mutation. Advances in computational technology especially in the areas of artificial intelligence and deep machine learning have further aided the discovery of potential biomarkers and molecular drug targets [34]. Integration of genomic testing of tissue specimens in standard of care practices is vital in strategizing the selection and sequence of targeted therapies to optimize clinical outcomes.

References

[1] Le Q-T, Giaccia AJ. Therapeutic exploitation of the physiological and molecular genetic alterations in head and neck cancer. Clin Cancer Res 2003;9(12):4287–95.

[2] Vermorken JB, Mesia R, Rivera F, Remenar E, Kawecki A, Rottey S, Erfan J, Zabolotnyy D, Kienzer HR, Cupissol D, Peyrade F, Benasso M, Vynnychenko I, De Raucourt D, Bokemeyer C, Schueler A, Amellal N, Hitt R. Platinum-based chemotherapy plus cetuximab in head and neck cancer. N Engl J Med 2008; 359(11):1116–27.

[3] Guo Y, Ahn M-J, Chan ATC, Wang C-H, Kang JH, Kim S-B, Bello MDG, Singh Arora R, Zhang Q, He X, Li P, Dechaphunkul A, Kumar V, Kamble K, Li W, Kandil A, Cohen EEW, Geng Y, Zografos L, Tang PZ. Afatinib versus methotrexate as second-line treatment for patients with recurrent and/or metastatic (R/M) head and neck squamous cell carcinoma (HNSCC) progressing on or after platinum-based therapy: LUX-head & neck 3 phase III trial. J Clin Oncol 2019;37(15_Suppl):6024.

[4] Day TA, Shirai K, O'Brien PE, Matheus MG, Godwin K, Sood AJ, Kompelli A, Vick JA, Martin D, Vitale-Cross L, Callejas-Varela JL, Wang Z, Wu X, Harismendy O, Molinolo AA, Lippman SM, Van Waes C, Szabo E, Gutkind JS. Inhibition of mTOR signaling and clinical activity of rapamycin in head and neck cancer in a window of opportunity trial. Clin Cancer Res 2019;25(4):1156–64.

[5] Dunn LA, Fury MG, Xiao H, Baxi SS, Sherman EJ, Korte S, Pfister C, Haque S, Katabi N, Ho AL, Pfister DG. A phase II study of temsirolimus added to low-dose weekly carboplatin and paclitaxel for patients with recurrent and/or metastatic (R/M) head and neck squamous cell carcinoma (HNSCC). Ann Oncol 2017;28(10):2533–8.

[6] Soulieres D, Faivre S, Mesía R, Remenár É, Li SH, Karpenko A, Dechaphunkul A, Ochsenreither S, Kiss LA, Lin JC, Nagarkar R, Tamás L, Kim SB, Erfán J, Alyasova A, Kasper S, Barone C, Turri S, Chakravartty A, Chol M, Aimone P, Hirawat S, Licitra L. Buparlisib and paclitaxel in patients with platinum-pretreated recurrent

or metastatic squamous cell carcinoma of the head and neck (BERIL-1): a randomised, double-blind, placebo-controlled phase 2 trial. Lancet Oncol 2017;18(3):323–35.

[7] Adkins D, Ley J, Neupane P, Worden F, Sacco AG, Palka K, Grilley-Olson JE, Maggiore R, Salama NN, Trinkaus K, Van Tine BA, Steuer CE, Saba NF, Oppelt P. Palbociclib and cetuximab in platinum-resistant and in cetuximab-resistant human papillomavirus-unrelated head and neck cancer: a multicentre, multigroup, phase 2 trial. Lancet Oncol 2019;2045(19):1–11.

[8] Seront E, Schmitz S, Papier M, Van Maanen A, Henry S, Lonchay C, Rottey S, Van Caloen G, Machiels JP. Phase 1 study evaluating the association of the cyclin-dependent kinase 4/6 inhibitor ribociclib and cetuximab in recurrent/metastatic p16-negative squamous cell carcinoma of the head and neck. Front Oncol 2019;9:155.

[9] Goncalves PH, Heilbrun LK, Barrett MT, Kummar S, Hansen AR, Siu LL, Piekarz RL, Sukari AW, Chao J, Pilat MJ, Smith DW, Casetta L, Boerner SA, Chen A, Lenkiewicz E, Malasi S, LoRusso PM. A phase 2 study of vorinostat in locally advanced, recurrent, or metastatic adenoid cystic carcinoma. Oncotarget 2017;8(20):32918–29.

[10] Rodriguez CP, Wu V, Voutsinas JM, Fromm JP, Pillarisetty VG, Lee S, Santana-Davila R, Goulart BHL, Baik CS, Chow LQM, Eaton KD, Martins RG. Phase I/II trial of pembrolizumab (P) and vorinostat (V) in recurrent metastatic head and neck squamous cell carcinomas (HN) and salivary gland cancer (SGC). J Clin Oncol 2018;36(15_suppl):6025.

[11] Locati LD, Perrone F, Cortelazzi B, Bergamini C, Bossi P, Civelli E, Morosi C, Lo Vullo S, Imbimbo M, Quattrone P, Dagrada GP, Granata R, Resteghini C, Mirabile A, Alfieri S, Orlandi E, Mariani L, Saibene G, Pilotti S, Licitra L. A phase II study of sorafenib in recurrent and/or metastatic salivary gland carcinomas: translational analyses and clinical impact. Eur J Cancer 2016;69:158–65.

[12] Tchekmedyian V, Sherman EJ, Dunn L, Tran C, Baxi S, Katabi N, Antonescu CR, Ostrovnaya I, Haque SS, Pfister DG, Ho AL. Phase II study of lenvatinib in patients with progressive, recurrent or metastatic adenoid cystic carcinoma. J Clin Oncol 2019;37(18):1529–37.

[13] Hong DS, Kummar S, Farago AF, Lassen UN, Berlin J, Schilder RJ, McDermott RS, Patel JD, Dowlati A, Doebele RC, Tan DS-W, Lee JJ, Nanda S, Childs BH, Ku N, Drilon AE, Hyman DM. Larotrectinib efficacy and safety in adult TRK fusion cancer patients. J Clin Oncol 2019;37(15_suppl):3122.

[14] Li BT, Shen R, Offin M, Buonocore DJ, Myers ML, Venkatesh A, Razavi P, Ginsberg MS, Ulaner GA, Solit DB, Hyman DM, Rudin CM, Gedvilaite E, Tsui D, Arcila ME, Kris MG, Weitsman G, Ng T, Scaltriti M, Ho AL. Ado-trastuzumab emtansine in patients with HER2 amplified salivary gland cancers (SGCs): results from a phase II basket trial. J Clin Oncol 2019;37(15_suppl):6001.

[15] Bonner JA, Harari PM, Giralt J, Azarnia N, Shin DM, Cohen RB, Jones CU, Sur R, Raben D, Jassem J, Ove R, Kies MS, Baselga J, Youssoufian H, Amellal N, Rowinsky EK, Ang KK. Radiotherapy plus cetuximab for squamous-cell carcinoma of the head and neck. N Engl J Med 2006;354(6):567–78.

[16] Gougis P, Bachelard CM, Kamal M, Gan HK, Borcoman E, Torossian N, Bièche I, Pharm D, Le Tourneau C. Clinical development of molecular targeted therapy in head and neck squamous cell carcinoma. JNCI Cancer Spectr 2019;3. https://doi.org/10.1093/jncics/pkz055.

[17] Temam S, Kawaguchi H, El-Naggar AK, Jelinek J, Tang H, Liu DD, Lang W, Issa J-P, Lee JJ, Mao L. Epidermal growth factor receptor copy number alterations correlate with poor clinical outcome in patients with head and neck squamous cancer. J Clin Oncol 2007;25:2164–70.

[18] US Food and Drug Administration. Drugs @ FDA, https://www.accessdata.fda.gov/scripts/cder/daf/. Accessed 25 August 2019.

[19] Wind S, Schnell D, Ebner T, Freiwald M, Stopfer P. Clinical pharmacokinetics and pharmacodynamics of afatinib. Clin Pharmacokinet 2017;56(3):235–50.

[20] Machiels JPH, Haddad RI, Fayette J, Licitra LF, Tahara M, Vermorken JB, Clement PM, Gauler T, Cupissol D, Grau JJ, Guigay J, Caponigro F, de Castro G, de Souza Viana L, Keilholz U, del Campo JM, Cong XJ, Ehrnrooth E, Cohen EEW. Afatinib versus methotrexate as second-line treatment in patients with recurrent or metastatic squamous-cell carcinoma of the head and neck progressing on or after platinum-based therapy (LUX-Head & Neck 1): an open-label, randomised phase 3 trial. Lancet Oncol 2015;16(5):583–94.

[21] Clement PM, Gauler T, Machiels JP, Haddad RI, Fayette J, Licitra LF, Tahara M, Cohen EEW, Cupissol D, Grau JJ, Guigay J, Caponigro F, de Castro G, de Souza VL, Keilholz U, Del Campo JM, Cong XJ, Ehrnrooth E, Vermorken JB. LUX-H&N 1 investigators. Afatinib versus methotrexate in older patients with second-line recurrent and/or metastatic head and neck squamous cell carcinoma: subgroup analysis of the LUX-Head & Neck 1 trial. Ann Oncol 2016;27(8):1585–93.

[22] cBioPortal for Cancer Genomics, http://www.cbioportal.org/. [Accessed 2 September 2019].

[23] Soulieres D, Licitra L, Mesía R, Remenar E, Li SH, Karpenko A, Chol M, Wang YA, Solovieff N, Bourdeau L, Sellami D, Faivre S. Molecular alterations and buparlisib efficacy in patients with squamous cell carcinoma of the head and neck: biomarker analysis from BERIL-1. Clin Cancer Res 2018;24(11):2505–16.

[24] Adlai Nortye, http://www.adlainortye.com/en_newsdetail.php?cid=69&id=111. [Accessed 4 September 2019].

[25] The Cancer Genome Atlas Network. Comprehensive genomic characterization of head and neck squamous cell carcinomas. Nature 2015;517(7536):576–82.

[26] Schvartsman G, Pinto NA, Bell D, Ferrarotto R. Salivary gland tumors: molecular characterization and therapeutic advances for metastatic disease. Head Neck 2019;41(1):239–47.

[27] Wagner VP, Martins MD, Guimaraes DM, Vasconcelos AC, Meurer L, Vargas PA, Fonseca FP, Squarize CH, Castilho RM. Reduced chromatin acetylation of malignant salivary gland tumors correlates with enhanced proliferation. J Oral Pathol Med 2017;46(9):792–7.

[28] Brogdon JL, Xu Y, Szabo SJ, An S, Buxton F, Cohen D, Huang Q. Histone deacetylase activities are required for innate immune cell control of Th1 but not Th2 effector cell function. Blood 2007;109(3):1123–30.

[29] Magner WJ, Kazim AL, Stewart C, Romano MA, Catalano G, Grande C, Keiser N, Santaniello F, Tomasi TB. Activation of MHC class I, II, and CD40 gene expression by histone deacetylase inhibitors. J Immunol 2000; 165(12):7017–24.

[30] Drilon A, Laetsch TW, Kummar S, Dubois SG, Lassen UN, Demetri GD, Nathenson M, Doebele RC, Farago AF, Pappo AS, Turpin B, Dowlati A, Brose MS, Mascarenhas L, Federman N, Berlin J, El-Deiry WS, Baik C, Deeken J, Boni V, Nagasubramanian R, Taylor M, Rudzinski ER, Meric-Bernstam F, Sohal DPS, Ma PC, Raez LE, Hechtman JF, Benayed R, Ladanyi M, Tuch BB, Ebata K, Cruickshank S, Ku NC, Cox MC, Hawkins DS, Hong DS, Hyman DM. Efficacy of larotrectinib in TRK fusion-positive cancers in adults and children. N Engl J Med 2018;378(8):731–9.

[31] Lange AM, Lo HW. Inhibiting TRK proteins in clinical cancer therapy. Cancer 2018;10(4):105.

[32] FDA approves an oncology drug that targets a key genetic driver of cancer, rather than a specific type of tumor, https://www.fda.gov/news-events/press-announcements/fda-approves-oncology-drug-targets-key-genetic-driver-cancer-rather-specific-type-tumor. [Accessed 8 September 2019].

[33] LoRusso PM, Weiss D, Guardino E, Girish S, Sliwkowski MX. Trastuzumab emtansine: a unique antibody-drug conjugate in development for human epidermal growth factor receptor 2-positive cancer. Clin Cancer Res 2011;17(20):6437–47.

[34] Shah P, Kendall F, Khozin S, Goosen R, Hu J, Laramie J, Ringel M, Schork N. Artificial intelligence and machine learning in clinical development: a translational perspective. NPJ Digit Med 2019;2:1–5.

CHAPTER

4

Novel drug delivery systems for head and neck cancers: Beyond the horizon

Yazeed M. Alhiyari[a], *Amanda Miller*[b], *Albert Y. Han*[a,c], *Maie A. St. John*[a,c,d,e]

[a]Department of Head and Neck Surgery, David Geffen School of Medicine at University of California Los Angeles, Los Angeles, CA, United States [b]David Geffen School of Medicine at University of California Los Angeles, Los Angeles, CA, United States [c]UCLA Head and Neck Cancer Program, UCLA Medical Center, Los Angeles, CA, United States [d]Department of Bioengineering, University of California Los Angeles, Los Angeles, CA, United States [e]UCLA Jonsson Comprehensive Cancer Center, Los Angeles, CA, United States

Abstract

Surgical resection of advanced head and neck squamous cell carcinoma (HNSCC) poses a unique clinical challenge due to nearby vital structures necessary for speech, swallowing, and breathing. Aggressive surgical resection severely impairs functioning and quality of life. Despite considerable advancements in the treatment of HNSCC, long-term survival has continued to be poor in the past 30 years. The standard of care for most HNSCC remains surgery followed by adjuvant radiation and/or chemotherapy. To improve patient survival and preserve critical functions that directly impact quality of life, innovative treatment modality and delivery mechanisms have been explored, with promising results. Several novel drug delivery modalities offer a new potential for patients with advanced HNSCC. These thoughtfully engineered materials can exhibit release of chemotherapeutics at variable kinetics, provide enhanced imaging techniques allowing for minimization of radiation exposure, and optimize immunomodulatory signals to combat the tumor microenvironment.

Abbreviations

aPD1 anti-programmed cell death 1
CDNs cyclic dinucleotides
EGFR epidermal growth factor receptor
HNSCC head and neck squamous cell carcinoma
HPV human papilloma virus
OPSCC oropharyngeal squamous cell carcinoma
PCL poly-e-caprolactone

https://doi.org/10.1016/B978-0-12-820679-9.00004-9

PLCL poly-DL-lactide-co-e-caprolactone
STING stimulator of interferon genes
TQ thymoquinone

Conflict of interest

No potential conflicts of interest were disclosed.

Introduction

Head and neck squamous cell carcinoma (HNSCC) causes more than 300,000 annual deaths worldwide and remains the sixth most common cancer. Head and neck cancer accounts for up to 4% of all cancer diagnoses in the United States [1]. This year alone, close to 70,000 people will develop head and neck cancer. A large majority of patients exhibit preexisting risk factors including a long smoking history or history of heavy alcohol use. Human papilloma virus (HPV) infection is also a significant contributor to about >90% of oropharyngeal squamous cell carcinoma (OPSCC) [2], and the number of people with oropharyngeal cancers associated with HPV infection is on the rise. Radiation exposure during treatment of a previous cancer, such as lymphomas in childhood or thyroid cancer, and immunosuppression are also risk factors for carcinogenesis of HNSCC.

OPSCC classically presents in patients in their 60s, but the age of diagnosis has been trending to a younger age [2]. Patients often present with a solitary neck mass that is painless. Visual inspection, physical examination, and cross-sectional imaging confirm the diagnosis; however, cytopathological swabs for in-office screenings (similar to a Pap-smear) are not yet available. Due to the lack of symptoms, patients are often diagnosed at an advanced stage that requires aggressive treatment that places critical neurovascular structures at risk.

Management of head and neck cancer involves a combination of treatment modalities often developed by a multidisciplinary team of specialists. Multiple clinicopathologic characteristics are incorporated at the time of treatment planning, including patient's age, medical comorbidities, subsite, histopathologic diagnosis, stage, grade, margin status, perineural invasion, and lymphovascular invasion. For patients with HNSCC the primary therapy often fails, requiring salvage treatment [3].

Surgery remains the mainstay treatment for HNSCC. However, there are severe impairments in both the functioning and quality of life following many resections. Due to the anatomical location of many HNSCCs, patients may be severely disfigured following surgery. This outcome has profound psychological and emotional effects, as well as permanently impaired fundamentally human functionalities including speaking, eating, and swallowing. Surgical planning, as well as reconstruction, necessitates careful consideration of oncologic sound surgery in the context of functional preservation. Often, cancers in the advanced stage are deemed unresectable due to encasement of critical vascular structures. In these cases, novel treatment plans including immunotherapy and local drug delivery can offer targeted therapy by minimizing systemic effects [4].

This chapter will explore exciting new developments in the realms of biomedical engineering, nanotechnology, and immunotherapy in advancing the treatment of late-stage head and neck cancer patients who have not been rescued by traditional and current treatment strategies. In the future, these technologies may be integrated into the mainstay treatment. These tools may also enhance our knowledge of oncogenesis, metastasis, and recurrence of HNSCC. Ultimately, it is of the utmost importance to further investigate these novel drug delivery modalities to significantly achieve an improvement of quality of life and cure rates over the long term.

Polymer films

Polymers may be synthesized in thin films or sheets that can be reshaped to match the surface of the tumor. In head and neck cancer management, a polymer drug delivery platform can provide a local, targeted therapy against the tumor, especially when critical structures are involved that result in positive margins and/or unresectability requiring nonsurgical treatment. The polymer components are completely biodegradable, and modulation of the chemical structure allows for optimization of the physical properties and release characteristics of the film. This modality is gaining in popularity, as the flexibility of polymer chemistry and drug modifications can create unique degradation and release kinetics tailored to a certain application.

The physical properties of a polymer film allow the surgeon to implant it directly into the tumor bed, and if necessary, peel, cut to size and shape, and reposition as needed. Furthermore, polymer layers can be combined to allow different concentrations of different chemicals to be released at different rates by carefully selecting the appropriate polymer backbone for each layer and each desired condition.

For example, a polymer created at UCLA, composed of a mixture of poly-e-caprolactone (PCL) and poly-DL-lactide-co-e-caprolactone (PLCL), was used to deliver various chemotherapeutics locally to debulked HNSCCs in a mouse model (Fig. 1) [6]. The breakdown of the polymer backbone into lactides and caprolactones is well known; both are used in FDA-approved devices with excellent biocompatibility.

PCL is highly versatile and affords the polymer film the function of encapsulating many drugs and small molecules. Its low glass transition temperature allows for high polymer chain flexibility at body temperature and affords higher drug diffusivity. PCL alone is not malleable enough, and PLCL is not strong enough for surgical placement and manipulation in vivo. However, the combination of each polymer component yields optimized flexibility and malleability for facile handling [5]. This polymer film can be used to slowly release cisplatin, a well-studied chemotherapeutic agent. Thus, the sustained release drug delivery system exhibited by this thoughtfully designed polymer layer showed improved antitumor efficacy and reduced systemic toxicity when tested in vivo and compared with systemically administered cisplatin [5].

With the controlled drug release kinetics of biocompatible polymers and meshes, a new set of potentially efficacious drugs to shrink tumors and prevent recurrence is now an active area of research. For example, some chemicals may hold significant promise in their ability to be toxic to cancer cells; however, their use in clinics or the operating room might have been previously hindered by the dosing required to achieve effective tumor control. This is especially true if these drug levels would be otherwise particularly toxic to all tissues, including those already damaged by traditional chemotherapy and radiation therapy treatment methods.

The applications for polymer use are not limited to current chemotherapeutics, as immune modulates such as anti-PD1 antibody, radio enhancers (e.g., gold nanoparticles), or other

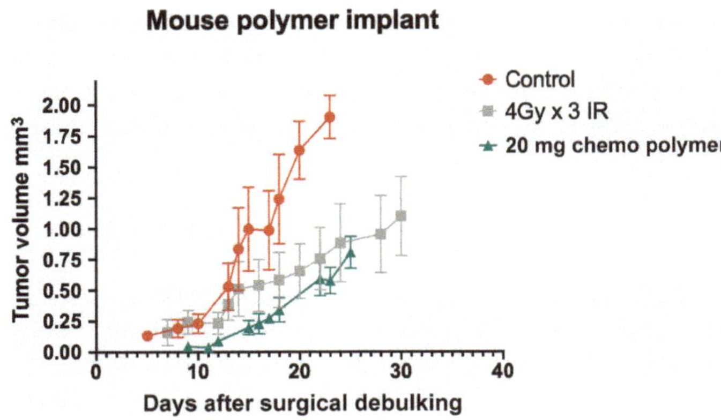

FIG. 1 Control polymer, *red line*, is PCL-PLCL vehicle control. *Green line* represents PCL-PLCL polymer embedded with chemotherapeutic TQ. The PCL-PLCL combinatorial polymer layers were placed surgically in mice with head and neck debulked tumors. The *gray line* represents fractionated radiotherapy 4Gy daily for 3days after tumor debulking. Here it is demonstrated that equivalent or superior tumor control can be achieved with a chemo/polymer combination to irradiation alone, and this is attributable to slow, extended, local release of the chemotherapeutic over time [5].

compounds may be embedded in the backbone material. Even compounds like a naturally occurring bioactive agent thymoquinone (TQ) found in black seed oil have long been claimed to exhibit antioxidant and antitumor effects when digested orally [7,8]. However, this agent's clinical use showed that high dosing may be required to see tumor-shrinking properties, and it can cause a wide range of side effects including mortality if introduced intravenously. Thus, by loading TQ and other molecules of its kind into polymer films that can securely wrap the tumor, systemic toxicity can be limited while enhancing the therapeutic power never previously harnessed.

Hydrogels and scaffolds

Hydrogel scaffolds have become a popular new method of bioengineering tissue regeneration, immune modulation, and cancer ablation. These materials hold the power to directly deliver drugs within the tumor environment while also restoring immune cell function and allowing for additional applications. Hydrogen scaffolds can release growth factors and drugs to direct the healing of surrounding tissues and bone regeneration following surgical resection of the tumor.

Polymeric hydrogels are of interest for their tunable physicochemical properties to be tailored directly to the needed application. These have the added advantage of being biocompatible and biodegradable, integrating well with the surrounding tissue. The controlled release of encapsulated drugs, proteins, and even cells can be achieved to provide many different effector functions, including the treatment of head and neck cancer while overcoming the challenges thereof.

It has been demonstrated in bone that natural products including fibrin, collagen, and gelatin, as well as polysaccharides (e.g., chitosan and alginate), can serve as foundations for the hydrogels to provide bone healing. Polymer backbones can be swapped and exchanged to ultimately effect mechanophysical properties and degradation characteristics of the drugs or biologics loaded in the gel [9].

Gel-like scaffolds can be injected, making for easy placement during surgery. Such scaffolds can hold a combination of chemo-immunotherapeutic drugs and deliver them locally. The polymeric network creates a gel that can be formed quickly inside the body. For example, anti-PD1 antibody and gemcitabine were combined and loaded into the gel for delivery within a tumor. Upon injection, the gel promotes the kinds of tumor characteristics that immunotherapies can identify. As the scaffold breaks down, it sequentially releases gemcitabine followed by anti-PD1. First causing cytotoxic cancer cell death, then stimulating an immune response effectively promoting better tumor control than single drug delivery regimens [10].

Many challenges exist with treating aggressive cancers as they are resistant to conventional therapies. However, the adaptability of hydrogels in allowing organic and inorganic compounds to mix and interact without spreading leads to unique treatment possibilities. One such challenge is dealing with multidrug-resistant cancers. Aetzi and colleagues demonstrated that embedding dark-gold nanoparticles modified by 5-fluorouracil in hydrogel can switch off the multidrug resistance protein expression in a breast cancer cell line [11]. The gold nanoparticles modified by 5-fluorouracil are also known as nanobeacons and release the drug only upon hybridization of DNA through a conformational change. The nanobeacons are labeled with a near-infrared dye and dark quencher at opposite ends, and the activity can be monitored via fluorescence [12]. With the power to reverse drug resistance, this drug delivery vehicle could significantly revolutionize outcomes in late-stage HNSCC.

Besides treating cancer, gels and scaffolds have found their way into regenerative medicine as well. Osteogenesis-inducing scaffolds have been employed to direct tissue engineering following mandibular defects as an alternative to autologous fibular free flap. The biomimetic is osteoconductive, osteoinductive, and supports osseointegration. Bone morphogenic protein 2 is released while stem cells are seeded into the scaffold [13], which can be placed into the defect following surgery. Fibrin-based scaffolds seeded with stem cells have been tested in animal models for wound healing in skin, vocal fold, and various other tissues to mitigate scar formation. These types of treatments may prove to be the future for patients with significant bone and/or tissue loss following their surgical treatment.

Decellularized scaffolds also hold promise in the treatment of head and neck cancer. They can modulate immune function and encourage healthy tissue growth while limiting tumor recurrence and synergizing with cancer immunotherapy. Biologic scaffolds are used to prime the immune microenvironment to shape tumor regression and improve the immunotherapy response [14].

Augmented radiation therapy and imaging

Another significant challenge when treating advanced head and neck cancer arises in the stages after salvage surgery. It is possible to debulk the tumor significantly, but there needs to be a treatment method for reducing cancerous tissue that is necessarily left behind as it may be attached to nearby structures such as the common carotid artery or internal jugular vein. It is currently the standard of care to provide radiation therapy for this purpose. A radiation oncologist may see the patient weekly for up to 3 months, delivering high-dose radiation. While this may be beneficial in reducing the cancerous tissue masses, it is also highly damaging to the delicate structures of the head and neck. Unfortunately, the field of radiation delivered must remain wide to include all areas that could harbor regional metastases. Structures might have artificially shifted during surgery, and often the cancer tissue that is left behind is not in its original location based on imaging prior to the operation. If the field of radiation is limited to precisely the location of cancer tissue, vital structures and healthy normal tissue can be protected as much as possible.

Delivering a lethal dose of radiation to cancer cells while sparing nearby normal tissues has always been a significant challenge in radiation therapy. We know that radiation has its unique clinical applications but is often limited by this problem. More modern technology such as interstitial brachytherapy attempts to solve this by surgically implanting radiation sources such as afterload catheters or seeds in and around the tumor left behind during surgery. These small devices deliver continuous slow-dose radiation to a limited volume but have been efficacious [15]. Even with this proposed method, there remains the requirement of health personnel to wear protective gear to prevent radiation exposure, as well as the requirement of the patient to remain in isolation away from family and loved ones. Lastly, there is always the possibility of the brachytherapy agents shifting from their original positions, limiting the radiation dose to the tumor bed as was intended.

One technique emerging in this field is the design of a radiopaque tracer polymer. The polymer will have properties that allow it to be seen outside of the background on standard CT imaging. There are a variety of materials that can give the polymer these properties. Gold and tantalum, for example, are metals with a high specific Z, which allow them to be dense and radiopaque on imaging. Calcium carbonate, similar to the components of bone, shows up as a dense structure against the background of human tissues while providing no biological side effects as it is degraded in the body (Fig. 2). If these polymers are placed exactly on the cancer tissue that is left behind, they can remain in the body for up to 3 months, lasting throughout the patient's radiation treatment. The radiation oncologist now has the added tool of being able to precisely localize the cancer tissue for radiation targeting, limiting the field of radiation as much as possible so that the surrounding healthy tissue can be spared the damage. As stated previously this style of polymer is unique in the fact that the length of time it resides in the body can be controlled based on the mixed ration of the polymer backbone. Furthermore, multiple layers with different components can be made in one polymer unit. Each layer can be engineered for different applications. A multilayered polymer could serve to be a marker for pre−/postoperative care, while also radiosensitizing the tumor and locally delivering chemotherapeutics and immune modulators.

Pretargeting is often employed to enhance imaging quality and involves pretreating with a targeting ligand and waiting for it to accumulate at the tumor site before imaging. This method can enhance specificity and reduce off-target radiation. For ease and speed, passive targeting without the need for a specific receptor can be used and may be superior in some situations. Goos et al. designed star-shaped copolymer nanoparticles conjugated with $Gd3+$ and radiolabeled with $89Zr$ [16]. These nanoparticles minimize the radiation dose to healthy tissue because they are able to accumulate at the site of the tumor. They enable superior imaging with enhanced contrast using PET combined with MRI.

Microneedle patches

Another intriguing drug delivery modality involves drug-loaded microneedles arranged in patches that can be placed locally on oral mucosa or skin. The microneedles painlessly

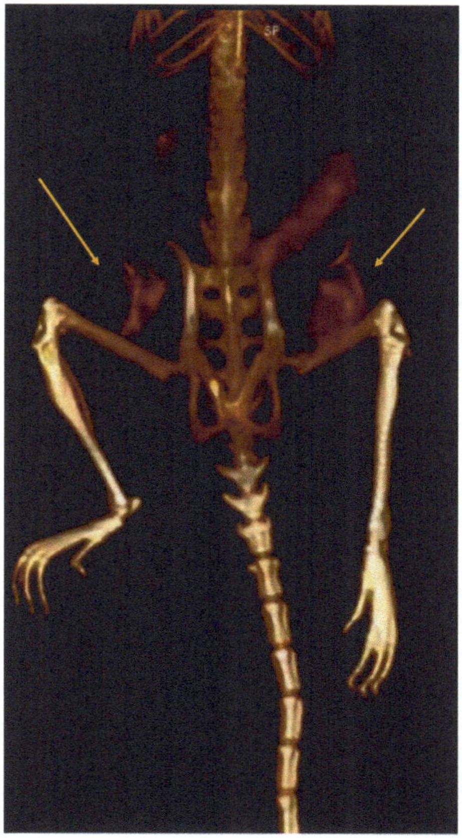

FIG. 2 Radiopaque polymer thin film containing calcium carbonate can be placed postoperatively and assist in localizing the remaining cancer tissue and limiting radiation field. The *arrows* show a calcium carbonate-embedded polymer surgically placed on either flank of the mouse in the bone-enhanced CT image. These polymers are designed to naturally dissolve over the time course of treatment and can be useful for pre/post-op evaluation.

pierce the surface epithelium and enable the local delivery of their contents to the underlying tumor. The chemical structure of the microneedles can also be modified such that small molecule drugs or therapeutic micro- and nanoparticles can be loaded within each needle, with the underlying patch layer remaining inert and quickly degrading, leaving the microneedles and their contents behind in the tissue [17].

This has recently been an incredibly popular technology that has many different clinical applications to explore. As such, it is not surprising that many different companies (Micron Biomedical, Nissha Co. and 3M) have begun manufacturing various versions of a patch coated in solid or loaded microneedles. Applications include long-acting contraception, skin care cosmetics, and alternative vaccines. Of course, there is additionally an important clinical unmet need in the treatment of head and neck cancers, especially those that are late-stage and unresectable. Wang and colleagues have synthesized a self-degradable microneedle patch with sustained delivery of anti-PD1 [18]. The microneedles are made of hyaluronic acid with dextran nanoparticles loaded with encapsulate anti-PD1 and glucose oxidase, allowing for release in a pH-sensitive manner. This patch shows efficacy in a single administration for the reduction of melanoma tumor growth and holds significant promise for use in head and neck cancer therapy as well.

Nanoparticles

Recent advancements in nanotechnology fabrication allowed for new avenues to promote local drug delivery through small particles. These particles can either behave as carriers for the therapeutic drug or act as the therapeutic agent themselves. Honing nanoparticles to the tumor microenvironment allows for minimization of systemic effects and greater efficacy in local tumor reduction. Nanoparticles loaded with various drugs can then be delivered directly to the tumor, largely bypassing systemic circulation and maximizing the therapeutic benefit [19,20]. These carrier-based drug delivery systems can exhibit controlled release of drugs while also enhancing the time the drug spends directly within the matrix of the tumor and increasing cellular uptake. Additionally, the large surface-to-volume ratio of nanoparticles maximizes the therapeutic effect of drugs [19].

Gold nanoparticles have shown significant clinical promise in several avenues. They can be used as a drug delivery platform, and can exhibit tumor growth inhibition and applications in photothermal therapy and imaging [21]. Kozakiewicz and colleagues showed that gold nanoparticles conjugated with antiepithelial growth factor receptor when incubated with HNSCC cell lines could inflict photothermal damage when stimulated with low-energy argon laser [22]. The use of argon laser may be most useful for applications in mucosal and cutaneous neoplasms. Another system developed by Lucky and colleagues employs the use of PEGylated titanium dioxide to increase tissue penetration of drugs using near-infrared technology. The nanoparticles utilize anti-EGFR antibodies to target the cancer cell surface and then via up-conversion, promote drug penetration and uptake [23].

Nanoparticles can be loaded with conventional chemotherapy agents. Zhou et al. created amphiphilic chitosan 160–240 nm nanoparticles, which can load paclitaxel and doxorubicin for drug delivery [24]. They showed sustained, controlled release in the tumor bed, which can be optimized by the backbone chemistry. Codelivery of two drugs is also possible with nanosystems. Cisplatin and paclitaxel were loaded onto nanostructured lipid carriers created from PEG-DSPE and conjugated to folate [25]. Folate enhanced the interaction with cells and uptake into the tumor. This approach showed superior killing of the FaDu cell line while exhibiting much less toxicity. Finally, Lang et al. used solvent evaporation to synthesize saracatinib-loaded self-assembling nanoparticles made of an amphiphilic polymer. Saracatinib inactivates Src with greater efficacy than systemic delivery in HNSCC [26].

Radiosensitizing drugs delivered to tumor cells via nanoparticles can be used to maximally prime cancer cells to receive the most damage from radiation therapy. Spherical gold nanoparticles coated with cisplatin and glucose for enhanced cellular uptake have been shown to simultaneously accomplish radiosensitization, drug delivery, and tumor imaging. These nanoparticles achieved better radiotherapy results than traditional approaches. Additionally, gold nanoparticles were found to provide a long-term tumor-targeted CT contrast agent to aid in imaging throughout and after treatment [27]. Nanoparticles hold great promise as they can be utilized in a wide range of applications from drug delivery to enhancing tumor imaging.

Immunomodulation

The advent of immunotherapy revolutionized management of late-stage cancer patients. Two anti-PD1 (aPD1) monoclonal antibodies, pembrolizumab and nivolumab, were recently approved as second-line treatment for metastatic HNSCC [28]. Similarly, cetuximab, erlotinib, and bevacizumab, epidermal growth factor receptor (EGFR) inhibitors, have led to improved patient survival. Unfortunately, suboptimal clinical response and possibility of recurrence warrant continue to be a challenge. Novel immunotherapeutic methods may play an important role in overcoming this challenge.

Similar to other systemically administered drugs and biomolecules, immune-active antibodies dissipate when they reach the desired location and therapeutic levels. Local and targeted administration would allow these rapidly degradable biomolecules to locate exactly within the tumor for maximal antitumor effect. To overcome this barrier, several approaches have been demonstrated in the literature. An innovative self-degradable microneedle patch has also been developed to exhibit sustained release of aPD1 [18]. The microneedle is composed of hyaluronic acid with integrated pH-sensitive dextran nanoparticles, encapsulating aPD1 and glucose oxidase. At physiologic pH, blood glucose is converted to gluconic acid, and this acidic environment promotes the dissociation of the nanoparticles with substantial release of aPD1. The microneedle patch described has shown significant promise in a melanoma mouse model. Further investigation could allow for the delivery of anti-CTLA4 antibody.

Currently under investigation is the ability to incorporate immunotherapy antibodies into the backbone of an implantable mesh. This mesh can be cut and placed directly on the surface of the tumor, including smaller pieces of tissue that had to be left behind after debulking surgery. The mesh would then slowly release aPD1 over several months to provide for maximum immune activation and tumor growth regression. This is an excellent alternative to systemic immunotherapy and would enable direct signaling of cells within the tumor immune microenvironment. Furthermore, the mesh can be combined with many of the other materials mentioned in the previous sections for a combinatorial approach to therapy.

Cyclic dinucleotides (CDNs), also known as stimulator of interferon genes (STING) agonists, are a powerful new class of immunotherapy drugs. These agents are currently being investigated in clinical trials, and it has been shown that CDNs need to be delivered via multiple injections and are ineffective for particularly aggressive tumors. As a solution to this, an injectable peptide hydrogel has been proposed, which would localize CDNs to cancer cells and release them slowly over time. Other groups have similarly found that the application of STING agonists in biomaterials to the resection cavity prevents recurrence and shows greater antitumor efficacy [29,30]. It may now become feasible with these kinds of biomaterials to employ novel immunotherapies that have been previously untapped.

Sophisticated altering of the anti-inflammatory tumor microenvironment that goes beyond administration of PD-1 antibodies has been proposed. The above described methods can be tailored to deliver cytokines and immunomodulatory drugs to the immune cells within the direct vicinity of debulked tumors. Characteristics of the tumor immune microenvironment ultimately direct local resistance of the antitumor immune response including evasion of apoptosis, induction of angiogenesis, deregulation of energy metabolism, and resistance to immune detection and destruction [31]. For this reason, future research should be directed at understanding the prognostic factors that can be identified in the tumor immune microenvironment and identifying cytokines that could boost immune response against tumors to eradicate remaining cancer cells and prevent recurrence.

HNSCC tumors exhibit local immunosuppression with depressed T cell-mediated responses and NK cell and antibody-dependent cytotoxic effects [32]. The role of various cytokines in tumor regression has been studied thoroughly and has contributed to a vast field of knowledge applicable to further optimizing drug delivery modalities for the purpose of altering immune cytokines. Thus far, the greatest challenge has been the lack of simple, effective protocols for delivery of these cytokines directly and locally to the tumor milieu.

The PCL-PLCL modular polymer previously described can release a combination of CCL21 and IL-7 in a slow and controlled manner. This approach can harness the property of CCL21 to promote chemotaxis of many immune effectors including dendritic cells, T-cells, and NK cells, as well as the ability of IL-7 to promote survival of immune cells even in the anti-inflammatory environment of the tumor. This polymer can be alternatively designed to load different sets of cytokines and immunomodulatory drugs as well (Fig. 3).

The GM-CSF recruitment factor and CPI antibodies can be released to locally recruit immune cells and activate dendritic cells to enhance antitumor responses. Exogenous T-cells

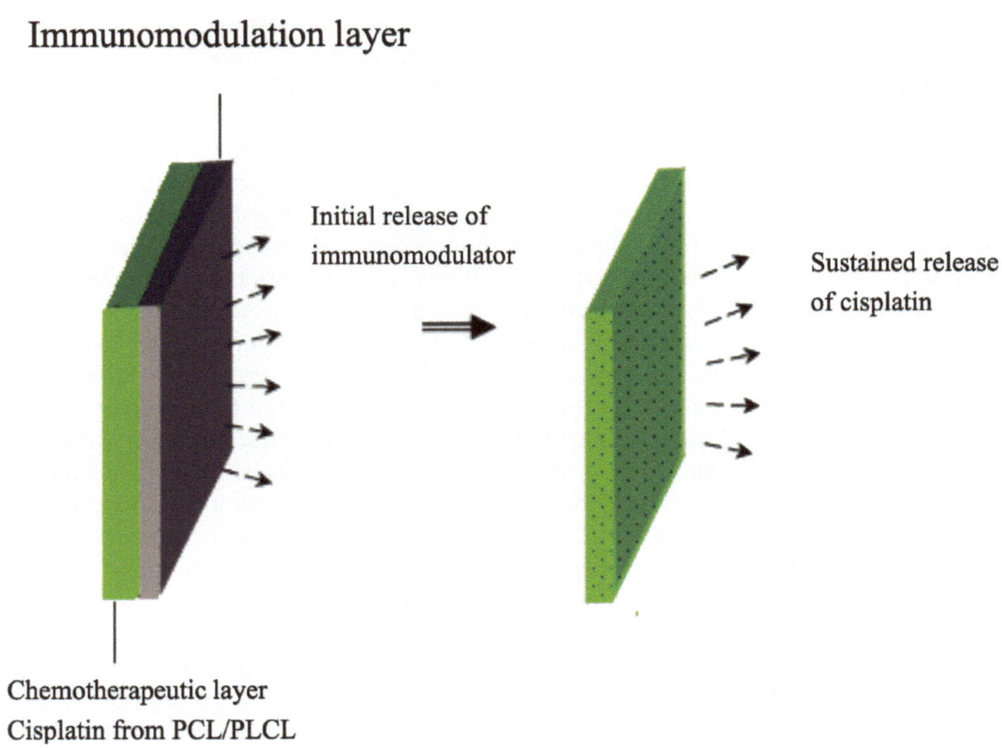

FIG. 3 Depicts the future of a hypothetical multilayer polymer backbone. Each layer can elute drugs, immunomodulators, or radiosensitizer, at different times or delivery rates, thus enhancing tumor killing and providing a more positive wound healing and anticancer environment [5]. Here the immunomodulator layer dissolves first (days to week) while the cisplatin layer dissolves over the course of the radiotherapy treatment (weeks to months).

encased in biopolymer implants can activate endogenous T-cells, allowing for expansion and tumor destruction. CCL17 chemokine recruits T-cells to the tumor, preventing metastasis and limiting tumor size. TLR7 or 8 and STING agonists prevent recurrence postoperatively when slowly released [33]. The flexible and versatile nature of localized immunomodulation by new biomaterials may completely revolutionize the management of head and neck cancers.

Conclusion

New emerging drug delivery systems and concepts hold a great promise for the treatment of head and neck cancers. Traditional therapeutic methods have limited long-term disease-free survival and erode posttreatment quality of life due to systemic toxicity of adjuvant therapy. Polymer sheets can be synthesized to surround the residual or unresectable tumor. Radiopaque tracing can be achieved with polymer layers and hydrogels, allowing for an ultra-narrow field of radiation and thereby preventing damage to local healthy structures. Microneedle patches placed directly on the skin and mucosa dissolve, dispelling their contents into the tumor. Nanoparticle-based technology allows for controlled, sustained release of drugs directly to the tumor, achieving superior tumor growth reduction at significantly lower concentrations. Finally, sophisticated modulation of the local tumor immune microenvironment can be achieved for maximal therapeutic effect activating resident immune cells.

Acknowledgments

We would like to thank members of the Department of Head and Neck Surgery at UCLA including Ashley Kita, Peter Pellionisz, and Penny Amescua for insight comments and administrative support.

References

[1] Siegel RL, Miller KD, Jemal A. Cancer statistics, 2017. CA Cancer J Clin 2017;67(1):7–30.
[2] Marur S, et al. HPV-associated head and neck cancer: a virus-related cancer epidemic. Lancet Oncol 2010;11(8):781–9.
[3] Matoscevic K, et al. Success of salvage treatment: a critical appraisal of salvage rates for different subsites of HNSCC. Otolaryngol Head Neck Surg 2014;151(3):454–61.
[4] Hanna GJ, et al. Rationale for neoadjuvant immunotherapy in head and neck squamous cell carcinoma. Oral Oncol 2017;73:65–9.
[5] Hu D, et al. A novel modular polymer platform for the treatment of head and neck squamous cell carcinoma in an animal model. Arch Otolaryngol Head Neck Surg 2012;138(4):412–7.
[6] Pellionisz PA, et al. Use of a novel polymer in an animal model of head and neck squamous cell carcinoma. Otolaryngol Head Neck Surg 2018;158(1):110–7.
[7] Asaduzzaman Khan M, et al. Thymoquinone, as an anticancer molecule: from basic research to clinical investigation. Oncotarget 2017;8(31):51907–19.
[8] Majdalawieh AF, Fayyad MW, Nasrallah GK. Anti-cancer properties and mechanisms of action of thymoquinone, the major active ingredient of *Nigella sativa*. Crit Rev Food Sci Nutr 2017;57(18):3911–28.
[9] Montalbano G, et al. Synthesis of bioinspired collagen/alginate/fibrin based hydrogels for soft tissue engineering. Korean J Couns Psychother 2018;91:236–46.
[10] Wang C, et al. In situ formed reactive oxygen species-responsive scaffold with gemcitabine and checkpoint inhibitor for combination therapy. Sci Transl Med 2018;10(429). eaan3682.

[11] Conde J, Oliva N, Artzi N. Implantable hydrogel embedded dark-gold nanoswitch as a theranostic probe to sense and overcome cancer multidrug resistance. Proc Natl Acad Sci U S A 2015;112(11):E1278–87.
[12] Bao C, et al. Bioresponsive antisense DNA gold nanobeacons as a hybrid in vivo theranostics platform for the inhibition of cancer cells and metastasis. Sci Rep 2015;5(1):12297.
[13] Nelms L, Palmer WJ. Tissue engineering in mandibular reconstruction: osteogenesis-inducing scaffolds. Plast Aesthet Res 2019;6:21. https://doi.org/10.20517/2347-9264.2019.40.
[14] Wolf MT, et al. The biologic scaffold immune microenvironment inhibits tumor formation and synergizes with immunotherapy. In: 42nd society for biomaterials annual meeting and exposition 2019: the pinnacle of biomaterials innovation and excellence. Society for Biomaterials; 2019.
[15] Skowronek J. Current status of brachytherapy in cancer treatment – short overview. J Contemp Brachyther 2017;9(6):581–9.
[16] Goos J, et al. Star polymers for PET/MRI: a passive pretargeting approach. J Nucl Med 2018;59(suppl 1):195.
[17] Kim Y-C, Park J-H, Prausnitz MR. Microneedles for drug and vaccine delivery. Adv Drug Deliv Rev 2012;64 (14):1547–68.
[18] Wang C, et al. Enhanced cancer immunotherapy by microneedle patch-assisted delivery of anti-PD1 antibody. Nano Lett 2016;16(4):2334–40.
[19] Panikkanvalappil SR, El-Sayed MA, El-Sayed IH. Advances in nanomedicine for head and neck cancer. In: Head and neck cancer. Springer; 2016. p. 827–44.
[20] Marcazzan S, et al. Nanomedicine, an emerging therapeutic strategy for oral cancer therapy. Oral Oncol 2018;76:1–7.
[21] Davidi ES, et al. Cisplatin-conjugated gold nanoparticles as a theranostic agent for head and neck cancer. Head Neck 2018;40(1):70–8.
[22] Kozakiewicz P, Grzybowska-Szatkowska L. Application of molecular targeted therapies in the treatment of head and neck squamous cell carcinoma. Oncol Lett 2018;15(5):7497–505.
[23] Lucky SS, et al. In vivo biocompatibility, biodistribution and therapeutic efficiency of titania coated upconversion nanoparticles for photodynamic therapy of solid oral cancers. Theranostics 2016;6(11):1844–65.
[24] Zhou H, et al. Synthesis and characterization of amphiphilic chitosan derivatives as a nano-carrier for paclitaxel delivery. J Control Release 2011;152(Suppl 1):e124–5.
[25] Yang J, Ju Z, Dong S. Cisplatin and paclitaxel co-delivered by folate-decorated lipid carriers for the treatment of head and neck cancer. Drug Deliv 2017;24(1):792–9.
[26] Lang L, et al. Combating head and neck cancer metastases by targeting Src using multifunctional nanoparticle-based saracatinib. J Hematol Oncol 2018;11(1):85.
[27] Tamar D, et al. All-in-one theranostic nanoagent for head and neck cancer treatment. In: Proc. SPIE; 2018.
[28] Yang B, et al. Progresses and perspectives of anti-PD-1/PD-L1 antibody therapy in head and neck cancers. Front Oncol 2018;8:563.
[29] Leach DG, et al. STINGel: controlled release of a cyclic dinucleotide for enhanced cancer immunotherapy. Biomaterials 2018;163:67–75.
[30] Gough MJ, Baird JR, Bell RB. Implantable biomaterials to provide local immunotherapy following surgical resection. Oncotarget 2018;9(102):37612–3.
[31] Peltanova B, Raudenska M, Masarik M. Effect of tumor microenvironment on pathogenesis of the head and neck squamous cell carcinoma: a systematic review. Mol Cancer 2019;18(1):63.
[32] Lin Y, Sharma S, John MS. CCL21 cancer immunotherapy. Cancers (Basel) 2014;6(2):1098–110.
[33] Leach DG, Young S, Hartgerink JD. Advances in immunotherapy delivery from implantable and injectable biomaterials. Acta Biomater 2019;88:15–31.

CHAPTER

5

HPV-positive oropharyngeal cancers: Today and tomorrow

Ameya A. Asarkar[a,b], *Cherie-Ann O. Nathan*[a,b]

[a]Department of Otolaryngology/Head Neck Surgery, Ochsner-LSU Health, Shreveport, LA, United States [b]Surgical Service, Otolaryngology Section, Overton Brooks Veterans Affairs Medical Center, Shreveport, LA, United States

Abstract

The incidence of oropharyngeal squamous cell carcinoma (OPSCC) has grown exponentially in the past few decades. The molecular and clinical attributes of human papilloma virus (HPV)-associated OPSCC have identified this disease as a distinct entity from HPV-negative OPSCC. The epidemic of HPV-positive OPSCC continues to grow, and identifying novel treatment algorithms is important.

Abbreviations

5-FU	5-fluorouracil
AJCC	American Joint Commission on Cancer
CAP	College of American Pathologists
CASP8	caspase-8
CDK	cyclin-dependent kinase
CI	confidence interval
cTNM	clinical tumor, nodes, metastasis
EGFR	epidermal growth factor receptor
FGFR3	fibroblast growth factor 3
HNSCC	head-neck squamous cell carcinoma
HPV	human papilloma virus
HR	hazard ratio
HR-HPV	high-risk human papilloma virus
IC	induction chemotherapy
IMRT	intensity-modulated radiotherapy treatment
ISH	in situ hybridization
MDADI	MD Anderson Dysphagia Inventory
NCCN	National Comprehensive Cancer Network
OPSCC	oropharyngeal squamous cell carcinoma
PD-1	programmed cell death-1

https://doi.org/10.1016/B978-0-12-820679-9.00005-0

PFS	progression-free survival
pTNM	pathological tumor, nodes, metastasis
QOL	quality of life
TCGA	The Cancer Genome Atlas
TNM	tumor, nodes, metastasis
TORS	transoral robotic surgery
UICC	Union for International Cancer Control
XRT	radiation therapy/radiotherapy

Conflict of interest

No potential conflicts of interest were disclosed.

Introduction

The temporal and causative relationship of tobacco use and alcohol consumption with head and neck cancers has been well established [1,2]. Syrjanen first suggested the causative association of HPV with head and neck cancers in 1983 [3]. Since then, research has been focused on elucidating the epidemiologic molecular and clinical behavior of HPV-associated cancers of the head and neck [4]. The incidence of HPV-positive OPSCC has dramatically increased over the past few decades, and HPV-positive OPSCC has been viewed as a distinct entity from the HPV-negative OPSCC caused by traditional risk factors such as alcohol and tobacco [5,6]. This chapter will focus on the epidemiology, ongoing molecular and clinical research, and the future directions to combat this cancer that is growing in epidemic proportions.

Epidemiology

The incidence of HPV-driven OPSCC has grown exponentially over the past few decades in the United States and in the developed world. About 3500 new cases of HPV-associated OPSCC are diagnosed in women, and about 15,500 are diagnosed in men each year in the United States [7]. During 1999–2015, there was a decrease in cervical carcinoma incidence rates by 1.6% per year, and the incidence of OPSCC increased by 2.7% per year among men and 0.8% per year among women [8]. Since 1990 the incidence of HPV-positive OPSCC has risen by more than 5% every year [9]. Between 1988 and 2004 the incidence of HPV-positive OPSCC has increased by 225% [9].

Patients with HPV-positive OPSCC are generally younger with a median age of 54 years and with minimal to no smoking history [10]. A combination of genetic and behavioral factors has been associated with HPV-positive OPSCC. This is a shift from the traditional demographics of older men with smoking history and alcohol abuse as the causative factors for OPSCC [11,12]. About 30% of patients with HPV-positive tumors are nonsmokers versus less than 5% in the HPV-negative groups [13]. The rapidly growing epidemic of HPV-positive OPSCC is a public health concern, and the number of HPV-positive OPSCC has already surpassed the number of cervical cancers [8,9].

This rising trend is also likely to be related to changing sexual practices. HPV is a known sexually transmitted virus. A National Health and Nutrition Examination Survey reported oral HPV infection in 6.9% of participants aged 14–69 years [14]. The increased number of lifetime oral sex partners (≥6 partners; odds ratio [OR] 3.4; 95% confidence interval [CI], 1.3–8.8; $P = .009$) and vaginal sex partners (≥26 partners; OR 3.1; 95% CI, 1.5–6.5; $P = .002$) increases the incidence of HPV-related OPSCC [15].

Various studies have shown that HPV-positive OPSCC has better population-based survival and prognosis than HPV-negative OPSCC. Ang et al. reported that the 3-year survival in HPV-positive patients was 82.4% versus 57.1% in HPV-negative patients ($P < .001$) [16]. After adjusting for age, race, stage, and tobacco exposure, HPV positivity had a risk reduction for death of 58% (hazard ratio [HR] 0.42; 95% confidence interval [CI] 0.27–0.66) [17]. These results are similar to many studies throughout the world [16]. Various studies have reported improved survival rates among HPV-positive nonsmokers compared with HPV-positive smokers even in recurrent tumors, further underpinning the importance of smoking cessation [18,19].

Current staging

The American Joint Committee on Cancer (AJCC) staging system introduced significant changes in the HPV-positive OPSCC with the eighth edition (Tables 1 and 2). As more data became available, it became apparent that the seventh edition was not able to discriminate and accurately predict survival differences between HPV-negative and HPV-positive patients [9,20]. The monumental rise of HPV-positive OPSCC thus warranted a separate clinical and pathological staging system to differentiate these two entities.

It was imperative to identify a universal, inexpensive, and reproducible test to differentiate HPV-positive from HPV-negative OPSCC. Although HPV-positive OPSCC is commonly found in patients who are nonsmokers, tobacco users are found in both groups of patients, making tobacco exposure not an ideal characteristic to differentiate between the two entities. HPV in situ hybridization (ISH) can be performed on tissue samples. However, the direct

TABLE 1 Clinical and pathologic T category for human papillomavirus-associated (p16-positive) oropharyngeal cancer, 8th edition, Staging manual.

T category	T criteria
T0	No primary identified
T1	Tumor 2 cm or smaller in greatest dimension
T2	Tumor larger than 2 cm but not larger than 4 cm in greatest dimension
T3	Tumor larger than 4 cm in greatest dimension or extension to lingual surface of epiglottis
T4	Moderately advanced local disease; tumor invades the larynx, extrinsic muscle of tongue, medial pterygoid, hard palate, or mandible or beyond[a]

[a] *Mucosal extension to lingual surface of epiglottis from primary tumors of the base of the tongue and vallecula does not constitute invasion of the larynx.*

This table is used with the permission of the American Joint Committee on Cancer (AJCC), Chicago, Illinois. The original source for this material is the AJCC Cancer Staging Manual, Eighth Edition (2017) published by Springer Science and Business Media LLC (springer.com). Amin MB, Edge SB, Greene FL, et al., eds. AJCC cancer staging manual. 8th ed. New York: Springer; 2017, with permission 2.

TABLE 2 Clinical N category human papillomavirus-associated (p16-positive) oropharyngeal cancer, 8th edition, Staging manual.

N category	N criteria
NX	Regional lymph nodes cannot be assessed
N0	No regional lymph node metastasis
N1	One or more ipsilateral lymph nodes, none larger than 6cm
N2	Contralateral or bilateral lymph nodes, none larger than 6cm
N3	Lymph node(s) larger than 6cm

This table is used with the permission of the American Joint Committee on Cancer (AJCC), Chicago, Illinois. The original source for this material is the AJCC Cancer Staging Manual, Eighth Edition (2017) published by Springer Science and Business Media LLC (springer.com). Amin MB, Edge SB, Greene FL, et al., eds. AJCC cancer staging manual. 8th ed. New York: Springer; 2017, with permission.

detection is expensive and not universally available, especially in developing countries, making it difficult for worldwide application. Another limitation with ISH is that the probes are stain specific. Non-HPV 16 types account for about 5%–10% of HPV-positive OPSCC [21].

The tumor suppressor protein p16 is a cyclin-dependent kinase 2A (CDK2A). p16 is an established biomarker for the function of HPV E7 oncoprotein and is a robust surrogate marker for HPV-associated carcinogenesis. Various studies have also identified p16 as an independent prognosticator for HPV-positive OPSCC [16,22]. Immunohistochemical staining for p16 is now recommended on all new patients with OPSCC by the College of American Pathologists (CAP) [23]. Specific criteria have been adopted to label OPSCC as high-risk HPV related (HR-HPV) [24,25]. The cutoff for p16 overexpression is diffuse (75%) tumor expression, with at least moderate (+2/3) staining intensity. Overexpression of p16 should be both intranuclear and cytoplasmic.

Two main differences between the "T" tumor staging of HR-HPV and p16 negative OPSCC are the absence of carcinoma in situ (Tis) and no distinction between T4a and T4b as in p16-negative OPSCC. The first change was necessary due to the nonaggressive pattern of p16-positive OPSCC and the lack of a distinct basement membrane in the epithelium. The second change was warranted as the survival curves of T4a and T4b categories as per the seventh edition AJCC converged and there was no statistical survival difference between the two groups [26].

With improved survival of the p16-positive OPSCC, most head and neck surgeons today offer transoral robotic surgery (TORS) with neck dissection for low-volume disease that is surgically amenable. Thus a separate pathologic classification for nodal disease was developed. Studies identified that neither size nor laterality of the lymph nodes conferred any survival benefit [27]. Further, extranodal extension and N3 nodal disease do not alter survival, and N3 disease behaves similar to N1 in p16-positive OPSCC [27]. Differences in survival are seen with >5 lymph nodes. Thus N1 includes 1–4 lymph nodes, and N2 includes more than 5 nodes (Table 3). Grouped TNM staging has also changed to combine both T and N staging into stages I, II, and III. Stage IV is only reserved for distant metastatic or M1 disease (Tables 4 and 5). This group inherently has poorer prognosis. This significant change in paradigm has prompted the down staging of almost 94% of patients [26].

TABLE 3 Pathologic N category human papillomavirus-associated (p16-positive) oropharyngeal cancer, 8th edition, Staging manual.

N category	N criteria
NX	Regional lymph nodes cannot be assessed
pN0	No regional lymph node metastasis
pN1	Metastasis in four or fewer lymph nodes
pN2	Metastasis in more than four lymph nodes

This table is used with the permission of the American Joint Committee on Cancer (AJCC), Chicago, Illinois. The original source for this material is the AJCC Cancer Staging Manual, Eighth Edition (2017) published by Springer Science and Business Media LLC (springer.com). Amin MB, Edge SB, Greene FL, et al., eds. AJCC cancer staging manual. 8th ed. New York: Springer; 2017, with permission.

TABLE 4 Anatomic stage and prognostic groups for clinical TNM grouping of human papillomavirus-associated (p16-positive) oropharyngeal cancer, 8th edition, Staging manual.

When T is	And N is	And M is	Then the stage group is
T0,T1, or T2	N0 or N1	M0	I
T0,T1, or T2	N2	M0	II
T3	N0, N1, or N2	M0	II
T0,T1,T2,T3, or T4	N3	M0	III
T4	N0, N1, N2, or N3	M0	III
Any T	Any N	M1	IV

This table is used with the permission of the American Joint Committee on Cancer (AJCC), Chicago, Illinois. The original source for this material is the AJCC Cancer Staging Manual, Eighth Edition (2017) published by Springer Science and Business Media LLC (springer.com). Amin MB, Edge SB, Greene FL, et al., eds. AJCC cancer staging manual. 8th ed. New York: Springer; 2017, with permission.

TABLE 5 Anatomic stage and prognostic groups for pathologic TNM grouping of human papillomavirus-associated (p16-positive) oropharyngeal cancer, 8th edition, Staging manual.

When T is	And N is	And M is	Then the stage group is
T0,T1, or T2	N0, N1	M0	I
T0, T1, or T2	N2	M0	II
T3 or T4	N0, N1	M0	II
T3 or T4	N2	M0	III
Any T	Any N	M1	IV

This table is used with the permission of the American Joint Committee on Cancer (AJCC), Chicago, Illinois. The original source for this material is the AJCC Cancer Staging Manual, Eighth Edition (2017) published by Springer Science and Business Media LLC (springer.com). Amin MB, Edge SB, Greene FL, et al., eds. AJCC cancer staging manual. 8th ed. New York: Springer; 2017, with permission.

Treatment options and outcomes

Historically the treatment algorithms for OPSCC have included radiotherapy (XRT) with or without chemotherapy and surgery reserved for salvage settings [28,29]. The main limitation of upfront surgery has been the associated morbidity involved as the access to these tumors was challenging. With the advent of minimally invasive techniques like TORS and transoral laser surgery (TLS), the indications have been expanded to include early T1, T2, and selected T3 tumors of the oropharynx [30–33]. Some clinicians have proposed less intense treatment (i.e., deintensification of the treatment) for HPV-related OPSCC given the significantly better prognosis for these tumors. However, the available data to date are retrospective, hence, insufficient to recommend altering the standard of care at this time. The National Comprehensive Cancer Network (NCCN) panel on head and neck cancers does not recommend deescalation of treatment outside the clinical trial setting at this point [34].

Nonsurgical treatment options

Radiation therapy

Radiation therapy has been the mainstay of treatment for OPSCC. The GORTEC group conducted one of the first OPSCC specific clinical trial. In this trial, participants were randomized into two groups, with one receiving radiotherapy with chemotherapy and the other receiving radiotherapy alone. Chemoradiotherapy added an overall survival benefit of 6.6% at 5 years [35]. Intensity-modulated radiotherapy treatment (IMRT) has become a valuable tool in a radiation oncologist's armamentarium. IMRT aims at reducing the radiation toxicity by reducing collateral damage to normal tissues and improving the accuracy of dose delivery. Reducing toxicity translates into minimizing xerostomia and swallowing dysfunction [36,37].

Systemic therapy with radiation therapy

The addition of chemotherapy to radiotherapy in locally advanced tumors has improved the outcomes in several studies [35,38–40]. A metaanalysis by Pignon et al. reported a hazard ratio (HR) of 0.81 ($P < .0001$) and an absolute benefit of 6.5% at 5 years among the concomitant chemoradiation trials [38]. Another metaanalysis of chemotherapy in head and neck cancer (MACH-NC) confirmed that treatment algorithms that include chemoradiation decreased the risk of death in 5872 patients compared with XRT alone (HR 0.88, 95%, CI 0.82–0.93). This translated into an absolute benefit of 8% with concurrent chemoradiation.

Platinum-based chemotherapy is the standard of care for regimens that warrant systemic therapy. In the MACH-NC metaanalysis, which compared different chemotherapy agents, platinum-based chemotherapy was superior to the nonplatinum-based regimens. Cisplatin (100 mg/m^2 every 3 weeks for 2–3 doses) is the most widely used chemotherapy agents. Other agents commonly used include carboplatin and 5-fluorouracil (5-FU) [29].

One of the early studies reported by Bonner et al. looked at the use of cetuximab as a radiosensitizing agent [41]. This trial established that cetuximab with radiotherapy improved locoregional control and reduced mortality. Patients receiving cetuximab plus radiotherapy had statistically significant prolonged progression-free survival (PFS) (HR 0.70, 95% CI 0.54–0.90, $P = .006$) [41]. This trial, however, did not compare cetuximab with cisplatin. Recently published results of RTOG 1016 compared radiation and cetuximab with radiation and cisplatin. Estimated 5-year survival rates were 84.6% (80.6–88.6) with cisplatin versus 77.9% (73.4–82.5) with cetuximab ($P = .02$). Progression-free survival was significantly worse with cetuximab compared with cisplatin (hazard ratio 1.72 [1.29–2.29]; one-sided log-rank $P = .0001$) with 5-year estimates of 78.4% (73.8–83.0) with cisplatin and 67.3% (62.4–72.2) with cetuximab [42]. Further the acute moderate to severe toxicity (77.4%, 95% CI 73–81.5 vs. 81.7%, 77.5–85.3; $P = .1586$) and late moderate to severe toxicity (16.5%, 95% CI 12.9–20.7 vs. 20.4%, 16.4–24.8; $P = .1904$) were similar in both the groups [42].

Surgical treatment options

Transoral robotic surgery (TORS)

Weinstein et al. reported the first application of transoral robotic surgery in head and neck cancer in 2006 [43]. Following this, in 2009, the US Food and drug Administration (FDA) approved the use of the da Vinci robot (Intuitive Surgical Inc., Sunnyvale, CA, USA) for use in head and neck cancers. The robot provides technical advantages including improved visualization with three-dimensional imaging, specifically avoids line-of-sight restrictions and tremor reduction, and allows for improved range of motion [44,45]. These advantages have led to increased utilization of TORS as the primary treatment modality, especially in T1 and T2 OPSCC. Various studies have reported excellent functional and oncologic outcome with TORS [45–49]. De Almeida et al. reported a large multiinstitutional study to validate the oncologic safety of TORS. The 3-year overall survival rate and disease-specific survival rates were 87.1% (95% CI, 81.4%–91.2%) and 92.5% (95% CI, 87.8%–95.5%), respectively [50]. Based on similar studies, deescalation of therapy for HPV-related OPSCC has been debated [51]. To date the data on TORS are promising and have the potential to improve the quality of life of patients, without compromising oncologic safety. However, the current evidence for deescalation of therapy, especially in HPV-related OPSCC, is limited to small retrospective studies, and prospective data from clinical trials are pending. Ongoing trials should be able to answer these questions.

Current status of deescalation strategies

There is increasing evidence that less intense treatment regimens could achieve comparable oncologic results with reduced toxicity [52–56]. Intuitively, deescalation seems to be an attractive option to improve the quality of life (QOL) of HPV-positive patients. Multiple ways in which this can be achieved are (1) minimally invasive surgery, (2) eliminating chemotherapy, (3) replacing chemotherapy agents with less toxic agents, (4) reducing dose of RT, and (5) induction chemotherapy followed by reduced RT dose.

Minimally invasive surgery

The most notable organ preservation strategies began with the Veterans Affairs Laryngeal Cancer study group's multicenter trial, which compared larynx preservation with total laryngectomy [57]. Since then, multiple trials conducted by the European Organization for Research and Treatment (EORTC) and Radiation Therapy Oncology Group (RTOG) have cemented the place of organ preservation strategies in the management of HNSCC [58,59]. Surgeries for OPSCC are cosmetically disfiguring due to the anatomical location of the tumors. Thus more traditional approaches of open surgery were replaced with nonsurgical organ-preserving radiotherapy with or without chemotherapy.

As mentioned before, with the advent of robotic surgery, TORS is now becoming the standard of care especially for early OPSCC. TORS has the potential to guide adjuvant therapy more accurately based on pathologic staging. This information could be used to calibrate the dose of radiation therapy or to omit it altogether. The proponents of deescalation for HPV-positive OPSCC use this rationale to deintensify therapy for patients. However, these strategies have not yet been proven in a randomized controlled trial (RCT). The current trial that could answer this dilemma is the ECOG-ACRIN (E3311) (NCT01898494) trial in which patients with stage III/IVa p16-positive OPSCC treated by transoral surgery and neck dissection are stratified into four arms based on their pathological results: (1) Arm A, patients staged as T1-T2/N0-N1 with negative margins receive no adjuvant therapy; (2) Arm B, patients with close margins, <1-mm extra capsular spread (ECS), 2–4 metastatic LN, perineural invasion, and/or lymphovascular invasion (low-dose IMRT, 50 Gy/25 Fractions); (3) Arm C, similar tumor characteristics and patients receive (standard-dose IMRT, 60 Gy/30 Fractions); and (4) Arm D, patients with positive margins, >1-mm ECs or >5 metastatic LN undergoes standard-dose IMRT (66 Gy/33Fractions) with weekly chemotherapy (cisplatin 40 mg/m^2). The primary end point is 2-year PFS. The investigators have completed accrual to the trial, and we await the 2-year follow-up results [60].

Adjuvant deescalation, extracapsular spread, p16+, and transoral (ADEPT) trial for oropharynx malignancy being conducted by the Washington University School of Medicine is a phase III trial (NCT01687413) to study the optimal intensity of adjuvant therapy. Patients enrolled in this trial have undergone surgical resection using minimally invasive techniques (with clear margins) and have had ECS in their lymph nodes. Four hundred ninety-six patients are randomized to receive either radiation alone (IMRT, 60 Gy/30 fractions) or radiation (IMRT, 60 Gy/30 fractions) and weekly cisplatin (40 mg/m^2) during therapy. The end points are disease-free survival and locoregional control [61].

The postoperative adjuvant treatment for HPV-positive tumor (PATHOS) trial (NCT02215265) is a multiinstitutional trial run in the United Kingdom. The estimated enrollment is 242 patients with HPV-positive cancer (T1-3, N0-2b) treated by transoral surgery and neck dissection. Patients are to be stratified into three groups based on their pathological results: Arm A—patients with no adverse histopathological features will not receive any adjuvant treatment. Arm B—T3 tumors (or T1/T2 tumor with additional high-risk factors), TNM eighth edition N0–N1 with evidence of perineural and/or vascular invasion or close margins (15 mm) are randomized between two postoperative radiation doses (60 Gy in 30 fractions over 6 weeks or 50 Gy in 25 fractions over 5 weeks). Arm C—included those with tumors of any T or any N stage with positive margins (<1 mm) around the primary tumor specimen, and/or lymph nodes with extracapsular spread are randomized to either radiation therapy with 60 Gy in 30 fractions over 6 weeks 60 Gy in 30 fractions over 6 weeks with concurrent cisplatin. The end point of this study is swallowing function measured by the MD Anderson Dysphagia Inventory (MDADI) score. Secondary outcomes include swallowing assessments, QOL questionnaires, acute and late toxicities, DFS, locoregional control, and distant metastases [62].

Replacing cisplatin with cetuximab

Cetuximab is a chimeric monoclonal antibody that targets epidermal growth factor receptor (EGFR). EGFR is involved in the activation of several oncogenic pathways and is overexpressed in up to 90% of HNSCC. The rationale that encourages the use of cetuximab in HPV-positive patients is mainly based on the study of Bonner et al. [41]. However, until recently, there was no direct comparison between the efficacy and toxicity profile of RT with cisplatin and RT with cetuximab in OPSCC.

This debate was put to rest with the results of RTOG 1016, which were published in 2018 by Gillison et al. RTOG 1016 was a randomized, multicenter, noninferiority trial at 182 healthcare centers in the United States and Canada. Patients were randomly assigned to receive either radiotherapy plus cetuximab or radiotherapy plus cisplatin. Patients received cetuximab at a loading dose of 400 mg/m^2 for 5–7 days before radiotherapy initiation, followed by cetuximab 250 mg/m^2 weekly for seven doses (total 2150 mg/m^2), or cisplatin 100 mg/m^2 on days 1 and 22 of radiotherapy (total 200 mg/m^2). All patients received accelerated IMRT (70 Gy in 35 fractions over 6 weeks at six fractions per week). The primary endpoint was overall survival. After median follow-up duration of 4.5 years, radiotherapy with cetuximab did not meet the noninferiority criteria for overall survival (hazard ratio [HR] 1.45, one-sided 95% upper CI 1.94; $P = .5056$ for noninferiority; one-sided log-rank $P = .0163$). The estimated 5-year overall survival was 77.9% (95% CI 73.4–82.5) in the cetuximab group versus 84.6% (80.6–88.6) in the cisplatin group. Progression-free survival was significantly lower in the cetuximab group compared with the cisplatin group (HR 1.72, 95% CI 1.29–2.29; $P = .0002$; 5-year progression-free survival 67%, 95% CI 62.4–72.2 vs. 78.4%, 73.8–83.0), and locoregional failure was significantly higher in the cetuximab group compared with the cisplatin group (HR 2.05, 95% CI 1.35–3.10; 5-year proportions 17.3%, 95% CI 13.7–21.4 vs. 9.9%, 6.9–13.6). Proportions of acute moderate to severe toxicity (77.4%, 95% CI 73.0–81.5 vs. 81.7%, 77.5–85.3; $P = .1586$) and late moderate to severe toxicity (16.5%, 95% CI 12.9–20.7 vs. 20.4%, 16.4–24.8; $P = .1904$) were similar between the cetuximab and cisplatin groups [42]. Thus this trial conclusively proved that in eligible patients, radiotherapy with cisplatin was still considered the standard of care.

Induction chemotherapy followed by reduced radiotherapy dose

The rationale behind this approach is that induction chemotherapy helps in chemo selecting the good responders to radiotherapy [63]. Studies have shown that HPV-positive tumors are more radiosensitive [64]. The ECOG 1308 trial is the first published study based on that concept [65]. In this trial, 80 patients received three cycles of induction chemotherapy (IC) with paclitaxel, cisplatin, and cetuximab. Patients with complete response received cetuximab and low-dose radiation at 54Gy, and patients with less than complete response received standard dose of 70 Gy. The IC regimen was well tolerated as 96% of patients received all planned cycles, without major treatment breaks. The progression-free survival at 2 years for complete responders treated with a decreased dose was 80% (95% CI 0.65–0.89). Patients with T4 (2-year PFS: 50%; 95% CI 0.11–0.80), N2c (2-year PFS: 73%: 95% CI 0.44–0.89), and smokers >10 pack years (2-year PFS: 65%; 95% CI 0.41–0.82) had poor outcomes compared with those with less than 10 pack years of tobacco history and less advanced tumor stages (<T4N2c) who had a 2-year PFS of 96% (95% CI 0.71–0.99). Patients treated with 54 Gy of radiation had reduced difficulty with swallowing solids (40% v 89%; $P = .011$) or impaired nutrition (10% vs. 44%; $P = .025$) at 1 year [63].

Chen et al. performed a single-arm phase II trial (NCT01716195), in which 44 patients with stage III/IV p16-positive OPSCC received two cycles of IC (paclitaxel and carboplatin given 21 days apart) followed by radiation combined with paclitaxel. The radiation dose was reduced to 54 Gy in complete or partial responders and to 60 Gy in nonresponders. Sixteen percent patients had treatment breaks for intolerable toxicity or social reasons. The 2-year locoregional control and progression-free survival were 95% (95% CI 80–99) and 92% (95% CI 77–97), respectively. The grade 3 or worse mucosal adverse events were reduced in 85% (95% CI 80–90) for patients treated with 54 Gy and 86% (95% CI 80–90) for those who received 60 Gy ($P = .47$) at 2 years. These outcomes, although not statistically significant, were achieved with 15%–20% decreased radiation doses [66].

The Quarterback study (NCT01706939) is an ongoing randomized phase III clinical trial comparing reduced and standard radiation therapy doses for locally advanced HPV-positive oropharynx cancer. Patients receive three cycles of docetaxel, cisplatin, and 5-FU (TPF) as induction chemotherapy. Patients with a complete or partial response are randomized (2:1 randomization) to reduced (56 Gy) or standard (70 Gy) dose radiotherapy with weekly carboplatin and cetuximab or carboplatin alone, respectively, as the second phase. Patients not meeting the response criteria are treated with standard-dose chemoradiation. Three hundred sixty-five patients, with HPV p16-positive oropharyngeal, nasopharyngeal, and carcinoma of unknown primary, are planned for accrual. The primary outcome measures include progression-free survival at 3 years. Toxicity, which will be assessed by symptom scores and QOL scores, will be studied as secondary outcome. Patients will be followed for 5 years. Preliminary outcomes were presented at the ASCO 2017 meeting. The 2-year PFS for patients receiving 56 Gy and standard dose were 87.5% and 83%, respectively [67,68].

The OPTIMA trial is investigating and deintensifying treatment based on response to induction chemotherapy in an exploratory pilot study. Nab-paclitaxel (paclitaxel albumin-stabilized nanoparticle formulation) and carboplatin will be used as an induction regimen followed by response-based therapy. This study will investigate the role of chemotherapy with high- and low-dose radiation and radiation alone in stage III or IV HPV-related OPSCC (NCT02258659) [69]. Paclitaxel and carboplatin used as induction therapies with concomitant paclitaxel in tumors of HPV-positive patients are another regimen under investigation (NCT02048020) [70].

The rationale behind these studies, where the response to a primary treatment determines the choice of the next treatment, is very relevant. Additionally, induction chemotherapy has the potential to reduce distant metastasis, as shown in prior studies [71]. Indeed the risk benefit ratio of reduced toxicity and cancer control is potentially narrow. While the concept of decreasing treatment intensity is attractive, physicians may be hesitant to change their practice especially when long-term survival outcomes of reduced treatment are still unknown. Treatment deintensification is a pragmatic approach in selected HPV-positive patients and must be proven in well-conducted clinical trials. Numerous studies exploring different strategies are currently ongoing but are still in the nascent stage. A balance between oncologic cure and toxicity must be maintained when deescalation is attempted.

Immunotherapy

Currently the literature on immunotherapy regimens specific to HPV-positive OPSCC is scarce. Platinum-based chemotherapy regimens and radiation therapy are the standard of care for treating OPSCC. The programmed cell death protein 1 (PD-1) inhibitors pembrolizumab and nivolumab are approved in recurrent or metastatic OPSCC that have failed to respond to platinum therapy, regardless of HPV status [72]. There are no data suggesting that standard immunotherapy regimens should change based on the HPV status.

The Eastern Cooperative Oncology Group (ECOG) 3161 trial will enroll patients with locally advanced intermediate-risk (stage II/III) HPV-positive oropharyngeal cancer (NCT03811015). Patients in the trial will be randomly assigned to either cisplatin and intensity-modulated radiation therapy followed by nivolumab or cisplatin and radiation followed by observation. This study will investigate both progression-free and overall survival [73]. The phase II/III NRG-HN005 trial will explore whether the use of immunotherapy can allow for safe deintensification of therapy in early-stage (stage I/II) HPV-positive OPSCC patients (NCT03952585). Deintensified radiation therapy with nivolumab will be compared with standard of care, that is, cisplatin and radiation. Progression-free survival and QOL indices will be investigated as primary outcomes in this trial [74].

Preventive strategies

Gardisil-4 (Merck & Co.), Cervarix (GlaxoSmithKline), and Gardisil-9 (Merck & Co.) are the three commercially approved vaccines for preventing HPV infection. Gardisil-4 is currently recommended by the Center for Disease Control (CDC) Advisory Committee on Immunization Practices (ACIP) for use in females between 9 and 26 years of age and males between 9 and 21 years of age, with the option to vaccinate up to age 26 for certain high-risk individuals. The first dose is recommended at 11–12 years and the second dose 6–12 months after the first dose [75]. Numerous large studies with over 10 years of follow-up have shown that these HPV vaccines are safe and have no reports of serious adverse effects [76–78]. Further, Herrero et al. observed that the efficacy of the vaccine was over 93% for preventing oral HPV infection when compared with control, suggesting that it has the potential to prevent HPV-related OPSCC [79]. Similar trials for OPSCC are difficult to conduct primarily due to the absence of a screening test for OPSCC and a lengthy latent period between HPV infection and cancer development.

Ongoing and future research

Research is ongoing to determine the molecular signatures of HPV-induced OPSCC and especially in patients with smoking history. This could potentially help further with risk stratification. FAT1, CASP8, NOTCH1, FGFR3, and KRAS are all overexpressed in smokers even in HPV-positive smokers [80]. Thus explaining that HPV-positive OPSCC is more aggressive in smokers than nonsmokers. *TP*53 mutations and CDKN2A alterations are infrequently seen in HPV-positive tumors [81]. The Cancer Genome Atlas Network (TCGA) study revealed mutations of PIK3CA, amplification of E2F1, and loss of TRAF3 in HPV-positive tumors [81]. Another study showed that EGFR correlated well with smoking status and EGFR overexpression could be used to predict survival in OPSCC [82]. However, the data available from these studies are limited, and molecular profiling is a topic of ongoing research.

Conclusions

HPV-related OPSCC is a distinct entity when compared with HPV-negative OPSCC. The inherently better prognosis that HPV-associated cancer patients have when compared with tobacco-related head and neck cancers has pushed investigators to look beyond the traditional treatment algorithms. The significantly better survival is well reflected with the changes in the AJCC eighth edition TNM staging system. Some question the utility of potentially toxic regimens for these cancers that have high cure rates. The concept of deescalation is exciting and promising with favorable outcomes in most of the early trials. However, long-term data are still evolving. Thus until there is conclusive evidence, the standard of care regimen includes surgery or radiation for early-stage disease and platinum-based chemoradiotherapy, either as a primary modality or in adjuvant settings, for advanced-stage disease. The continued development of minimally invasive surgical techniques such as TORS and TLM offers an alternate approach to treatment of these tumors while reducing the toxicity and improving quality of life. Several studies have shown the oncologic safety of these procedures when patients are selected appropriately. With the expanding indications and technological advancements, better quality of life for patients can be achieved.

Immunotherapy is an exciting exploratory modality, and as new drugs get FDA approval, over the next few years, we will see a rise in the practice of personalized medicine based on patient and tumor factors. Research into molecular signatures of HPV-related tumors is ongoing, and this will help us optimize treatment at an individual level.

References

[1] Argiris A, Karamouzis MV, Raben D, Ferris RL. Head and neck cancer. Lancet 2008;371(9625):1695–709.

[2] Jethwa AR, Khariwala SS. Tobacco-related carcinogenesis in head and neck cancer. Cancer Metastasis Rev 2017;36(3):411–23. https://doi.org/10.1007/s10555-017-9689-6.

[3] Martín-Hernán F, Sánchez-Hernández J-G, Cano J, Campo J, del Romero J. Oral cancer, HPV infection and evidence of sexual transmission. Med Oral Patol Oral Cir Bucal 2013;18(3):e439–44.

[4] Ndiaye C, Mena M, Alemany L, et al. HPV DNA, E6/E7 mRNA, and p16INK4a detection in head and neck cancers: a systematic review and meta-analysis. Lancet Oncol 2014;15:1319–31.

[5] Shaikh MH, McMillan NAJ, Johnson NW. HPV-associated head and neck cancers in the Asia Pacific: a critical literature review & meta-analysis. Cancer Epidemiol 2015;39:923–38.

[6] D'Souza G, Kreimer AR, Viscidi R, et al. Case-control study of human papillomavirus and oropharyngeal cancer. N Engl J Med 2007;356:1944–56.

[7] Viens LJ, Henley SJ, Watson M, Markowitz LE, Thomas CC, Thompson TD, Razzaghi H, Saraiya M, Centers for Disease Control and Prevention (CDC). Human papillomavirus–associated cancers—United States, 2008–2012. MMWR Morb Mortal Wkly Rep 2016;65(26):661–6.

[8] Van Dyne EA, Henley SJ, Saraiya M, Thomas CC, Markowitz LE, Benard VB. Trends in human papillomavirus–associated cancers—United States, 1999–2015. MMWR Morb Mortal Wkly Rep 2018;67:918–24.

[9] Chaturvedi AK, Engels EA, Pfeiffer RM, et al. Human papillomavirus and rising oropharyngeal cancer incidence in the United States. J Clin Oncol 2011;29:4294–301.

[10] Elrefaey S, Massaro MA, Chiocca S, Chiesa F, Ansarin M. HPV in oropharyngeal cancer: the basics to know in clinical practice. Acta Otorhinolaryngol Ital 2014;34:299–309.

[11] Deschler DG, Richmon JD, Khariwala SS, Ferris RL, Wang MB. The "new" head and neck cancer patient—young, nonsmoker, nondrinker, and HPV positive: evaluation. Otolaryngol Head Neck Surg 2014;151:375–80.

[12] Gallagher ST, Deal AM, Ballard D, Mayer DK. Oropharyngeal cancer and HPV: measuring knowledge and impact among survivors of head and neck cancer. Clin J Oncol Nurs 2017;21:321–30.

[13] Hong AM, Martin A, Chatfield M, et al. Human papillomavirus, smoking status and outcomes in tonsillar squamous cell carcinoma. Int J Cancer 2013;132(12):2748–54.

[14] Gillison ML, Broutian T, Pickard RK, et al. Prevalence of oral HPV infection in the United States, 2009–2010. JAMA 2012;307(7):693–703.

[15] D'Souza G, Kreimer AR, Viscidi R, et al. Case-control study of human papillomavirus and oropharyngeal cancer. N Engl J Med 2007;356(19):1944–56.

[16] Ang KK, Harris J, Wheeler R, et al. Human papillomavirus and survival of patients with oropharyngeal cancer. N Engl J Med 2010;363:24–35.

[17] Dayyani F, Etzel CJ, Liu M, et al. Meta-analysis of the impact of human papillomavirus (HPV) on cancer risk and overall survival in head and neck squamous cell carcinomas (HNSCC). Head Neck Oncol 2010;2:15.

[18] Maxwell JH, Kumar B, Feng FY, et al. Tobacco use in human papillomavirus-positive advanced oropharynx cancer patients related to increased risk of distant metastases and tumor recurrence. Clin Cancer Res 2010;16:1226–35.

[19] Hafkamp HC, Manni JJ, Haesevoets A, et al. Marked differences in survival rate between smokers and nonsmokers with HPV 16-associated tonsillar carcinomas. Int J Cancer 2008;122:2656–64.

[20] Edge SB, Byrd DR, Compton CC, Fritz AG, Greene FL, Trotti A, editors. AJCC cancer staging manual. 7th ed. New York: Springer; 2010.

[21] Kreimer AR, Clifford GM, Boyle P, Franceschi S. Human papillomavirus types in head and neck squamous cell carcinomas carcinomas worldwide: a systematic review. Cancer Epidemiol Biomarkers Prev 2005;14:467–75.

[22] Rischin D, Young RJ, Fisher R, et al. Prognostic significance of p16INK4A and human papillomavirus in patients with oropharyngeal cancer treated on TROG 02.02 phase III trial. J Clin Oncol 2010;28:4142–8.

[23] Lewis Jr JS, Beadle B, Bishop JA, et al. Human papillomavirus testing in head and neck carcinomas: guideline from the college of American Pathologists. Arch Pathol Lab Med 2018;142:559–97.

[24] Jordan RC, Lingen MW, Perez-Ordonez B, et al. Validation of methods for oropharyngeal cancer HPV status determination in US cooperative group trials. Am J Surg Pathol 2012;36:945–54.

[25] Schlecht NF, Brandwein-Gensler M, Nuovo GJ, et al. A comparison of clinically utilized human papillomavirus detection methods in head and neck cancer. Mod Pathol 2011;24:1295–305.

[26] Husain ZA, Chen T, Corso CD, et al. A comparison of prognostic ability of staging systems for human papilloma virus related oropharyngeal squamous cell carcinoma. JAMA Oncol 2017;3(3):358–65.
[27] Sinha P, Kallogjeri D, Gay H, et al. High metastatic node number, not extracapsular spread of N classification is a node related prognosticator in transorally resected, neck-dissected p16 positive oropharynx cancer. Oral Oncol 2015;51(5):514–20.
[28] Denis F, Garaud P, Bardet E, et al. Final results of the 94-01 French Head and Neck Oncology and Radiotherapy Group randomized trial comparing radiotherapy alone with concomitant radiochemotherapy in advanced-stage oropharynx carcinoma. J Clin Oncol 2004;22:69–76.
[29] Bourhis J, Calais G, Lapeyre M, et al. Concomitant radiochemotherapy or accelerated radiotherapy: analysis of two randomized trials of the French Head and Neck Cancer Group (GORTEC). Semin Oncol 2004;31:822–6.
[30] Weinstein GS, O'Malley Jr BW, Magnuson JS, et al. Transoral robotic surgery: a multicenter study to assess feasibility, safety, and surgical margins. Laryngoscope 2012;122:1701–7.
[31] Li RJ, Richmon JD. Transoral endoscopic surgery: new surgical techniques for oropharyngeal cancer. Otolaryngol Clin North Am 2012;45:823–44.
[32] Hinni ML, Zarka MA, Hoxworth JM. Margin mapping in transoral surgery for head and neck cancer. Laryngoscope 2013;123:1190–8.
[33] Haughey BH, Hinni ML, Salassa JR, et al. Transoral laser microsurgery as primary treatment for advanced-stage oropharyngeal cancer: a United States multicenter study. Head Neck 2011;33:1683–94.
[34] https://www.nccn.org/professionals/physician_gls/pdf/head-and-neck.pdf.
[35] Calais G, Alfonsi M, Bardet E, et al. Randomized trial of radiation therapy versus concomitant chemotherapy and radiation therapy for advanced-stage oropharynx carcinoma. J Natl Cancer Inst 1999;91(24):2081–6.
[36] Nutting CM, Morden JP, Harrington KJ, et al. Parotid-sparing intensity modulated versusconventional radiotherapy in head and neck cancer (PARSPORT): a phase 3 multicentre randomised controlled trial. Lancet Oncol 2011;12(2):127–36.
[37] Petkar I, Rooney K, Roe JW, et al. DARS: a phase III randomised multicentre study of dysphagia optimised intensity-modulated radiotherapy (Do-IMRT) versus standard intensity-modulated radiotherapy (S-IMRT) in head and neck cancer. BMC Cancer 2016;16(1):770.
[38] Pignon JP, Bourhis J, Domenge C, et al. Chemotherapy added to locoregional treatment for head and neck squamous-cell carcinoma: three meta-analyses of updated individual data. MACH-NC Collaborative Group. Meta-Analysis of Chemotherapy on Head and Neck Cancer. Lancet 2000;355(9208):949–55.
[39] Bird T, De Felice F, Michaelidou A, et al. Outcomes of intensity-modulated radiotherapy as primary treatment for oropharyngeal squamous cell carcinoma—a European single institution analysis. Clin Otolaryngol 2017; 42(1):115–22.
[40] Argiris A. Update on chemoradiotherapy for head and neck cancer. Curr Opin Oncol 2002;14:323–9.
[41] Bonner JA, Harari PM, Giralt J, Cohen RB, Jones CU, Sur RK, et al. Radiotherapy plus cetuximab for locoregionally advanced head and neck cancer: 5-year survival data from a phase 3 randomised trial, and relation between cetuximab-induced rash and survival. Lancet Oncol 2010;11(1):21–8.
[42] Gillison ML, Trotti AM, Harris J, et al. Radiotherapy plus cetuximab or cisplatin in human papilloma virus—positive oropharyngeal cancer (NRG Oncology RTOG 1016): a randomised, multicenter, non-inferiority trial. Lancet 2019;393(1016):40–50.
[43] Weinstein GS, O'Malley BW, Snyder W, et al. Transoral robotic surgery: radical tonsillectomy. Arch Otolaryngol Head Neck Surg 2007;133(12):1220–6.
[44] McLeod IK, Melder PC. Da Vinci robot-assisted excision of a vallecular cyst: a case report. Ear Nose Throat J 2005;84(3):170–2.
[45] Cracchiolo JR, Roman BR, Kutler DI, Kuhel WI, Cohen MA. Adoption of transoral robotic surgery compared with other surgical modalities for treatment of oropharyngeal squamous cell carcinoma. J Surg Oncol 2016;114(4):405–11.
[46] Hutcheson KA, Holsinger FC, Kupferman ME, et al. Functional outcomes after TORS for oropharyngeal cancer: a systematic review. Eur Arch Otorhinolaryngol 2015;272(2):463–71.
[47] Leonhardt FD, Quon H, Abrahão M, O'Malley Jr BW, Weinstein GS. Transoral robotic surgery for oropharyngeal carcinoma and its impact on patient-reported quality of life and function. Head Neck 2012; 32(2):146–54.

[48] Hurtuk AM, Marcinow A, Agrawal A, Old M, Teknos TN, Ozer E. Quality-of-life outcomes in transoral robotic surgery. Otolaryngol Head Neck Surg 2012;146(1):68–73.
[49] Moore EJ, Hinni ML, Olsen KD, Price DL, Laborde RR, Inman JC. Cost considerations in the treatment of oropharyngeal squamous cell carcinoma. Otolaryngol Head Neck Surg 2012;146(6):946–51.
[50] de Almeida JR, Li R, Magneuson JS, et al. Oncologic outcomes after transoral robotic surgery: a multiinstitutional study. JAMA Otolarungol Head Neck Surg 2015;141(12):1043–51.
[51] Chen AM, Zahra T, Daly ME, et al. Definitive radiation therapy without chemotherapy for human papilloma virus—related oropharyngeal carcinomas. Head Neck 2013;35:1652–6.
[52] Vermorken JB, Psyrri A, Mesia R, et al. Impact of tumor HPV status on outcome in patients with recurrent and/or metastatic squamous cell carcinoma of the head and neck receiving chemotherapy with or without cetuximab: retrospective analysis of the phase III EXTREME trial. Ann Oncol 2014;25:801–7.
[53] Rosenthal DI, Harari PM, Giralt J, et al. Association of human papillomavirus and p16 status with outcomes in the IMCL-9815 phase III registration trial for patients with locoregionally advanced oropharyngeal squamous cell carcinoma of the head and neck treated with radiotherapy with or without cetuximab. J Clin Oncol 2016;34(12):1300–8.
[54] Seiwert T, Fayette J, Cupissol D, et al. A randomized, openlabel, phase II study of afatinib (BIBW 2992) versus cetuximab in recurrent or metastatic squamous cell carcinoma of the head and neck: final data. In: Presented at the multidisciplinary head and neck cancer symposium, Phoenix, AZ, 26 January; 2012.
[55] Dowthwaite SA, Franklin JH, Palma DA, et al. The role of transoral robotic surgery in the management of oropharyngeal cancer: a review of the literature. ISRN Oncol 2012;2012:945162.
[56] O'Sullivan B, Huang SH, Siu LL, et al. Deintensification candidate subgroups in human papillomavirus-related oropharyngeal cancer according to minimal risk of distant metastasis. J Clin Oncol 2013;31:543–50.
[57] The Department of Veterans Affairs Laryngeal Cancer Study Group. Induction chemotherapy plus radiation compared with surgery plus radiation in patients with advanced laryngeal cancer. N Engl J Med 1991;324: 1685–90.
[58] Lefebvre JL, Chevalier D, Luboinski B, et al. Larynx preservation in pyriform sinus cancer: preliminary results of a European Organization for Research and Treatment of Cancer phase III trial. EORTC Head and Neck Cancer Cooperative Group. J Natl Cancer Inst 1996;88:890–9.
[59] Forastiere AA, Goepfert H, Manor M, et al. Concurrent chemoradiotherapy and radiotherapy for organ preservation in advanced laryngeal cancer. N Engl J Med 2003;349:2091–8.
[60] https://www.cancer.gov/about-cancer/treatment/clinicaltrials/search/v?id=NCT01898494&r=1.
[61] https://clinicaltrials.gov/ct2/show/NCT01687413.
[62] https://clinicaltrials.gov/ct2/show/NCT02215265.
[63] Ensley JF, Jacobs JR, Weaver A. Correlation between response to cis-platinum-combination chemotherapy and subsequent radiotherapy in previously untreated patients with advanced squamous cell cancers of the head and neck. Cancer 1984;54:811–4.
[64] Mirghani H, Amen F, Tao Y, Deutsch E, Levy A. Increased radiosensitivity of HPV-positive head and neck cancers: molecular basis and therapeutic perspectives. Cancer Treat Rev 2015;41:844–52.
[65] Marur S, Li S, Cmelak AJ, Gillison ML. E1308: phase II trial of induction chemotherapy followed by reduced-dose radiation and weekly cetuximab in patients with HPV-associated resectable squamous cell carcinoma of the oropharynx—ECOG-ACRIN cancer research group. J Clin Oncol 2016;10(33):836–45.
[66] Chen AM, Felix C, Wang PC. Reduced-dose radiotherapy for human papillomavirus-associated squamous-cell carcinoma of the oropharynx: a single-arm, phase 2 study. Lancet Oncol 2017;18:803–11.
[67] https://clinicaltrials.gov/ct2/show/NCT01706939.
[68] Rainey H, Roy E, Selkridge I. Standard of care vs reduced-dose chemoradiation after induction chemotherapy in HPV+oropharyngeal carcinoma patients. J Clin Oncol 2017;35(1):6069.
[69] https://clinicaltrials.gov/ct2/show/NCT02258659.
[70] https://clinicaltrials.gov/ct2/show/NCT02048020.
[71] O'Sullivan B, Huang SH, Siu LL. Deintensification candidate subgroups in human papillomavirus-related oropharyngeal cancer according to minimal risk of distant metastasis. J Clin Oncol 2013;31:543–50.
[72] Chow LQ, Haddad R, Gupta S, et al. Antitumor activity of pembrolizumab in biomarker-unselected patients with recurrent and/or metastatic head and neck squamous cell carcinoma: results from the phase Ib KEYNOTE-012 expansion cohort. J Clin Oncol 2016;34:3838–45.

[73] https://ecog-acrin.org/clinical-trials/ea3161-educational-materials.
[74] https://clinicaltrials.gov/ct2/show/NCT03952585.
[75] Markowitz LE, et al. Human papillomavirus vaccination: recommendations of the Advisory Committee on Immunization Practices (ACIP). MMWR Recomm Rep 2014;63:1–30.
[76] Naud PS, et al. Sustained efficacy, immunogenicity, and safety of the HPV-16/18 AS04-adjuvanted vaccine: final analysis of a long-term follow-up study up to 9.4 years post-vaccination. Hum Vaccin Immunother 2014;10:2147–62.
[77] Scheller NM, Pasternak B, Svanstrom H, Hviid A. Quadrivalent human papillomavirus vaccine and the risk of venous thromboembolism. JAMA 2014;312:187–8.
[78] Stokley S, et al. Human papillomavirus vaccination coverage among adolescents, 2007-2013, and postlicensure vaccine safety monitoring, 2006-2014—United States. MMWR Morb Mortal Wkly Rep 2014;63:620–4.
[79] Herrero R, et al. Reduced prevalence of oral human papillomavirus (HPV) 4 years after bivalent HPV vaccination in a randomized clinical trial in Costa Rica. PLoS One 2013;8.
[80] Zevallos JP, Yim E, Brennan P, et al. Molecular profile of human papillomavirus–positive oropharyngeal squamous cell carcinoma stratified by smoking status. Int J Radiat Oncol Biol Phys 2016;94(4):864.
[81] Cancer Genome Atlas Network. Comprehensive genomic characterization of head and neck squamous cell carcinoma. Nature 2015;517:576–82.
[82] Sivarajah S, Kostiuk M, Lindsay C, et al. EGFR as a biomarker of smoking status and survival in oropharyngeal squamous cell carcinoma. J Otolaryngol Head Neck Surg 2019;48:1.

CHAPTER

6

Treatment resistance and a glimmer of hope, the promise of new agents

Vlad C. Sandulache[a], *Jeffrey N. Myers*[b]

[a]Bobby R. Alford Department of Otolaryngology Head and Neck Surgery, Baylor College of Medicine, Houston, TX, United States [b]Department of Head and Neck Surgery, The University of Texas MD Anderson Cancer Center, Houston, TX, United States

Abstract

Treatment resistance, whether intrinsic or acquired, is a critical factor limiting the utility of nonsurgical treatment regimens for head and neck cancers and other solid tumors. In this chapter, we discuss existing evidence regarding both mechanisms of resistance and their impact on relative effectiveness of conventional chemotherapy, targeted agents, and radiation. Although largely driven by the intrinsic tumor biology, more specifically individual genomic and epigenetic events, treatment resistance often involves wholesale reprogramming of transcriptional and translational cellular profiles and alterations in cellular energetic and metabolomic constraints. Such broad-based alterations generate a complex scientific and translational problem and can result in the development of unpredictable cross-resistance patterns. Recent data arising from preclinical and clinical studies of targeted agents provide some potentially useful insights and strategies for minimizing the potential for development of resistant clones.

Abbreviations

AJCC	American Joint Committee on Cancer
EMT	epithelial mesenchymal transition
FDG	fluorodeoxyglucose
HNSCC	head and neck squamous cell carcinoma
HPV	human papillomavirus
LINAC	linear accelerator
MRI	magnetic resonance imaging
NAD	nicotinamide adenine dinucleotide
PARP	poly ADP ribose polymerase
PDX	patient derived xenograft
PET	positron emission tomography

https://doi.org/10.1016/B978-0-12-820679-9.00006-2

Conflict of interest

No potential conflicts of interest were disclosed.

Introduction

Development of resistance to cancer treatment is a major source of treatment failure for over half a million Americans afflicted with a solid tumor each year. Multiple factors have been shown to drive tumor resistance, and yet its clinical behavior remains as unpredictable as the biological mechanisms arising in individual cells and tumors.

Most physicians are exposed to the concept of sensitivity and resistance in the context of microbiology, as these both are routinely tested ex vivo using bacterial cultures obtained from patient specimens. Over time, however, these relatively straightforward and simple concepts have increasingly been applied to the field of cancer, and often tumors or cancer cells are referred to as "sensitive" or respectively "resistant" to cancer treatment. Although simple from a grammatical perspective, the use of this term in the context of cancer belies significant complexity and, thus, can potentially negatively impact our planned remedies. Unlike bacteria, cancers are almost never tested ex vivo for relative sensitivity or resistance to specific targeted or nontargeted agents and/or radiation. Therefore relative sensitivity is imputed prior to treatment, most often empirically, less frequently using predictive biomarkers (i.e., mutational signature). In the modern era, high-resolution imaging obtained during, or rapidly following, treatment completion can also inform clinicians regarding the speed of treatment response and, thus, potentially relative sensitivity or resistance to treatment.

Even without delving into the concept of cancer stem cells, it is both biologically clear and experimentally proven that solid tumors are highly heterogeneous and consist of myriad clones with varying genomic, epigenetic, and gene expression patterns and, therefore, different phenotypes, including varying proliferative rates, and differential sensitivity and/or resistance to tumoricidal agents. Although this has been known for decades and studied at a preclinical level, more recently single-cell genomic analysis has demonstrated the phenomenon conclusively. Furthermore, solid tumors are geographically heterogeneous secondary to variable vascularity and diverse neural, stromal, and immune microenvironmental distribution throughout a given tumor and therefore variable nutrient and oxygen availability. Since radiation is a critical component of head and neck cancer treatment and has a spatial component, treatment response is, therefore, impacted by the dimensional constraints of the tumor. Whether and how much a tumor responds to treatment is a function of clonal extinction and repopulation that depends on both the aforementioned biologic makeup of tumor cells and their geographic distribution within the tumor mass. This is critically important, because if we simply use gross tumor regression as a surrogate for treatment sensitivity, we may make inappropriate predictions regarding the relative risk of recurrence from residual resistant clones.

This chapter sets out to define treatment resistance, discuss existing means of detecting its presence in solid tumors, and outline some of the mechanisms utilized by cancer cells to overcome the tumoricidal effects of conventional chemotherapy, targeted agents, and radiation.

Treatment resistance: Defining the problem

As we revert to thinking about tumor resistance in a manner similar to bacterial antibiotic resistance, two potential means of development of resistance become apparent. The first is intrinsic resistance, whereby a tumor simply fails to respond to chemotherapy, targeted agents or radiation, because of intrinsic tumor biology. As will be discussed later, complete treatment resistance is likely biologically rare and clinically difficult to ascertain. However, we do know that cells can easily manifest the opposite phenomenon, namely, a dramatic intrinsic sensitivity to conventional chemotherapy and radiation. Examples of this phenomenon can be encountered in patients with Ataxia telangiectasia, Fanconi anemia, and Li-Fraumeni syndrome that manifest with significantly higher rates of acquired malignancies despite equivalent carcinogen exposure primarily due to loss to tumor suppressor function and decreased surveillance of DNA damage and activation of DNA repair pathways [1–3].

Similarly, some tumors are intrinsically sensitive to specific treatments. For example, in the clinic, BRAF mutant tumors such as melanoma and thyroid cancer are exquisitely sensitive to BRAF inhibitors. In the preclinical setting, *TP53* mutational status has been linked to intrinsic differential response to cytotoxic chemotherapy and radiation [4–7]. PI3-kinase mutants also demonstrate preferential response to PI3-kinase inhibitors [8]. To what degree cell line and murine model data can be extrapolated to human patterns of intrinsic resistance however remains unclear. Within the human disease spectrum, human papillomavirus (HPV)–associated tumors, particularly in the oropharynx, have consistently demonstrated superior treatment response to both chemotherapy and radiation, disease free and overall survival compared with their HPV-negative counterparts. This effect is so robust it has been incorporated into the 8th edition of the *AJCC Staging Manual* [9–12].

The degree to which tumors fail to respond at all remains unclear, but there are significant patient data that demonstrate the phenomenon. Patients with intermediate and high-risk oropharyngeal SCC demonstrate a high rate of in-field locoregional recurrence and a nontrivial rate of incomplete response despite delivery of potentially curative doses of chemoradiation (Fig. 1) [10, 13]. Failure to respond can be seen relatively clearly in the setting of induction chemotherapy, where anywhere from 20% to 40% of tumors will fail to demonstrate any clinical or imaging-based response depending on regimen and disease site [14, 15]. The relative time frame, however, would favor the latter mechanism. Unfortunately, most clinical trials utilize multiagent chemotherapy regimens (with or without targeted agents), so it is difficult to identify precisely what drives tumor response [16, 17].

Treatment resistance in cancer can also be acquired, wherein treatment induces properties in cells that render them resistant to further treatment (Fig. 2). To evaluate and characterize this phenomenon preclinically, investigators often utilize prolonged exposure of tumor cells to individual treatments of increasing intensity. Here the experience for cisplatin is instructive with respect to magnitude of the phenomenon of acquired resistance and its biological complexities. Since cisplatin is often utilized in the treatment of head and neck cancer, many efforts have focused on elucidating mechanisms of response to this compound [18]. The acquired resistance to cisplatin using established human cell lines

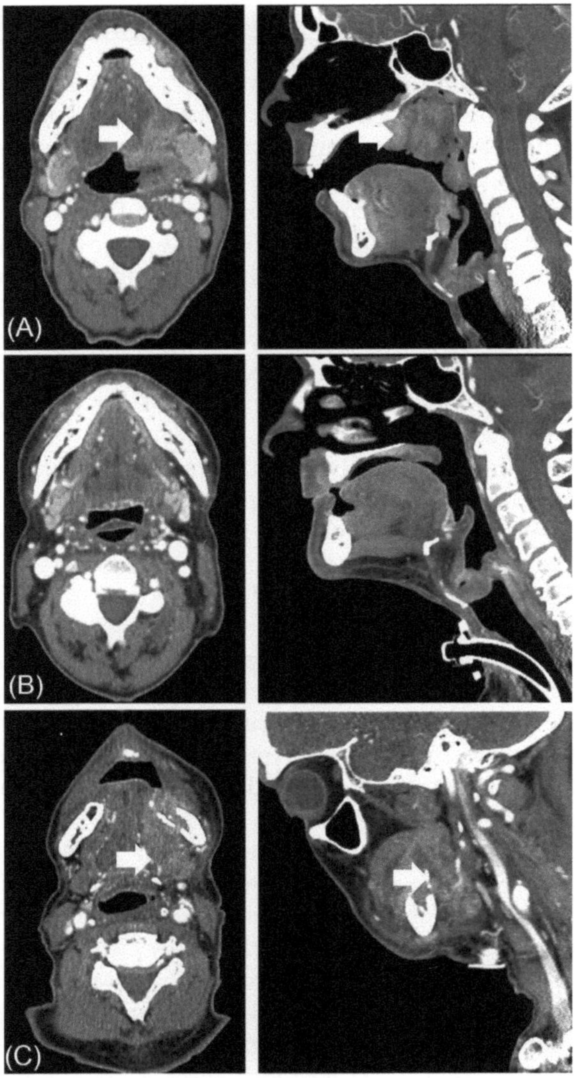

FIG. 1 Local recurrence likely indicating treatment resistance. (A) A 51-year-old male smoker presented with a T4N2bM0 SCCA of the oropharynx (p16-) *(white arrow)* and received curative intent chemoradiation (cisplatin) resulting in a complete response (B). (C) The patient developed local recurrence 18 months later *(white arrow)*, which was treated surgically. Patient initiated reirradiation but recurred locally prior to the midradiation point.

is accompanied by broad shifts in gene expression including genes involved in basic metabolism, cell cycle progression and survival, and modulation of distinct signaling cascades such as PI3kinase-Akt. More recently, similar analyses have been performed using PDX models that can increase tumor heterogeneity and potentially increase fidelity to the clinical scenario being tested [19].

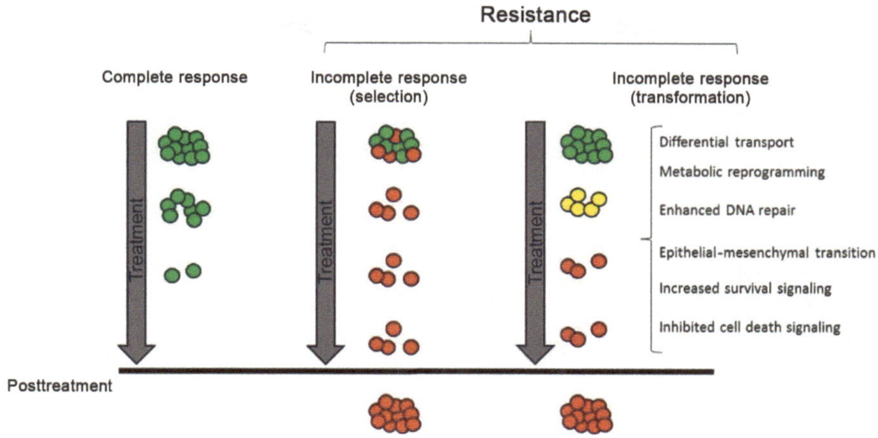

FIG. 2 Putative mechanisms of response and resistance. Tumors made up of sensitive cells *(green)* demonstrate progressive clonal extinction during treatment, with no residual, viable tumor cells following completion of treatment and persistent clinical response. Tumors made up of sensitive *(green)* and resistant *(red)* cells demonstrate clonal extinction of the sensitive cells, but persistence of resistant cells, resulting in repopulation and recurrence following treatment completion. Alternatively, sensitive cells *(green)* can acquire resistance *(yellow)* through transcriptional, translational, and metabolomic reprogramming during treatment and become resistant *(red)*. Cells with intrinsic or acquired resistance can repopulate the tumor volume resulting in clinical recurrence following treatment completion.

Sometimes the mechanisms underlying individual drug resistance are as simple as they are in bacteria, for example, differential drug transport [20]. In the case of cisplatin, this can be accomplished through upregulation of energy-dependent transporters, either specific to metal ions (i.e., high-affinity copper transporter 1) or nonspecific, such as those which can drive multidrug resistance through upregulation of P-glycoprotein, that can pump cisplatin out of cells. Alternatively, since cisplatin is a highly reactive molecule, essentially a free radical, resistance can be acquired through the enhanced generation of molecules designed to scavenge free radicals [21]. Although glutathione is the most common such molecule, there are multiple mechanisms for enhancing reducing equivalent pools including increased generation of NADH and NADPH, production of thiol containing proteins such as thioredoxin, and altered enzymatic activity shifting amino acid flux into glutathione generation and recovery.

Singular mechanisms present a relatively straightforward problem from a clinical perspective assuming that nontoxic methods exist to overcome them (i.e., copper ions and regenerators of reducing equivalents). More complex mechanisms, however, pose essentially impossible problems from a translational perspective. Such is the case when acquisition of cisplatin resistance is associated with shifts in epithelial-mesenchymal transition (EMT) markers, which represent wholesale reprogramming of cellular activity; this is further accentuated when these shifts vary from cellular background to cellular background depending on their permissive genotype [22]. In some cases, shifts in EMT may result in multidrug resistance, whereas in other cases, the event may be limited to a single compound that further complicates translational solutions [23]. Perhaps one of the most problematic mechanisms of cisplatin resistance is differential activation or regulation of DNA repair mechanisms [24].

The acquisition of enhanced single-strand or double-strand break repair capacity through continued cisplatin exposure such as differential BRCA, Rad 51, or KIF4A activity, has the potential to diminish cisplatin effectiveness. Furthermore, this acquired resistance based on enhance DNA repair can also potentially impact entire classes of conventional cytotoxic chemotherapy and external beam radiation all of which depend on the accumulation of DNA damage to kill tumor cells.

Resolving intrinsic from acquired resistance is not simply a semantic difficulty, but has real-world implications for therapeutic regimen design and implementation. The frequency of individual drug administration, combination of cytotoxic and targeted agents with or without radiation, and sequencing of drug regimens all can be substantially impacted by the mechanisms discussed in the preceding text and are discussed in more detail in the succeeding text.

Treatment resistance: Detecting the problem

Whether resistance is intrinsic or acquired, it is critical to be able to identify its emergence rapidly and consistently. Current clinical paradigms rely on clinical or imaging-based response (i.e., changes in gross tumor volume) and disease recurrence to infer relative treatment sensitivity. However, this standard approach is neither sensitive enough to improve clinical regimens nor specific enough to inform translational research efforts.

Two technological advances have helped ameliorate this current limitation. *The first consists of improvements in imaging techniques.* Positron emission tomography (PET) combined with computed tomography (CT) has now become a standard tool in the diagnosis of head and neck malignancies, particularly head and neck squamous cell carcinoma (HNSCC). PET tracers have routinely been investigated for utility in detecting variable tumor vascularity and hypoxia, with reasonable success in preclinical models and clinical trials, although clinical utility has not yet been demonstrated. In addition to making pretreatment predictions, utilization of in-treatment anatomic and metabolic imaging can begin to provide real-time response measurements prior to treatment completion, potentially allowing for the development of adaptive treatment regimens [25]. Imaging-based measurements of treatment response are attractive because they are noninvasive and iterative, in addition to being patient specific, thus allowing for potential development of precision oncology approaches. However, given the complexity and expense of modern imaging techniques, continued development is required prior to broad-based inclusion into clinical algorithms.

Prospective studies of [(18)F]-misonidazole ((18)F-FMISO) positron emission tomography (PET) scans in the setting of concurrent HNSCC chemoradiation regimens have demonstrated high rates of pretreatment hypoxia, associated with treatment resistance, in both primary tumors and nodal metastases, which resolved in nearly 50% of patients during treatment delivery [26, 27]. In-treatment resolution of hypoxia suggests the potential for PET-driven deescalation in responsive tumors that regain a normoxic signature [28]. One critical aspect of this type of approach is the need for reproducible measurements of hypoxic and normoxic conditions across variable tumor volumes, differential tumor biology, and individual patient metabolism. Although promising initial results exist, reproducibility will require testing across larger patient populations and variable tumor characteristics [29]. Combining

FMISO and FDG-PET can allow for dual measurement of tumor hypoxia and cytotoxic effects of chemoradiation regimens, to inform about treatment efficacy and drive overall dose modifications [30]. Utilization of FDG-PET alone has demonstrated promise in predicting ultimate clinical response with relatively robust data from several groups, both within the spectrum of HNSCC and other solid tumors. Gensheimer et al. [31] evaluated mid-radiotherapy PET-CT and found that although no PET metrics were associated with overall survival, some were positively associated with local recurrence. Perhaps most importantly, early out-of-field progression captured on mid-treatment imaging was universally associated with poor overall survival. A larger study of NSCLC tumors demonstrated utility for mid-treatment FDG-PET in modulating the overall radiation dose [32]. In this case, mid-treatment PET metrics were predictive of overall survival and deescalated dosing based on mid-treatment PET (i.e., metabolic tumor volume (MTV)) actually demonstrated improved survival compared with conventional approaches. It is important to note, however, that in such a circumstance, survival data represent not only an aggregate measure of deescalation effectiveness but also identification of radiation responders. Identifying the actual driver of survival effect would require more extensive prospective testing. Potential benefit was also found by Carpenter et al. in cervical cancer [33] and esophageal cancer by Kim et al. [34] In all cases, the most predictive PET parameter was change in the metabolic tumor volume, which likely provides a relatively accurate measure of chemoradiation cytotoxicity.

Magnetic resonance imaging (MRI) has excellent soft tissue resolution, and a spatial resolution that is superior to PET-CT or contrast enhanced CT (CECT). Multiparametric MRI has been under investigation by multiple groups for the detection and prediction of both normal tissue toxicity and tumor response. Because of its excellent spatial resolution and soft tissue definition, MRI can provide real-time, in-treatment information regarding both soft tissue events such as inflammation and alterations in vascularity, with implications for both normal tissue toxicity and tumor biology and response to conventional and targeted agents [35–38]. The utilization of diffusion weighted sequences has demonstrated utility in predicting radiation response in cervical nodal basins, with nonresponders demonstrating higher pretreatment diffusion coefficient values (D) [39, 40]. Combining FDG-PET with multiparametric MRI can further increase the complexity associated with data acquisition and interpretation. However, this approach leverages two distinct, powerful predictive platforms in a manner that may become clinically useful with continued advances in PET-CT simulation and utilization of MRI-LINAC platforms for radiation delivery [41]. An important consideration for these types of studies is the need to accurately and consistently characterize changes in FDG-PET (i.e., MTV) and MRI characteristics (i.e., diffusion coefficient (D), apparent diffusion coefficient (ADC) across the entire course of radiation), since the magnitude of parameter changes and the relative velocity can vary between tumors and patients and as a function of treatment time within each individual tumor (i.e., first half of treatment vs second half of treatment) [42]. Nevertheless, despite these biological and technological complexities, utilization of real-time imaging has the potential to revolutionize radiation-based treatments and potential effectiveness, especially by leveraging newer advances in artificial intelligence.

Despite advances in data acquisition and analysis, both anatomic and metabolic imaging platforms are fundamentally limited by spatial resolution. As a result, both PET and MRI are limited to providing information about bulk changes in tumor volume or activity during, or immediately following treatment. This limitation can be counterbalanced by a *second technological advance*, namely, improved detection of genomic and epigenetic alterations using small amounts of tissue, circulating tumor cells or cell free (cf) DNA. Multiple groups have demonstrated the utility of these tools in the context of head and neck malignancies, most robustly in the context of anaplastic or poorly differentiated thyroid carcinoma and melanoma. Among these, cfDNA has become robust enough to incorporated into multiple commercial platforms with increasing coverage of genomic and epigenetic events that makes it ideal for repeated, real-time interrogation of shifts in tumor burden and/or biology [43]. The utilization of CTC data is more complicated for technical reasons [44] primarily related to CTC isolation methods and subsequent down-stream analysis [45]. Although both cfDNA and CTC can provide real-time information regarding shifts in tumor biology during treatment, they suffer from sampling bias by not paralleling shifts in the intratumor cellular distribution in a representative manner. As such, they reproduce the problem inherent to solid tumor biopsy, whether via fine-needle aspiration, core biopsy, or partial excisional biopsy, namely, representation bias. However, whereas biopsies are used to establish the presence or absence of cancer, cfDNA and CTC data would ideally not only reveal the presence or absence of malignancy but also allow us to track shits in tumor biology. Fortunately, the minimally invasive nature of obtaining cfDNA and CTC data provides a significant clinical advantage compared with solid tumor invasive biopsies, namely, the potential for tolerable iterative sampling. Because repetitive sampling is feasible in most patients, cfDNA and CTA hold significant potential for clinical utility as diagnostic tools and guides for treatment modulation.

Single-cell RNA sequencing provides a partial solution to the problem of representation bias by allowing us to deconvolute cell populations in solid tumors including fibroblasts, tumors cells (and potentially tumor subpopulation), and immunocytes contextualizing data from solid tumors and potentially cfDNA and CTC data. Transient shifts in tumor biology, namely, epithelial to mesenchymal transition (EMT) measured in such a fashion can be analyzed in the context of treatment response [46]. This is particularly relevant for epithelial malignancies such as HNSCC. Using single-cell analysis, investigators have been able to identify and characterize shifts in EMT and integrate it into predictive analyses of clinical outcomes such as metastasis. More direct analysis can also inform regarding potential intrinsic (pretreatment) or acquired (mid- and posttreatment) treatment resistance [47]. Pretreatment single-cell analysis of solid tumor biopsies can potentially identify resistant clones for targeted agents, conventional chemotherapy or radiation using predictive algorithms. In-treatment or posttreatment single-cell analysis of biopsy or CTCs can potentially be used to track clonal expansion secondary to the development of resistance through shifts in multigene signatures such as EMT, or single genes or transcription factors (i.e., SOX2 and SOX9).

For the first time in several generations of scientists and clinicians, we stand on the threshold of a revolution in cancer treatment, more specifically in identifying and predicting response to cancer treatment in a personalized fashion. By combining in-treatment imaging with in-treatment and posttreatment interrogation of circulating tumor DNA it may be possible in the future to expedite evaluation of treatment response and monitor development or expansion of resistant clones faster than clinical progression or clinical imaging might demonstrate. However, this excitement must be tempered by the realization that our basic understanding of treatment resistance remains rudimentary and that tumor biology generates complex challenges that require a multidisciplinary approach and understanding of molecular biology, biochemistry, and metabolism.

Biology is destiny

To overcome resistance, we need to understand what drives it. To a large degree it is tumor biology. A fraction of treatment resistance will be driven by geographical and spatial constraints, which will be discussed in the radiation section in the succeeding text, but largely the intrinsic biology of each tumor cell will dictate the presence or emergence of treatment resistance. For some cancers of the head and neck region, the last two decades have shown new paths of leveraging tumor biology toward translational end points. Melanoma of the head and neck has benefited equally to melanoma of other sites with respect to both targeted agents (i.e., BRAF $\pm$ MEKi) and immunotherapy [48–52]. Anaplastic thyroid carcinoma, long considered a uniformly fatal disease for which treatment was often considered futile, has also shown promising results with targeted strategies driven by genomic and epigenetic changes, in particular the combination of BRAF and MEK inhibitors [53, 54].

In contrast the vast majority of head and neck cancers, namely HNSCC, have not experienced such positive developments. It has now been a decade since the genome of HNSCC was first unraveled with parallel publications by two large multiinstitutional teams [55, 56]. As first described in those initial studies, the biology of HNSCC is a desert of targetable or actionable biological events. First, HNSCC biology is largely defined by tumor suppressor events such as loss of p53 function and mutations in *NOTCH1*. As in other cancers, tumor suppressor events cannot be readily targeted directly. Second the frequency of nontumor suppressor events is rare.

TP53 mutation is one of the most common genomic events associated with cancer development across solid tumors. Dozens of books have been written about the functions of the p53 protein, and hundreds of articles have addressed the role of mutational events on tumorigenesis. Despite this, the precise interaction between *TP53* mutations, tumorigenesis, and treatment response within the spectrum of HNSCC remains unclear [7]. In general, mutations in *TP53* are thought to confer resistance to conventional treatment through multiple overlapping mechanisms, including, but not limited to, metabolic reprogramming, alterations in DNA repair pathways, modulation of programmed cell death cascades, and differential regulation of multiple protein signaling cascades. However, not all mutations appear to behave in a consistent manner. Classification schemes have been described including disruptive versus nondisruptive mutations, high-risk versus low-risk mutations, but no one classification has been widely adopted as reflective of effects on treatment response. In large part, this reflects still conflicting preclinical and clinical data [4–6, 57]. As a result, treatment selection based on *TP53* mutational status or p53 functional status is not yet feasible. Reversal of p53 mutation is theoretically possible and has been evaluated both in preclinical studies and limited early phase clinical trials, though promising clinical results have yet to materialize [58]. Restoration of full protein function, however difficult, may represent the best avenue for modulating treatment response in a somewhat predictable fashion but has been limited to date by challenges of gene delivery throughout an entire tumor [59].

Similar difficulties are associated with genomic alterations in other tumor suppressors such as *NOTCH1*, made more difficult by the complex effects of this protein on cellular signaling [60]. A few genomic alterations within HNSCC are directly targetable, the most specific example being PIK3CA mutations [60, 61]. As in other disease sites, mutant tumor cells demonstrate enhanced sensitivity to targeted agents, although whether this can be leveraged in the clinical setting remains unclear due to specific agent toxicity profiles at this time. The scarcity of directly actionable mutations within HNSCC has led investigators to posit a role for more indirect approaches predicated on identifying the Achilles heel of certain genomic events such as *TP53* mutations. Identification of altered mechanisms of cell death, loss of metabolic flexibility, and alteration in programmed and nonprogrammed cell death pathways has led to a variety of approaches targeting this common event to restore sensitivity to treatment, but have not yet materialized in the translational space [4, 62–65]. One additional layer of complexity associated with HNSCC biology is driven by the heterogeneity of the disease itself. Even outside of the binary divide generated by association with the human papillomavirus (HPV), there is significant evidence that HNSCC biology varies from site to site (i.e., oral cavity vs larynx) and that these differences may in fact have significant therapeutic implications [66].

Radiation resistance

Radiation is a mainstay treatment for advanced-stage head and neck cancers, including HNSCC, either as definitive treatment or in the adjuvant setting following surgical removal of the tumor. As such, relative radiation response is a critical driver of oncologic outcomes in HNSCC. A very broad topic, radiation resistance, and or sensitivity is an exceedingly complex subject to both study and understand. Unlike systemic agents, where response is largely a function of achievable plasma and intratumor dose, radiation response is driven by multiple overlapping characteristics of the treatment modality that are not related to tumor biology.

First, different radiation particles can generate variable antitumor activity. Neutrons have demonstrated exquisite efficacy in head and neck tumors, including in less radio-responsive tumors such as salivary gland malignancies, albeit at the cost of significant normal tissue toxicity and thus morbidity including significant acute and chronic xerostomia [67, 68]. Protons and more recently carbon ions have been investigated in the clinical setting to expand indications outside the traditional scope of rare tumors and pediatric malignancies, although the impact of particle choice on both efficacy and toxicity remains unclear [69, 70]. Most institutions within the United States have adopted intensity modulated radiation therapy (IMRT) techniques using photons as a function of decreased normal tissue toxicity and increased effectiveness compared with traditional radiation techniques [71–74]. However, even within the spectrum of a single-particle modality, fractionation can still play a critically important if not yet completely understood role within certain constraints. Attempts to improve response with variable fractionation and additional selected boosting have not yielded consistent clinical results to date [9, 75]. However, it is clear that single fraction dose delivered and the number of fractions can change the relative efficacy and potentially increase effectiveness in less responsive tumors [76].

Total dose delivered is thought to be constrained superiorly by normal tissue toxicity, developed empirically over the last 50 years. However, it is now clear that some tumors, especially those associated with HPV demonstrate equivalent oncologic outcomes at lower total dose levels [77, 78]. Reductions in total dose of over 10%–15% have now been reported in small trials, in very select patient cohorts, with maintenance of oncologic equivalency to traditional regimens. Unfortunately, many of these small studies add highly variable chemotherapy regimens to the radiation regimen, making it extremely difficult to separate the radiation response from the total chemoradiation response.

How to improve radiation response and address potential resistance therefore remains an exceedingly difficult proposition. *First*, preclinical studies that expand outside of basic studies of DNA damage and repair must cope with particle choice and reproduce fractionation regimens that mimic clinical practice. This is difficult and is rarely done in animal models. *Second*, there is still some disagreement over what primarily drives relative radiation effectiveness. Even if there is agreement that reactive oxygen species (ROS) generate a majority of DNA damage and antitumor activity, precisely what aspects of ROS generation are most important to maximize (i.e., peak ROS burst, duration of ROS generation, area over the curve) remains unclear. Preclinical models therefore must be developed and optimized to address subtle nuances of ROS generation and neutralization, with consideration for particle choice and fractionation. Moreover, these models must then incorporate temporal aspects of cytotoxic chemotherapy administration to generate data more compatible with the human experience. The impact of altered DNA repair remains unclear within HNSCC, particularly as it relates to intrinsic resistance. Although theoretically an excellent means of achieving radiosensitization, PARP inhibitors have demonstrated very limited efficacy in the context of HNSCC to date [79–81]. Exacerbation of DNA damage through enhanced metabolic generation of ROS stress has been investigated including in the context of clinical trials of metformin, but robust data regarding clinical efficacy have not yet emerged [4, 82–85].

Imaging-based modulation of radiation fields, fractionation, and total dose offers the most promise of any available approaches. Since radiation is already intrinsically linked to imaging modalities such as contrast enhanced CT, PET-CT it is ideally suited for adaptive regimens, with selected boosting. This is already integrated in some fractionation schemes and can be further developed. Whether utilizing multiparametric MRI or hyperpolarized MRI to measure tumor viability, vascularity, and metabolic state or using PET derivatives to measure and estimate hypoxic fraction, there is a technological avenue to both measuring radiation sensitivity and boosting radiation dose within selected tumor volumes that continues to escape us with respect to conventional and targeted agents, where all existing approaches based on predictive algorithms of relative tumor sensitivity and/or resistance.

Cytotoxic chemotherapy resistance: Translational implications

To say that we have a very poor understanding of conventional chemotherapy response and resistance within the spectrum of head and neck cancer is at the same time controversial and demonstrably true. Although thousands of articles have been written on intrinsic and acquired resistance to common chemotherapeutic agents such as cisplatin, including hundreds within the context of HNSCC, to date, we simply do not know what drives response and what we can do to modulate clinical effectiveness in a consistent manner. What we do know is that some tumors can and do demonstrate a blunted response to doses of cisplatin, which are very effective in a majority of patients, either in increasing the effectiveness of concomitant radiation or in reducing the potential for distant metastasis (Fig. 3).

Multiple clinical trials, including two recently completed trials in HPV-associated oropharyngeal cancer, have confirmed the primacy of platinum agents, namely, cisplatin in the

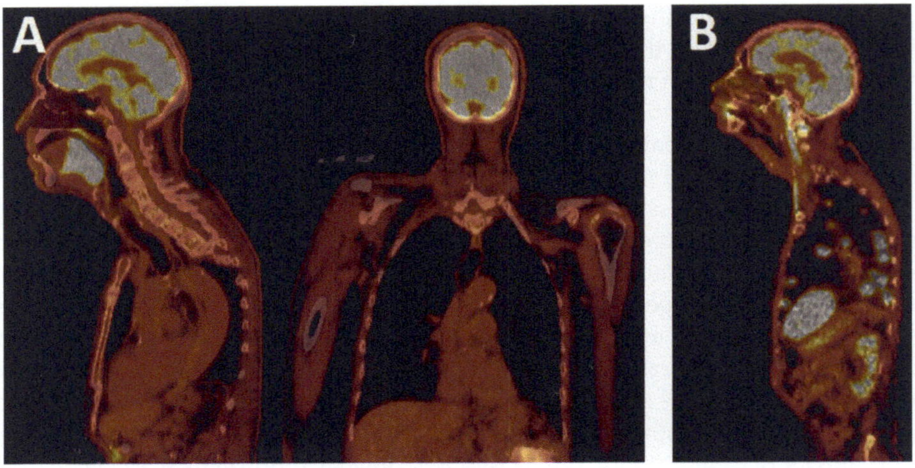

FIG. 3 In-treatment progression. (A) A 75-year-old male presented with pT4aN3bcM0 HPV-oropharyngeal squamous cell carcinoma. The patient underwent surgical resection and free flap reconstruction (microscopic negative margins). Adjuvant chemoradiation was initiated within 4 weeks of surgery; after four cycles of weekly cisplatin and 58Gy of radiation patient developed hemoptysis. (B) FDG-PET demonstrated widespread lung metastasis. The patient was converted to palliative immunotherapy.

treatment of advanced-stage disease [86, 87]. Although we know much about how cisplatin triggers cell death in HNSCC cells and tumors and multiple mechanisms of response and resistance have been described, the complexities discussed earlier in this chapter are reflected in a lack of conformity in cisplatin clinical use. Put simply, at the present time, there is no cisplatin administration regimen that is consistently employed across institutions in the United States in the context of HNSCC. Analyses of clinical datasets suggest that total or cumulative cisplatin dose is associated with improved outcomes; how to best achieve this remains unclear [88, 89]. High-dose regimens deliver higher doses of cisplatin at a lower frequency. Often, this has been associated with greater rates of acute toxicity, specifically nephrotoxicity and neutropenia. Conversely, low-dose cisplatin regimens deliver lower doses of cisplatin on a weekly basis over the entire course of radiation. Although acute toxicity from an individual dose may be lower, some studies have shown a greater potential for cumulative toxicity and potentially treatment interruptions. As a result, some studies have demonstrated higher clinical effectiveness for high-dose cisplatin regimens. This is in contrast to other analyses that demonstrate a more favorable toxicity profile for the low-dose regimen [90]. The lack of consensus on cisplatin administration has generated sufficient frustration to where it is expected that cisplatin regimens will be tested head to head in a prospective fashion in the near future. Unfortunately, much like most other clinical trials, any prospective trial will be conducted in highly functional and motivated individuals, which will continue to leave open the question of tolerability in the broader patient population outside of ECOG 0-1. It is precisely this variability in tolerability that drives the conflicting systematic review and metaanalysis results reported over the last decade in the HNSCC literature [91, 92].

As frustrating as the debate over cisplatin regimens has been for clinicians and patients alike, much of it frankly skirts some of the critical lessons learned from preclinical models.

First, cumulative cisplatin dose is not in fact an intuitive driver of cisplatin effectiveness in a cellular model, particularly if the sustained dose is low enough to allow for rapid active export and neutralization of intracellular cisplatin molecules by free radical scavengers. It is also not compatible with our understanding of DNA damage mechanisms, which suggest that low-dose DNA repair can be sustained by cancer cells without initiation of programmed or stochastic cell death. *Second*, cumulative dose ignores the concomitant administration of radiation and any potential interactions between cisplatin and radiation at a cellular level, whether through maximization of ROS generation or sustained DNA damage in an active cell cycle program. Low-dose, continuous administration of cisplatin is in fact precisely the mechanism used in the laboratory to generate acquired cisplatin resistance either through direct individual mechanisms or broader cellular reprogramming (i.e., EMT). This would therefore be quite counterproductive in the clinical setting, particularly in an induction chemotherapy approach. Unfortunately, these are simply extrapolations from preclinical data, since we lack real-time information from clinical trials that directly compare cisplatin regimens, acquired via metabolic imaging or real-time interrogation of tumors under active treatment via cfDNA or CTC coupled to single-cell RNA profiling [93].

Although cisplatin has been in clinical use for nearly half a century in the United States and around the world, our understanding of cisplatin plasma and tumor pharmacokinetics is poor, or at the very least provides very little translationally relevant data to guide trial design. We have some insight into cisplatin plasma kinetics, particularly as it relates to organ toxicity, but much less so in the context of actual solid tumor processing of cisplatin-induced stress and DNA damage [94–96]. Even less is known about other platinum derivatives, which are employed in clinical practice in HNSCC, such as carboplatin [97] or other commonly utilized cytotoxic agents such as docetaxel [98]. This is particularly problematic since there is in fact no level I evidence that cisplatin is in fact superior to other cytotoxic agents in the setting of HNSCC and recent clinical trials and retrospective clinical data suggest the opposite may be true [99, 100].

The translational implications of intrinsic or acquired resistance, driven in part by tumor biology and potentially in part by vascular constraints, are clear at this point and fairly dramatic. *First*, how do we choose among cytotoxic agents, and within each cytotoxic agent how do we dose the drug, not only with respect to both overall time frame but also critically with respect to administration of radiation fractions? How do we approach potentially aggressive tumor biology? Do we overcome it with higher drug doses and combinatorial strategies, or do we develop predictive algorithms that identify a single optimal compound from the available choices (i.e., cisplatin vs docetaxel)? *Second*, how do we integrate complex regimens within the concept of acquired resistance? Do we vary induction compounds from subsequent compounds to be used in the concurrent setting or during treatment consolidation? Do we exclude DNA damaging agents from the induction phase of treatment to avoid the possibility of enhanced DNA repair mechanisms which can alter radiation effectiveness? These options do exist even within the relatively limited spectrum of cytotoxic chemotherapy previously utilized in solid tumors; the primary driver of our uncertainty is the lack of robust preclinical models and prospectively acquired clinical data points. *Third*, what is the primary driver of chemotherapy effectiveness? Is it generation of free radicals and subsequent DNA damage? If so, metabolic targeting of ROS scavenging pathways and close temporal integration with individual radiation fractions may be able to generate critical improvements in effectiveness without significant alterations in total dose and thus toxicity. Alternatively, if the primary driver of chemotherapy effectiveness is to induce replicative stress through generation of paradoxical cell cycle progression and arrest signals, we must utilize a different approach. Since replicative stress secondary to treatment is largely a function of intrinsic biology, it will be critical to develop adjuvants that are biologically specific and nontoxic and can overcome biological constraints (see next section regarding Wee1-kinase inhibitors and *TP53* mutant HNSCC).

None of these questions can be answered without a wholesale transformation of our approach to HNSCC biology. Comprehensive, genomically characterized cell lines and PDX models must be integrated into high-throughput genomic and chemical screening approaches, paralleled by real-time tumor read-outs in the form of either metabolic imaging or direct metabolic interrogation to understand the dynamic interactions among the intrinsic tumor biology, extrinsic chemotherapy-generated stress, and acquisition of potential escape mechanisms.

Targeted agent response and resistance: Lessons for the future

Targeted agents are a dual edge sword. For the optimistic scientist, they represent an elegant means of targeting specific genomic (mutations and translocations) or epigenetic events. This view has been substantiated by the dramatic and specific effects of imatinib mesylate against chronic myelogenous leukemia (CML) tumors, which exhibited the BCR-ABL fusion event [101, 102]. It is further supported by the dramatic initial effects of BRAF inhibitors in the setting of *BRAF* mutant melanoma [48].

For the optimistic clinician, targeted agents represent a departure from the significant toxicity of conventional agents such as nephrotoxicity and ototoxicity for cisplatin. These results were supported by the introduction of cetuximab into clinical practice. The initial study in HNSCC demonstrated increased effectiveness compared with radiation alone at the cost of minimal reported toxicity [103]. Unfortunately the toxicity profile of targeted agents remains a critical issue both in the setting of clinical trials and clinical practice more broadly. RTOG1016 and De-ESCALaTE HPV demonstrated not only inferiority for cetuximab compared with cisplatin but also a concerning toxicity profile, not superior to cytotoxic chemotherapy [86, 87]. To some degree, these disappointing results were partially predictable based on preclinical and retrospective studies that failed to detect a predictive biological signature for cetuximab effectiveness in HNSCC. Although multiple genomic and epigenetic alterations have been associated with cetuximab response, including high EGFR, PTEN, EREG, and AREG expression, none has survived validation across multiple patient cohorts with sufficient predictive robustness to enter clinical practice [104, 105]. As a result of this work and other relatively disappointing results in the clinical trial setting, cetuximab has now been relegated to a secondary option for cisplatin-ineligible patients. Whether or not combination with immune-oncology approaches can change this in the future remains to be seen [106].

One of the lessons drawn from the cetuximab experience, especially when juxtaposed onto success stories such as BRAF inhibitors and imatinib, is the critical need for a biological rationale and more specifically the need to identify true biological dependencies within

HNSCC. It has become clear that small, quantitative differences in gene expression or protein levels between tumor and normal tissue do not generate a clinically relevant therapeutic index required for efficacy of the targeted agent. What is required is qualitative changes in cellular signaling and associated cellular phenotypes that can be targeted. One example of this approach is combining distinct genomic alterations (i.e., *TP53* mutations) with targeted inhibition of p53 mediated events [63]. The addition of Wee-1 kinase inhibition to cisplatin in the setting of mutant *TP53* abrogates the G2 blockade common to these cells resulting in significant resensitization to cisplatin in preclinical models. This effect is equally prominent in the setting of radiation-induced G2 blockade; as a result Wee-1 kinase targeting generates a preferential chemoradiosensitizing effect precisely in the most aggressive HNSCC tumor cells, namely, those with mutant *TP53* [107, 108]. Whether this approach, predicated on a synergistic interaction between a genomic event and a targeted agent, with promising resensitizing potential will materialize remains to be seen in active and planned clinical trials.

Unfortunately, for the pessimistic statistician and scientist, targeted agents are a losing proposition. In a race between millions of clones and billions of individual tumor cells within each solid tumor and scientific research, discovery, and clinical development, the numbers favor the tumor by several orders of magnitude. Evolution of tumors, whether through acquisition of secondary events, or through a simpler process of clonal selection, extirpation, and expansion, is not only theoretically possible but also has been demonstrated in the clinical setting, most clearly in the context of BRAF mutant melanoma, which can evolve in response to prolonged BRAF targeting [109, 110]. Acknowledging this reality is important, because it must shape our approach to combinatorial strategies for both conventional chemotherapy and targeted agents.

An additional, nonoverlapping difficulty is that of multidrug resistance. There exists a built-in assumption within oncology practice that conditioned or acquired resistance can be overcome by switching to another drug. This assumption is quite likely false and has been demonstrated as such in the context of microbiology and acquisition of antibiotic resistance. Complex preclinical analyses have shown that acquisition of cisplatin resistance dramatically shifts the genomic, epigenetic, transcriptional, and metabolic landscape of solid tumors including HNSCC, in ways that can impact the effectiveness of targeted compounds, for example, those targeting the Pi3K-Akt-mTOR pathway [111]. It has long been known that cisplatin exposure can trigger acquisition of cisplatin resistance through both simple and complex mechanisms (as discussed in the preceding text). Within the context of more complex mechanisms such as EMT, there is significant capacity for development of cross-resistance to both other cytotoxic agents and targeted agents.

As a result, future approaches to addressing the potential of intrinsic and acquired resistance must include several components. *First*, targeting of multiple pathways simultaneously is likely to be more effective than targeting of a single pathway, as long as the toxicity profile remains within reasonable clinical parameters. One good example of this is BRAF-MEK inhibition that has demonstrated excellent outcomes in melanoma with quite durable responses, almost on par with immunotherapy [49–51]. This efficacy has led to utilization of this combinatorial strategy in anaplastic thyroid carcinoma, one of the most aggressive head and neck solid tumor. However, tumors have shown the ability to adapt even to combinatorial approaches through upregulation of secondary signaling cascades or reactivation of presumably blocked pathways such as ERK1/2 [112–114]. As such, tumors maintain a competitive advantage over the clinician. Tumors can continually activate secondary and tertiary pathways as a method of therapeutic escape; some of these may be theoretically targetable [115]. In contrast, clinical addition of additional targeted agents is limited by increasing and potentially unsustainable toxicity [52, 114].

Second, future approaches should ideally leverage driver events in tumor biology, with a relatively fixed profile (i.e., mutations, translocations, and deletions) with the goal of achieving synthetic lethality between targeted agents and individual biological events, or potentially broader synthetic cytotoxicity as in the case of Wee-1 kinase inhibitors. Although likely to be rare, the combination of targeted agent hypersensitivity secondary to a biological event (i.e., CDKN2a/p16 copy number loss) and a tolerable class of targeted agents (i.e., checkpoint kinase inhibitors) can lead to potentially dramatic effects in a subset of HNSCC tumors, which although limited in scope, can be potentially curative for a subset of patients [62]. *Third,* future approaches should leverage combinatorial strategies that take advantage of the fact that we now have three relatively distinct tools at our disposal: conventional and cytotoxic chemotherapy, targeted agents, and immunotherapy. It may therefore be possible to combine these superclasses in selected cases to maximize efficacy, while maintaining tolerable toxicity if the toxicity profile is largely not overlapping [116]. Excitement about this approach must be tempered, however, by a realization of just how limited our understanding is of the mechanistic underpinnings of many of the agents we consider to be so-called targeted and our still limited understanding of tumor immunology, particularly for more rare diseases such as HNSCC [117, 118].

Finally, it is quite possible that small but substantial gains can be made in clinical efficacy through reevaluation of a relatively old concept, namely, metronomic chemotherapy. This approach, introduced almost two decades ago, continues to be utilized in some disease models and may in fact hold great promise, particularly as it relates to combinatorial strategies that rely on interactions between superclasses of systemic agents [119–121]. The possibility of modulating the tumor immune microenvironment using low, tolerable doses of conventional cytotoxic agents such as cisplatin and thus greatly potentiate the antitumor effectiveness of checkpoint inhibition is as exciting as the notion of modulating cytotoxic escape using low-dose, targeted tyrosine kinase inhibition.

Acknowledgments

There is no dedicated funding relevant to the current work.

References

[1] Correa H. Li-Fraumeni syndrome. J Pediatr Genet 2016;5:84–8.
[2] Fukunaga H, Taki Y, Prise KM. Diversity of ATM gene variants: a population-based genome data analysis for precision medicine. Hum Genomics 2019;13:38.
[3] van Os NJ, Roeleveld N, Weemaes CM, et al. Health risks for ataxia-telangiectasia mutated heterozygotes: a systematic review, meta-analysis and evidence-based guideline. Clin Genet 2016;90:105–17.
[4] Skinner HD, Sandulache VC, Ow TJ, et al. TP53 disruptive mutations lead to head and neck cancer treatment failure through inhibition of radiation-induced senescence. Clin Cancer Res 2012;18:290–300.
[5] Osman AA, Neskey DM, Katsonis P, et al. Evolutionary action score of TP53 coding variants is predictive of platinum response in head and neck cancer patients. Cancer Res 2015;75:1205–15.

[6] Neskey DM, Osman AA, Ow TJ, et al. Evolutionary action score of TP53 identifies high-risk mutations associated with decreased survival and increased distant metastases in head and neck cancer. Cancer Res 2015;75:1527–36.

[7] Sano D, Xie TX, Ow TJ, et al. Disruptive TP53 mutation is associated with aggressive disease characteristics in an orthotopic murine model of oral tongue cancer. Clin Cancer Res 2011;17:6658–70.

[8] Lehmann BD, Bauer JA, Schafer JM, et al. PIK3CA mutations in androgen receptor-positive triple negative breast cancer confer sensitivity to the combination of PI3K and androgen receptor inhibitors. Breast Cancer Res 2014;16:406.

[9] Ang KK, Harris J, Wheeler R, et al. Human papillomavirus and survival of patients with oropharyngeal cancer. N Engl J Med 2010;363:24–35.

[10] Sandulache VC, Hamblin J, Lai S, et al. Oropharyngeal squamous cell carcinoma in the veteran population: association with traditional carcinogen exposure and poor clinical outcomes. Head Neck 2015;37:1246–53.

[11] Fakhry C, Zhang Q, Gillison ML, et al. Validation of NRG oncology/RTOG-0129 risk groups for HPV-positive and HPV-negative oropharyngeal squamous cell cancer: implications for risk-based therapeutic intensity trials. Cancer 2019;125:2027–38.

[12] Lydiatt WM, Patel SG, O'Sullivan B, et al. Head and Neck cancers-major changes in the American Joint Committee on cancer eighth edition cancer staging manual. CA Cancer J Clin 2017;67:122–37.

[13] Sandulache VC, Vandelaar LJ, Skinner HD, et al. Salvage total laryngectomy after external-beam radiotherapy: a 20-year experience. Head Neck 2016;38(Suppl. 1):E1962–8.

[14] Zhong LP, Zhang CP, Ren GX, et al. Randomized phase III trial of induction chemotherapy with docetaxel, cisplatin, and fluorouracil followed by surgery versus up-front surgery in locally advanced resectable oral squamous cell carcinoma. J Clin Oncol 2013;31:744–51.

[15] Department of Veterans Affairs Laryngeal Cancer Study G, Wolf GT, Fisher SG, et al. Induction chemotherapy plus radiation compared with surgery plus radiation in patients with advanced laryngeal cancer. N Engl J Med 1991;324:1685–90.

[16] Cohen EE, Karrison TG, Kocherginsky M, et al. Phase III randomized trial of induction chemotherapy in patients with N2 or N3 locally advanced head and neck cancer. J Clin Oncol 2014;32:2735–43.

[17] Haddad R, O'Neill A, Rabinowits G, et al. Induction chemotherapy followed by concurrent chemoradiotherapy (sequential chemoradiotherapy) versus concurrent chemoradiotherapy alone in locally advanced head and neck cancer (PARADIGM): a randomised phase 3 trial. Lancet Oncol 2013;14:257–64.

[18] Wang Y, Liu L, Chen Z. Transcriptome profiling of cervical cancer cells acquired resistance to cisplatin by deep sequencing. Artif Cells Nanomed Biotechnol 2019;47:2820–9.

[19] Ricci F, Brunelli L, Affatato R, et al. Overcoming platinum-acquired resistance in ovarian cancer patient-derived xenografts. Ther Adv Med Oncol 2019;11.

[20] Liang ZD, Long Y, Tsai WB, et al. Mechanistic basis for overcoming platinum resistance using copper chelating agents. Mol Cancer Ther 2012;11:2483–94.

[21] Chen HH, Song IS, Hossain A, et al. Elevated glutathione levels confer cellular sensitization to cisplatin toxicity by up-regulation of copper transporter hCtr1. Mol Pharmacol 2008;74:697–704.

[22] Mezencev R, Matyunina LV, Wagner GT, McDonald JF. Acquired resistance of pancreatic cancer cells to cisplatin is multifactorial with cell context-dependent involvement of resistance genes. Cancer Gene Ther 2016;23:446–53.

[23] Arumugam T, Ramachandran V, Fournier KF, et al. Epithelial to mesenchymal transition contributes to drug resistance in pancreatic cancer. Cancer Res 2009;69:5820–8.

[24] Wan Q, Shen Y, Zhao H, et al. Impaired DNA double-strand breaks repair by kinesin family member 4A inhibition renders human H1299 non-small-cell lung cancer cells sensitive to cisplatin. J Cell Physiol 2019;234:10360–71.

[25] Min M, Lin P, Liney G, et al. A review of the predictive role of functional imaging in patients with mucosal primary head and neck cancer treated with radiation therapy. J Med Imaging Radiat Oncol 2017;61:99–123.

[26] Lee N, Nehmeh S, Schoder H, et al. Prospective trial incorporating pre-/mid-treatment [18F]-misonidazole positron emission tomography for head-and-neck cancer patients undergoing concurrent chemoradiotherapy. Int J Radiat Oncol Biol Phys 2009;75:101–8.

[27] Lee N, Schoder H, Beattie B, et al. Strategy of using intratreatment hypoxia imaging to selectively and safely guide radiation dose de-escalation concurrent with chemotherapy for locoregionally advanced human papillomavirus-related oropharyngeal carcinoma. Int J Radiat Oncol Biol Phys 2016;96:9–17.

[28] Nehmeh SA, Lee NY, Schroder H, et al. Reproducibility of intratumor distribution of (18)F-fluoromisonidazole in head and neck cancer. Int J Radiat Oncol Biol Phys 2008;70:235–42.
[29] Okamoto S, Shiga T, Yasuda K, et al. High reproducibility of tumor hypoxia evaluated by 18F-fluoromisonidazole PET for head and neck cancer. J Nucl Med 2013;54:201–7.
[30] Okamoto S, Shiga T, Yasuda K, et al. The reoxygenation of hypoxia and the reduction of glucose metabolism in head and neck cancer by fractionated radiotherapy with intensity-modulated radiation therapy. Eur J Nucl Med Mol Imaging 2016;43:2147–54.
[31] Gensheimer MF, Hong JC, Chang-Halpenny C, et al. Mid-radiotherapy PET/CT for prognostication and detection of early progression in patients with stage III non-small cell lung cancer. Radiother Oncol 2017;125:338–43.
[32] Kong FS, Li L, Wang W, et al. Greater reduction in mid-treatment FDG-PET volume may be associated with worse survival in non-small cell lung cancer. Radiother Oncol 2019;132:241–9.
[33] Carpenter DJ, Jacobs CD, Wong TZ, Craciunescu O, Chino JP. Changes on midchemoradiation therapy fluorodeoxyglucose positron emission tomography for cervical cancer are associated with prognosis. Int J Radiat Oncol Biol Phys 2019;105:356–66.
[34] Kim N, Cho H, Yun M, Park KR, Lee CG. Prognostic values of mid-radiotherapy (18)F-FDG PET/CT in patients with esophageal cancer. Radiat Oncol 2019;14:27.
[35] Joint H, Neck radiotherapy MRIDC. Dynamic contrast-enhanced MRI detects acute radiotherapy-induced alterations in mandibular microvasculature: prospective assessment of imaging biomarkers of normal tissue injury. Sci Rep 2016;6:29864.
[36] Joint H, Neck radiotherapy MRIDC. A multi-institutional comparison of dynamic contrast-enhanced magnetic resonance imaging parameter calculations. Sci Rep 2017;7:11185.
[37] Joint H, Neck radiotherapy MRIDC. Dynamic contrast-enhanced magnetic resonance imaging for head and neck cancers. Sci Data 2018;5:180008.
[38] Joint H, Neck radiotherapy MRIDC. A prospective longitudinal assessment of MRI signal intensity kinetics of non-target muscles in patients with advanced stage oropharyngeal cancer in relationship to radiotherapy dose and post-treatment radiation-associated dysphagia: preliminary findings from a randomized trial. Radiother Oncol 2019;130:46–55.
[39] Ding Y, Hazle JD, Mohamed AS, et al. Intravoxel incoherent motion imaging kinetics during chemoradiotherapy for human papillomavirus-associated squamous cell carcinoma of the oropharynx: preliminary results from a prospective pilot study. NMR Biomed 2015;28:1645–54.
[40] Marzi S, Piludu F, Sanguineti G, et al. The prediction of the treatment response of cervical nodes using intravoxel incoherent motion diffusion-weighted imaging. Eur J Radiol 2017;92:93–102.
[41] Bowen SR, Yuh WTC, Hippe DS, et al. Tumor radiomic heterogeneity: multiparametric functional imaging to characterize variability and predict response following cervical cancer radiation therapy. J Magn Reson Imaging 2018;47:1388–96.
[42] Min M, Lee MT, Lin P, et al. Assessment of serial multi-parametric functional MRI (diffusion-weighted imaging and R2*) with (18)F-FDG-PET in patients with head and neck cancer treated with radiation therapy. Br J Radiol 2016;89:20150530.
[43] Sandulache VC, Williams MD, Lai SY, et al. Real-time genomic characterization utilizing circulating cell-free DNA in patients with anaplastic thyroid carcinoma. Thyroid 2017;27:81–7.
[44] Payne K, Brooks J, Spruce R, et al. Circulating tumour cell biomarkers in head and neck cancer: current progress and future prospects. Cancers (Basel) 2019;11:1115–28.
[45] Davis-Marcisak EF, Sherman TD, Orugunta P, et al. Differential variation analysis enables detection of tumor heterogeneity using single-cell RNA-sequencing data. Cancer Res 2019;79:5102–12.
[46] Puram SV, Tirosh I, Parikh AS, et al. Single-cell transcriptomic analysis of primary and metastatic tumor ecosystems in head and neck cancer. Cell 2017;171:1611–24. e1624.
[47] Sharma A, Cao EY, Kumar V, et al. Longitudinal single-cell RNA sequencing of patient-derived primary cells reveals drug-induced infidelity in stem cell hierarchy. Nat Commun 2018;9:4931.
[48] Chapman PB, Hauschild A, Robert C, et al. Improved survival with vemurafenib in melanoma with BRAF V600E mutation. N Engl J Med 2011;364:2507–16.
[49] Flaherty KT, Infante JR, Daud A, et al. Combined BRAF and MEK inhibition in melanoma with BRAF V600 mutations. N Engl J Med 2012;367:1694–703.

[50] Long GV, Stroyakovskiy D, Gogas H, et al. Combined BRAF and MEK inhibition versus BRAF inhibition alone in melanoma. N Engl J Med 2014;371:1877–88.
[51] Liu M, Yang X, Liu J, et al. Efficacy and safety of BRAF inhibition alone versus combined BRAF and MEK inhibition in melanoma: a meta-analysis of randomized controlled trials. Oncotarget 2017;8: 32258–69.
[52] Broman KK, Dossett LA, Sun J, Eroglu Z, Zager JS. Update on BRAF and MEK inhibition for treatment of melanoma in metastatic, unresectable, and adjuvant settings. Expert Opin Drug Saf 2019;18:381–92.
[53] Cabanillas ME, Ferrarotto R, Garden AS, et al. Neoadjuvant BRAF- and immune-directed therapy for anaplastic thyroid carcinoma. Thyroid 2018;28:945–51.
[54] Subbiah V, Kreitman RJ, Wainberg ZA, et al. Dabrafenib and Trametinib treatment in patients with locally advanced or metastatic BRAF V600-mutant anaplastic thyroid cancer. J Clin Oncol 2018;36:7–13.
[55] Agrawal N, Frederick MJ, Pickering CR, et al. Exome sequencing of head and neck squamous cell carcinoma reveals inactivating mutations in NOTCH1. Science 2011;333:1154–7.
[56] Stransky N, Egloff AM, Tward AD, et al. The mutational landscape of head and neck squamous cell carcinoma. Science 2011;333:1157–60.
[57] Poeta ML, Manola J, Goldwasser MA, et al. TP53 mutations and survival in squamous-cell carcinoma of the head and neck. N Engl J Med 2007;357:2552–61.
[58] Shangary S, Wang S. Small-molecule inhibitors of the MDM2-p53 protein-protein interaction to reactivate p53 function: a novel approach for cancer therapy. Annu Rev Pharmacol Toxicol 2009;49:223–41.
[59] Clayman GL, El-Naggar AK, Lippman SM, et al. Adenovirus-mediated p53 gene transfer in patients with advanced recurrent head and neck squamous cell carcinoma. J Clin Oncol 1998;16(6):2221–32.
[60] Pickering CR, Zhang J, Yoo SY, et al. Integrative genomic characterization of oral squamous cell carcinoma identifies frequent somatic drivers. Cancer Discov 2013;3:770–81.
[61] Chen X, Cao Y, Sedhom W, et al. Distinct roles of PIK3CA in the enrichment and maintenance of cancer stem cells in head and neck squamous cell carcinoma. Mol Oncol 2020;14(1):139–58.
[62] Gadhikar MA, Zhang J, Shen L, et al. CDKN2A/p16 deletion in head and neck cancer cells is associated with CDK2 activation, replication stress, and vulnerability to CHK1 inhibition. Cancer Res 2018;78:781–97.
[63] Osman AA, Monroe MM, Ortega Alves MV, et al. Wee-1 kinase inhibition overcomes cisplatin resistance associated with high-risk TP53 mutations in head and neck cancer through mitotic arrest followed by senescence. Mol Cancer Ther 2015;14:608–19.
[64] Sandulache VC, Ow TJ, Pickering CR, et al. Glucose, not glutamine, is the dominant energy source required for proliferation and survival of head and neck squamous carcinoma cells. Cancer 2011;117:2926–38.
[65] Sandulache VC, Skinner HD, Ow TJ, et al. Individualizing antimetabolic treatment strategies for head and neck squamous cell carcinoma based on TP53 mutational status. Cancer 2012;118:711–21.
[66] Reddy RB, Khora SS, Suresh A. Molecular prognosticators in clinically and pathologically distinct cohorts of head and neck squamous cell carcinoma—a meta-analysis approach. PLoS One 2019;14.
[67] Davis C, Sikes J, Namaranian P, Laramore G, Dillon JK. Neutron beam radiation therapy: an overview of treatment and oral complications when treating salivary gland malignancies. J Oral Maxillofac Surg 2016;74:830–5.
[68] Liao JJ, Parvathaneni U, Laramore GE, et al. Fast neutron radiotherapy for primary mucosal melanomas of the head and neck. Head Neck 2014;36:1162–7.
[69] Yang J, Gao J, Qiu X, et al. Intensity-modulated proton and carbon-ion radiation therapy in the management of head and neck sarcomas. Cancer Med 2019;8:4574–86.
[70] Holliday EB, Kocak-Uzel E, Feng L, et al. Dosimetric advantages of intensity-modulated proton therapy for oropharyngeal cancer compared with intensity-modulated radiation: a case-matched control analysis. Med Dosim 2016;41:189–94.
[71] Nutting CM, Morden JP, Harrington KJ, et al. Parotid-sparing intensity modulated versus conventional radiotherapy in head and neck cancer (PARSPORT): a phase 3 multicentre randomised controlled trial. Lancet Oncol 2011;12:127–36.
[72] Xiang M, Holsinger FC, Colevas AD, Chen MM, Le QT, Beadle BM. Survival of patients with head and neck cancer treated with definitive radiotherapy and concurrent cisplatin or concurrent cetuximab: a surveillance, epidemiology, and end results-medicare analysis. Cancer 2018;124:4486–94.
[73] Beadle BM, Liao KP, Elting LS, et al. Improved survival using intensity-modulated radiation therapy in head and neck cancers: a SEER-medicare analysis. Cancer 2014;120:702–10.

[74] Beadle BM, Liao KP, Giordano SH, et al. Reduced feeding tube duration with intensity-modulated radiation therapy for head and neck cancer: a surveillance, epidemiology, and end results-medicare analysis. Cancer 2017;123:283–93.
[75] Jiang L, Zhang Y, Yang Z, Liang F, Wu J, Wang R. A comparison of clinical outcomes between simultaneous integrated boost (SIB) versus sequential boost (SEQ) intensity modulated radiation therapy (IMRT) for head and neck cancer: a meta-analysis. Medicine (Baltimore) 2019;98.
[76] Guadagnolo BA, Zagars GK. Adjuvant radiation therapy for high-risk nodal metastases from cutaneous melanoma. Lancet Oncol 2009;10:409–16.
[77] Hegde JV, Shaverdian N, Daly ME, et al. Patient-reported quality-of-life outcomes after de-escalated chemoradiation for human papillomavirus-positive oropharyngeal carcinoma: findings from a phase 2 trial. Cancer 2018;124:521–9.
[78] Pearlstein KA, Wang K, Amdur RJ, et al. Quality of life for patients with favorable-risk HPV-associated oropharyngeal cancer after de-intensified chemoradiotherapy. Int J Radiat Oncol Biol Phys 2019;103:646–53.
[79] Lee TW, Wong WW, Dickson BD, et al. Radiosensitization of head and neck squamous cell carcinoma lines by DNA-PK inhibitors is more effective than PARP-1 inhibition and is enhanced by SLFN11 and hypoxia. Int J Radiat Biol 2019;95(12):1597–612.
[80] Parsels LA, Karnak D, Parsels JD, et al. PARP1 trapping and DNA replication stress enhance radiosensitization with combined WEE1 and PARP inhibitors. Mol Cancer Res 2018;16:222–32.
[81] de Haan R, van Werkhoven E, van den Heuvel MM, et al. Study protocols of three parallel phase 1 trials combining radical radiotherapy with the PARP inhibitor olaparib. BMC Cancer 2019;19:901.
[82] Sharma PK, Dwarakanath BS, Varshney R. Radiosensitization by 2-deoxy-D-glucose and 6-aminonicotinamide involves activation of redox sensitive ASK1-JNK/p38MAPK signaling in head and neck cancer cells. Free Radic Biol Med 2012;53:1500–13.
[83] Simons AL, Ahmad IM, Mattson DM, Dornfeld KJ, Spitz DR. 2-Deoxy-D-glucose combined with cisplatin enhances cytotoxicity via metabolic oxidative stress in human head and neck cancer cells. Cancer Res 2007;67:3364–70.
[84] Simons AL, Fath MA, Mattson DM, et al. Enhanced response of human head and neck cancer xenograft tumors to cisplatin combined with 2-deoxy-D-glucose correlates with increased 18F-FDG uptake as determined by PET imaging. Int J Radiat Oncol Biol Phys 2007;69:1222–30.
[85] Simons AL, Parsons AD, Foster KA, Orcutt KP, Fath MA, Spitz DR. Inhibition of glutathione and thioredoxin metabolism enhances sensitivity to perifosine in head and neck cancer cells. J Oncol 2009;2009:519563.
[86] Gillison ML, Trotti AM, Harris J, et al. Radiotherapy plus cetuximab or cisplatin in human papillomavirus-positive oropharyngeal cancer (NRG Oncology RTOG 1016): a randomised, multicentre, non-inferiority trial. Lancet 2019;393:40–50.
[87] Mehanna H, Robinson M, Hartley A, et al. Radiotherapy plus cisplatin or cetuximab in low-risk human papillomavirus-positive oropharyngeal cancer (De-ESCALaTE HPV): an open-label randomised controlled phase 3 trial. Lancet 2019;393:51–60.
[88] Strojan P, Vermorken JB, Beitler JJ, et al. Cumulative cisplatin dose in concurrent chemoradiotherapy for head and neck cancer: a systematic review. Head Neck 2016;38(Suppl 1):E2151–8.
[89] Szturz P, Wouters K, Kiyota N, et al. Low-dose vs. high-dose cisplatin: lessons learned from 59 chemoradiotherapy trials in head and neck cancer. Front Oncol 2019;9:86.
[90] Szturz P, Wouters K, Kiyota N, et al. Weekly low-dose versus three-weekly high-dose cisplatin for concurrent chemoradiation in locoregionally advanced non-nasopharyngeal head and neck cancer: a systematic review and meta-analysis of aggregate data. Oncologist 2017;22:1056–66.
[91] Mackiewicz J, Rybarczyk-Kasiuchnicz A, Lasinska I, et al. The comparison of acute toxicity in 2 treatment courses: three-weekly and weekly cisplatin treatment administered with radiotherapy in patients with head and neck squamous cell carcinoma. Medicine (Baltimore) 2017;96.
[92] Mohamed A, Twardy B, Zordok MA, et al. Concurrent chemoradiotherapy with weekly versus triweekly cisplatin in locally advanced squamous cell carcinoma of the head and neck: comparative analysis. Head Neck 2019;41:1490–8.
[93] Noronha V, Joshi A, Patil VM, et al. Once-a-week versus once-every-3-weeks cisplatin chemoradiation for locally advanced head and neck cancer: a phase III randomized noninferiority trial. J Clin Oncol 2018;36:1064–72.

[94] Stewart DJ, Mikhael NZ, Nanji AA, et al. Renal and hepatic concentrations of platinum: relationship to cisplatin time, dose, and nephrotoxicity. J Clin Oncol 1985;3:1251–6.
[95] Gandara DR, DeGregorio MW, Wold H, et al. High-dose cisplatin in hypertonic saline: reduced toxicity of a modified dose schedule and correlation with plasma pharmacokinetics. A Northern California Oncology Group Pilot Study in non-small-cell lung cancer. J Clin Oncol 1986;4:1787–93.
[96] Gandara DR, Perez EA, Phillips WA, Lawrence HJ, DeGregorio M. Evaluation of cisplatin dose intensity: current status and future prospects. Anticancer Res 1989;9:1121–8.
[97] Xiang M, Colevas AD, Holsinger FC, Le QX, Beadle BM. Survival after definitive chemoradiotherapy with concurrent cisplatin or carboplatin for head and neck cancer. J Natl Compr Canc Netw 2019;17:1065–73.
[98] Liao JF, Zhang Q, Du XJ, et al. Concurrent chemoradiotherapy with weekly docetaxel versus cisplatin in the treatment of locoregionally advanced nasopharyngeal carcinoma: a propensity score-matched analysis. Cancer Commun (Lond) 2019;39:40.
[99] Liberato NL, Rognoni C, Rubrichi S, et al. Adding docetaxel to cisplatin and fluorouracil in patients with unresectable head and neck cancer: a cost-utility analysis. Ann Oncol 2012;23:1825–32.
[100] Vermorken JB, Remenar E, van Herpen C, et al. Cisplatin, fluorouracil, and docetaxel in unresectable head and neck cancer. N Engl J Med 2007;357:1695–704.
[101] Deininger MW, Goldman JM, Lydon N, Melo JV. The tyrosine kinase inhibitor CGP57148B selectively inhibits the growth of BCR-ABL-positive cells. Blood 1997;90:3691–8.
[102] Druker BJ, Tamura S, Buchdunger E, et al. Effects of a selective inhibitor of the Abl tyrosine kinase on the growth of Bcr-Abl positive cells. Nat Med 1996;2:561–6.
[103] Bonner JA, Harari PM, Giralt J, et al. Radiotherapy plus cetuximab for squamous-cell carcinoma of the head and neck. N Engl J Med 2006;354:567–78.
[104] Eze N, Lee JW, Yang DH, et al. PTEN loss is associated with resistance to cetuximab in patients with head and neck squamous cell carcinoma. Oral Oncol 2019;91:69–78.
[105] Kogashiwa Y, Inoue H, Kuba K, et al. Prognostic role of epiregulin/amphiregulin expression in recurrent/metastatic head and neck cancer treated with cetuximab. Head Neck 2018;40:2424–31.
[106] Taberna M, Oliva M, Mesia R. Cetuximab-containing combinations in locally advanced and recurrent or metastatic head and neck squamous cell carcinoma. Front Oncol 2019;9:383.
[107] Bridges KA, Hirai H, Buser CA, et al. MK-1775, a novel Wee1 kinase inhibitor, radiosensitizes p53-defective human tumor cells. Clin Cancer Res 2011;17:5638–48.
[108] Busch CJ, Kroger MS, Jensen J, et al. G2-checkpoint targeting and radiosensitization of HPV/p16-positive HNSCC cells through the inhibition of Chk1 and Wee1. Radiother Oncol 2017;122:260–6.
[109] Shi H, Hugo W, Kong X, et al. Acquired resistance and clonal evolution in melanoma during BRAF inhibitor therapy. Cancer Discov 2014;4:80–93.
[110] Van Allen EM, Wagle N, Sucker A, et al. The genetic landscape of clinical resistance to RAF inhibition in metastatic melanoma. Cancer Discov 2014;4:94–109.
[111] Niehr F, Eder T, Pilz T, et al. Multilayered omics-based analysis of a head and neck cancer model of cisplatin resistance reveals intratumoral heterogeneity and treatment-induced clonal selection. Clin Cancer Res 2018;24:158–68.
[112] Wang VE, Xue JY, Frederick DT, et al. Adaptive resistance to dual BRAF/MEK inhibition in BRAF-driven tumors through autocrine FGFR pathway activation. Clin Cancer Res 2019;25(23):7202–17.
[113] Sanchez IM, Purwin TJ, Chervoneva I, et al. In vivo ERK1/2 reporter predictively models response and resistance to combined BRAF and MEK inhibitors in melanoma. Mol Cancer Ther 2019;18:1637–48.
[114] Rossi A, Roberto M, Panebianco M, Botticelli A, Mazzuca F, Marchetti P. Drug resistance of BRAF-mutant melanoma: review of up-to-date mechanisms of action and promising targeted agents. Eur J Pharmacol 2019;862:172621.
[115] De Pauw I, Lardon F, Van den Bossche J, et al. Simultaneous targeting of EGFR, HER2, and HER4 by afatinib overcomes intrinsic and acquired cetuximab resistance in head and neck squamous cell carcinoma cell lines. Mol Oncol 2018;12:830–54.
[116] Ribas A, Lawrence D, Atkinson V, et al. Combined BRAF and MEK inhibition with PD-1 blockade immunotherapy in BRAF-mutant melanoma. Nat Med 2019;25:936–40.
[117] De Pauw I, Wouters A, Van den Bossche J, et al. Dual targeting of epidermal growth factor receptor and HER3 by MEHD7945A as monotherapy or in combination with cisplatin partially overcomes

cetuximab resistance in head and neck squamous cell carcinoma cell lines. Cancer Biother Radiopharm 2017;32:229–38.

[118] Lin A, Giuliano CJ, Palladino A, et al. Off-target toxicity is a common mechanism of action of cancer drugs undergoing clinical trials. Sci Transl Med 2019;11(509):eaaw8412.

[119] Gasparini G. Metronomic scheduling: the future of chemotherapy? Lancet Oncol 2001;2:733–40.

[120] Peereboom DM, Alban TJ, Grabowski MM, et al. Metronomic capecitabine as an immune modulator in glioblastoma patients reduces myeloid-derived suppressor cells. JCI Insight 2019;4:e130748.

[121] Longton E, Schmit K, Fransolet M, Clement F, Michiels C. Appropriate sequence for afatinib and cisplatin combination improves anticancer activity in head and neck squamous cell carcinoma. Front Oncol 2018;8:432.

CHAPTER

7

Immunosurveillance and mechanisms of immune escape in head and neck cancer

Jean-Nicolas Gallant, Young J. Kim

Department of Otolaryngology—Head and Neck Surgery, Vanderbilt University Medical Center, Nashville, TN, United States

Abstract

In this chapter, we provide a brief introduction to tumor immunology, the concept of immunosurveillance, and mechanisms of immune escape in head and neck cancer. We focus on head and neck squamous cell carcinoma (HNSCC), using this predominant subtype of head and neck cancer as an illustrative example. We detail the mechanisms that this malignancy uses to create an immunosuppressive tumor microenvironment to fuel tumor growth, including disruption of antigen presentation via the major histocompatibility complex class I (MHC-I), dysregulation of the tumor microenvironment cytokine balance, imbalance of the immune checkpoints and costimulatory molecules, and the recruitment of immunosuppressive cells. Throughout, we specifically focus on mechanisms of tumor escape from natural killer (NK) and T cells, the primary effector immune cell types that are being leveraged for clinical therapy.

Abbreviations

APC	antigen-presenting cell
APM	antigen-presenting machinery
CAF	cancer-associated fibroblast
CC	chemokine
CD	cluster of differentiation
CLT	calreticulin
CNX	calnexin
CTL	cytotoxic T lymphocyte ($CD8^+$ T cell)
CTLA	CTL-associated protein
GITR	glucocorticoid-induced TNFR-related protein
HNSCC	head and neck squamous cell carcinoma
IFN	interferon

https://doi.org/10.1016/B978-0-12-820679-9.00007-4

IL	interleukin
KIR	killer immunoglobulin receptor
LAG	lymphocyte-activating gene
LMP	latent membrane protein
MDSC	myeloid-derived suppressor cell
MHC	major histocompatibility complex
NK	natural killer
PD	programmed death
PG	prostaglandin
PVR	poliovirus receptor
TAM	tumor-associated macrophage
TAP	antigen peptide transporter
TCR	T-cell receptor
TGF	transforming growth factor
TIGIT	T-cell Ig and ITIM domain
TIM	T-cell immunoglobulin and mucin domain-containing protein
TME	tumor microenvironment
TPN	TAP-binding protein
Treg	regulatory T cell

Conflict of interest

YJK serves on advisory boards for Dracen, Aduro, Novartis, Sanofi, and Takeda.

Immunosurveillance via NK cells and CTLs

Great advances in our understanding of cancer cell biology have been made during the past 100 years. Decades of experimentation culminated in what are known as the hallmarks of cancer, the key factors that determine cancer cell transformation [1]. However, there is more to a tumor than the cancer cells that comprise its bulk, including extracellular matrix, endothelial cells, fibroblasts, signaling molecules, and immune cells. The latter have long been suspected to play a key role in tumor growth, even if they could not be directly visualized. Dating back to ancient Egypt, tumors have been known to regress during a febrile illness. However, this phenomenon could not directly be observed until the 18th century, with the advent of the microscope and histology. At that time, two German physicians, Busch and Fehleisen, independently observed regression of tumors in patients with coincidental erysipelas, a rash caused by infection with the bacterium *Streptococcus pyogenes*. An American surgeon, Coley [2], followed up on his own independent observation of cancer regression during erysipelas by injecting heat-inactivated *S. pyogenes* ("Coley's toxins") into patients with inoperable cancers—reporting remissions in some cases. Around the same time, at the turn of the 20th century, Ehrlich [3] predicted that the immune system actively and continually repressed the growth of cancers. However, so little was known about the composition and function of the immune system at the time, that it was not possible to test this surveillance hypothesis. As immunologists gained a better understanding of transplant and tumor immunobiology and genetics over the next 50 years, Burnet [4] and Thomas [5] revisited Ehrlich's prediction and put together the first clear articulation of the tumor immunosurveillance hypothesis that thymus-dependent (T) cells constantly surveyed the host for and eliminated transformed cells. We've come a long way in the ensuing 75 years and now recognize that immunosurveillance of cancer is a complex coordinated process involving both the innate and adaptive immune systems; thymus-dependent cells do play a central role, as demonstrated by the clinical efficacy of immune checkpoint inhibitors, but so do a host of other immune cells.

Both the innate and adaptive arms of the immune system play an important role in immunosurveillance [6]. The innate immune system includes dendritic cells, natural killer (NK) cells, macrophages, neutrophils, eosinophils, basophils, and mast cells. These innate immune cells provide first-line protection against disease, including cancer. Innate immunity does not require prior stimulation by antigen, is nonspecific, and diminishes in efficacy after repeated exposures to the same disease or antigenic challenge. The adaptive immune system includes B and T lymphocytes. These B and T cells require formal presentation by antigen-presenting cells (APCs) for their activation and create specific and durable humoral (B) and cell-mediated (T) responses, respectively. The complexity of these cellular relationships is exponentially complicated with molecular signaling processes that are both intrinsic and extrinsic to the immune cells such as the complement system, defensins, and pattern recognition receptors. These multiple dimensions of complexity help to explain why tumor immunology is still in its relative infancy; the interconnectedness of all these pieces all makes it difficult to dissect out discrete conclusions. Hence, this chapter will focus on two key effector cells—NK and ($CD8^+$) T cells (as representative of the innate and adaptive immune systems, respectively)—in head and neck cancer immunosurveillance.

NK cells surveil for cancer cells in two ways: they sense either the absence of major histocompatibility complex class I (MHC-I) proteins or the presence of one or more stress-associated proteins on the surfaces of cancer cells. NK cells use a specific receptor, killer immunoglobulin receptor (KIR) molecules, to recognize MHC-I molecules displayed by potential target cells. Tumor cells and virally infected cells tend to downregulate these MHC-I molecules, and they become "foreign" cells in the eyes of the NK cells. Binding of the target cell MHC-I molecules to KIR causes this receptor to release inhibitory signals into the NK cell that prevent it from launching an attack on the MHC-I positive cell. NK cells also display a variety of activating or inhibitory receptors. For example, NKG2D recognizes stress-associated cell proteins displayed by target cells and signals for strong activation of an NK cell's cytotoxic response. On the other hand, NKG2A binds to other cellular molecules and transduces an inhibitory signaling to suppress NK cytokine secretion and cytotoxicity. The balance of inhibitory and activating signals is key to NK cell function [7].

$CD8^+$ T cells, often referred to as cytotoxic T lymphocytes (CTLs), are the primary adaptive immune effectors in antitumor immune responses. CTLs are activated after creating an immune synapse between their T-cell receptors (TCRs) and cognate peptide-loaded MHC-I (on an APC) in a secondary lymphoid organ. After this first signal (signal 1, the TCR: MHC-I immune synapse formation), signal 2 (costimulation, e.g., CD80/CD86-CD28 signaling) and signal 3 (cytokine signaling, e.g., IL-2 binding), naïve CTLs expand and give rise to effector CTLs. These effector cells subsequently surveil for cancer by circulating through the bloodstream or tissues, recognizing their cognate peptide-MHC-I complexes, and exerting cytolytic functions (secreting perforins, granzymes, and various cytokines) to lyse-infected cells. Almost all nucleated cells in the body, including tumor cells, express MHC-I molecules on their surface and, hence, can potentially present peptides to CTLs. A fundamental tenet of tumor immunology is that cancer cells display antigens on their MHC-I molecules that differentiate them from their nontransformed counterparts. Such "tumor antigens" include differentiation, mutational (neo–), overexpressed, viral, and/or cancer/testis antigens [8].

From immunosurveillance to immunoediting to immune escape

Transformed cells that acquire malignant or potentially malignant changes are usually recognized and killed by NK and CTLs via the surveillance mechanisms described in the preceding text. This killing occurs because cancer cells typically display either danger signals or tumor antigens resulting from their transformation. This display of antigen leads to destruction by NK and CTL cells, respectively. However, some malignant cells are able to escape immunosurveillance by manipulating their own characteristics (e.g., inhibition of MHC-I molecules on cell surface) and those of other immune cells in their microenvironment. The idea that the immune system not only protects the host against tumor formation but also be coopted to increase tumorigenicity is the basis of the cancer immunoediting hypothesis, which stresses the dual host-protective and tumor-promoting actions of multiple immune cells on tumors.

Cancer immunoediting consists of three phases: elimination, equilibrium, and escape (Fig. 1). The elimination phase refers to the initial destruction of tumor cells by the innate immune system (NK cells and macrophages), followed by presentation of the tumor antigens to the adaptive immune systems (thereby creating tumor-specific CTLs). The elimination phase has not yet been directly observed in vivo, but its existence has been inferred from extensive murine experimentation that demonstrated the earlier onset or greater penetrance of neoplasia in mice lacking certain immune cell subsets, recognition molecules, effector pathways, or cytokines and by studies comparing tumor initiation, growth, and metastases in wild-type versus immunodeficient mice [9].

Certain tumor clones may survive the elimination phase and enter the equilibrium phase, in which the adaptive immune system prevents tumor cell outgrowth, but a small niche of viable tumor cells survives. In equilibrium the immune system maintains residual tumor cells in a functional state of dormancy. These latent tumor cells may reside in patients for decades before resuming growth as either recurrent primary tumors or distant metastases. Experiments have suggested that adaptive immunity, but not innate immunity, is responsible for maintaining the occult tumor cells in equilibrium. This observation mechanistically distinguishes equilibrium from elimination because the latter displays an obligate requirement for both innate and adaptive immunity [10]. It is during the equilibrium phase that tumor cells are selected ("edited") by the immune system for survival. The end result is the generation of poorly immunogenic tumor clones that become invisible to the elimination mechanisms of the immune system and promote cancer progression.

In the escape phase, tumor cells that have acquired the ability to circumvent immune recognition and/or destruction emerge as progressively growing, clinically visible tumors. Progression from equilibrium to the escape phase can occur because the tumor cell population changes in response to the immune system's editing functions and/or because the host immune system changes in response to the tumor cells themselves. In short, tumor cell escape can occur through multiple mechanisms. In the following section, we focus on four different mechanisms in head and neck squamous cell carcinoma (HNSCC): disruption of antigen

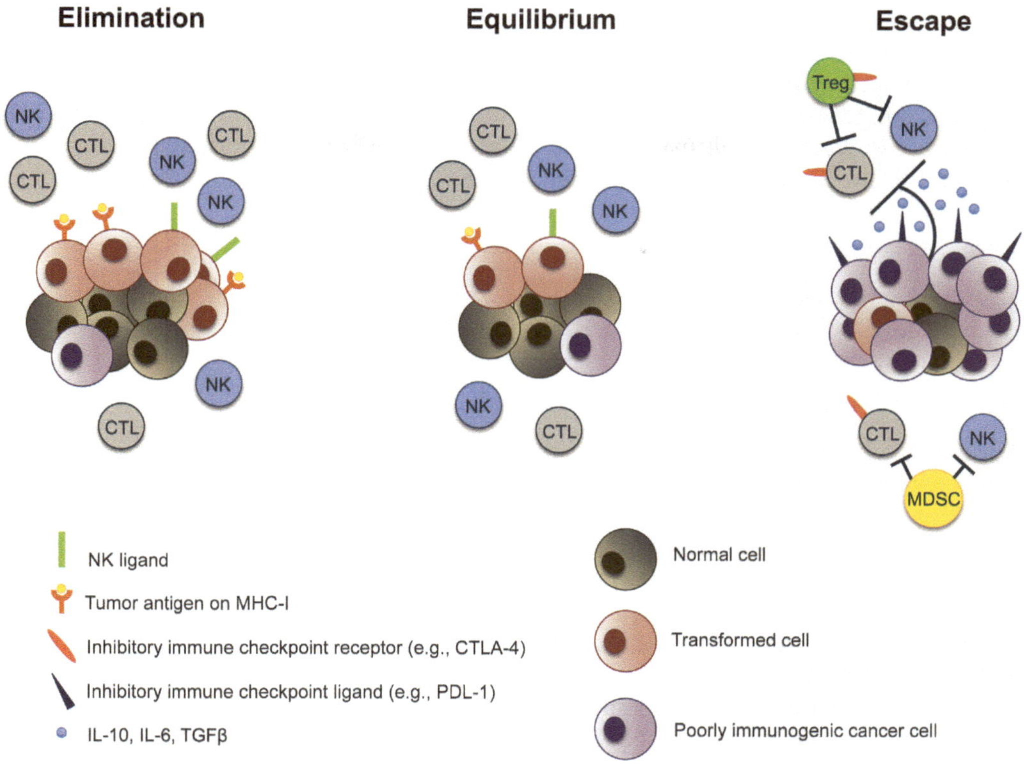

FIG. 1 The three phases of immunoediting in head and neck cancer. According to the immunoediting hypothesis, clinically apparent tumors arise in a stepwise fashion in the progression from immunosurveillance and elimination, to equilibrium and evolution, to escape and tumor outgrowth. The elimination phase refers to the initial destruction of tumor cells by the innate immune system, followed by presentation of the tumor antigens to the adaptive immune systems. Tumor cells during this phase of immunoediting are immunogenic and display both MHC-I-peptide complexes and NK ligands. As a result, immune infiltrates are bountiful during the elimination phase. Certain tumor clones may survive the elimination phase and enter the equilibrium phase, in which the immune system prevents tumor cell outgrowth, but a small niche of viable tumor cells survives due to natural selection. The end result is the generation of poorly immunogenic tumor clones that become invisible to the elimination mechanisms of the immune system and promote cancer progression. These poorly immunogenic cancer cells may then enter the escape phase, in which their outgrowth is no longer blocked by the immune system. Poorly immunogenic cells cause clinically apparent disease. *NK*, natural killer; *CTL*, cytotoxic T lymphocyte; *Treg*, regulatory T cell; *MDSC*, myeloid-derived suppressor cell.

presentation via MHC-I, dysregulation of the tumor microenvironment cytokine balance, recruitment of immunosuppressive cells, and imbalance of immune checkpoints.

We use HNSCC as the representative head and neck cancer because it is the most common head and neck cancer—an otherwise disparate set of pathologies united by anatomic location. HNSCC is the most common histologic subtype in 90% of head and neck cancers. Every year, there are more than 60,000 new cases of HNSCC diagnosed in the United States. Inflammatory insults, such as chronic alcohol and tobacco exposure, are common etiologic factors in HNSCC. Additionally, poor dental hygiene and human papillomavirus (HPV) infection are known to cause HNSCC (particularly oropharyngeal cancers). While the etiology of HNSCC carcinogenesis is variable, a unifying biological paradigm that partly explains these clinical risk factors indicates that chronic inflammation is the underlying biological process that shapes the genetic changes of the transformed epithelial cells. The fact that viruses and inflammatory environmental insults cause HNSCC speak to the importance of the immune systems in this cancer. In fact, once detected, HNSCC is one of the most immunosuppressive cancers for reasons outlined in the remainder of this chapter.

Mechanisms of immune escape in HNSCC

Disruption of antigen presentation via MHC-I

Almost all cell types, including cancer cells, routinely divert a portion of their recently synthesized proteins to the MHC-I antigen-presenting machinery (APM). Successful APM requires cleavage of proteins to appropriate size, loading of the antigen onto the MHC-I molecule, and trafficking of the complex to the cell surface. This process is accomplished by the proteasome, specific peptide transporters, and select chaperone proteins. All of these components play an important role in the generation of antigenic peptides, their translocation into the endoplasmic reticulum, and loading of the β2-microglobulin-associated MHC-I.

In HNSCC, defects in nearly all of the APM components have been observed, ultimately leading to disrupted antigen presentation and cancer cell immune escape. For one, critical subunits of the proteasome are IFNγ inducible, and loss of response to IFNγ (which commonly occurs in HNSCC) may disrupt this process [11]. Alterations in the expression of proteasome subunits (LMP2, LMP7, and LMP10) [12], peptide transporters (TAP1 and TAP2) [13], and chaperone proteins (CNX, CLT, TPN, and ERp57) [14] also have been observed in HNSCC. This downregulation has been seen occurring in both primary HNSCC tumors and metastatic lesions. Transcription of HLA-I (the gene that encodes MHC-I) and B2M (the gene that encodes β2-microglobulin) may be suppressed due to hypermethylation of DNA CpG islands and histone hypoacetylation. Moreover, loss of heterozygosity at chromosome 6p21 (where HLA-I is located) has been documented in about 1/3 of HNSCC and is associated with clinical reduction in disease-free interval and overall survival [15]. Taken together, these mechanisms lead to a partial loss of antigenic peptide presentation via MHC-I in over 50% of HNSCC [16].

However, the complete loss of MHC-I peptide display would be a potent-stimulating signal for NK cell tumor killing. Therefore tumor-mediated MHC-I dysfunction is nuanced to maintain immune evasion. This explains why some tumor cells will selectively suppress the expression of only one of the six key MHC-I molecules that are normally expressed concomitantly by cells throughout the body. Moreover, there is typical deficiency and downregulation of APM components without a significant difference in MHC-I expression itself. This nuanced suppression may block presentation of a particular tumor antigen, thereby allowing the tumor cell to evade attack by antigen-specific CTLs without provoking an attack by NK cells, since only a small proportion of the total population of cell surface MHC-I molecules would be lost.

Dysregulation of the tumor microenvironment cytokine balance

Cytokines are extracellular regulatory factors of the immune system. These small peptides/proteins include chemokines (CCs), interferons (IFNs), interleukins (ILs), and tumor necrosis factors (TNFs). Cytokines are produced by a broad range of cells, including most immune cells, and endothelial cells and fibroblasts. Cytokines can be grouped in a variety of different ways, including structurally, functionally, and immunologically. With regard to tumor immunology, type 1 cytokines are those that favor cell-mediated (CTL) anticancer responses (e.g., interferons and interleukin-2). Type 2 cytokines (e.g., IL-10 and TGF-β [transforming growth factor beta]) tend to promote tumor inflammation and antibody production. Measuring the balance of these cytokines in tumors is difficult. Since human solid tumor tissues are known to be infiltrated by various kinds of immune cells, it has to be distinguished directly between tumor-derived cytokines and cytokines produced by tumor-stimulated immune cells, respectively. In this section, we focus on the cytokines that have been shown to be produced by HNSCC cells themselves. These include the multifunctional cytokine TGF-β, the potently immunosuppressive cytokine IL-10, and the proinflammatory cytokine IL-6.

TGF-β is a multifunctional cytokine involved in immune signaling and stem cell function. TGF-β functions as a tumor suppressor in normal epithelial cells or in the early stages of oncogenesis. However, as a tumor develops, TGF-β becomes a potent tumor promoter in the epithelium at a later stage and even increases the production of TGF-β, supporting tumor progression and metastasis. It has been shown that HNSCC overexpress TGF-β, which suppresses NK and CTL and is a key cytokine in the differentiation of regulatory T cells (Tregs and immunosuppressive cells; more on these later) [17]. TGF-β is known to inhibit the proliferation and function of T and B lymphocytes and the function of macrophages. Under certain conditions, TGF-β can even induce apoptosis of dendritic cells and macrophages—the two key antigen-presenting cells of the immune system. In addition, TGF-β, which was portrayed earlier as being toxic for certain types of immunocytes, is also thought to help induce other types of T cells to transdifferentiate into Tregs, both in the thymus and in peripheral tissues [18].

IL-10 is a potent immunosuppressive cytokine produced by multiple cell types. IL-10 can lead to downregulation of most Type 1 cytokines and costimulatory molecules. The cytokine leads to a net balance of humoral over cellular immunity. It has been shown that HNSCC overexpresses IL-10 [19]. Interestingly the release of IL-10 by human cancer cells is mimicked by Epstein–Barr virus (EBV), which has acquired and remodeled a cellular IL-10 gene. By forcing virus-infected cells to release an IL-10-like immunosuppressive cytokine, Epstein-Barr virus protects these cells from direct attack by cytotoxic immune cells, thereby ensuring a safe cellular haven for itself over extended periods of time [20].

IL-6 is a potent proinflammatory cytokine that is predominantly produced by granulocytes. Traditionally, IL-6 is produced by various cells such as T lymphocytes, macrophages, or monocytes and involved in distinct cellular processes like cell differentiation, the production of acute phase proteins in the liver, the proliferation of B lymphocytes, and the production of neutrophils. IL-6 has been shown to have pro- and antiinflammatory properties, and thus dysregulation of IL-6 cytokine signaling often contributes to various kinds of cancer playing a central role as a differentiation and growth factor of tumor cells. The cytokine has been correlated with recurrence and survival in head and neck cancers. Elevated levels of IL-6 is frequently well detected in the serum of patients with HNSCC but not in benign lesions, and the constitutive expression of IL-6 has been found to be responsible for downregulation of the costimulatory molecule CD80 [21].

HNSCC is also known to deregulate many other cytokines. Of note are IL-1, CCL22, and PGE2. IL-1 seems to play a crucial role in the regulation of cytokine production by tumor and resident tissue cells and has been shown to be elevated in HNSCC cells [22]. HNSCC has also been shown to release CCL22 to recruit Tregs; the latter display the cognate receptor, termed CCR4, on their surface [23]. Once present within a tumor mass, the Tregs can suppress the actions of CTLs (more on this later). Prostaglandin E_2 (PGE_2) is a prosurvival, proangiogenic molecule that is produced by many cancers, including HNSCC. High-expression levels of PGE_2 have been detected in HNSCC and in its surrounding mucosa and were shown to have a significant impact on tumor growth. PGE_2 is regularly produced by cells of HNSCC and interferes with monocyte functions such as migration and adherence to endothelial cells [24].

Imbalance of immune checkpoints

As briefly discussed the responses of T cells to the antigens being displayed by APCs can also be tweaked in various ways and usually proceed through two phases. Before CTLs can surveil for cancer, they must be activated and matured. During this first phase the initial encounter of T cell with APCs induces various responses (signal 1) that are intensified by the encounter between the CD28 cell surface receptor of the T cells and complementary ligands displayed by the APCs (signal 2). Subsequently, during the second phase of the response, several other cell surface T-cell receptors, termed checkpoint molecules, mediate critical feedback loops to ensure that T-cell activation is finely tuned and that any resulting immune responses are limited (signal 3). In the healthy state, effector functions of the immune system must be held in check to prevent damage to self (autoimmunity) or prolonged activation. Checkpoint molecules keep the immune system in check, hence their name. Primarily surface molecules on immune (mainly T) cells and this class of receptors respond to binding of their ligands by quelling the activity of local immune cells. Characteristics of the local inflammatory environment, including cytokines and cross talks, among various immune cell subsets, can promote upregulation of these receptors or ligands on hematopoietic cells, immune cells, and nonimmune cells such as epithelia (including tumor cells). Coaptation of immune checkpoint pathways is an important route of immune escape in HNSCC. Mapping of the complicated immune checkpoint signaling within the tumor microenvironment (TME) has yielded many therapeutic targets (see the other chapter in this text), and logic follows that intervening at multiple intersections on this highway simultaneously with combination therapy has the potential for improved treatment efficacy compared with monotherapy. Multiple immune checkpoint receptors have been described, and expression patterns and cell distribution are varied, giving these molecules diverse and tunable contributions to immunomodulation. There are both stimulatory and inhibitory immune checkpoints (Fig. 2). The best studied of these molecules are the so-called inhibitory immune checkpoints, cytotoxic T lymphocyte-associated protein 4 (CTLA4) and programmed death 1 (PD-1), which are found on T cells.

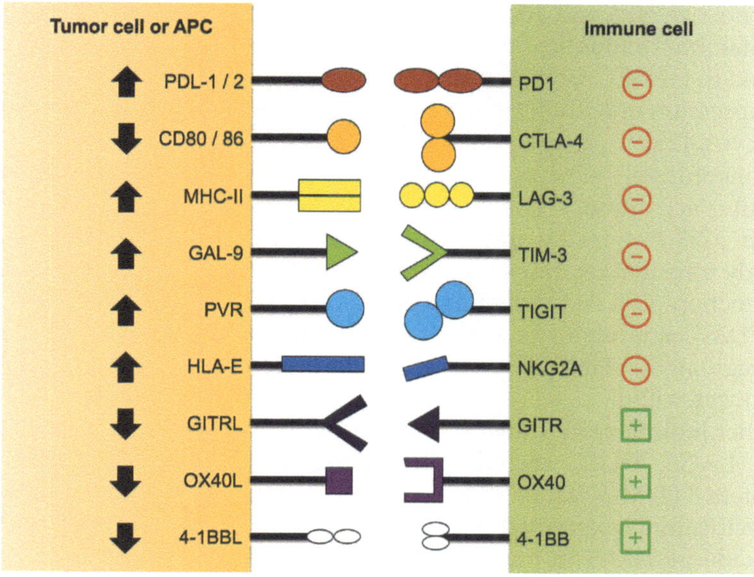

FIG. 2 Costimulatory and inhibitory pathways in head and neck cancer. Costimulatory and inhibitory molecules expressed on immune cells (mainly T cells) and tumors cells and APCs. These molecules, often called checkpoint proteins, are sometimes overexpressed on head and neck cancer cells *(as indicated by arrows)*. Some molecules, like CD80, have dual inhibitory and stimulatory roles (e.g., CD80:CD28 provides "signal 2") and are often seemingly paradoxically downregulated on tumor cells. The effects of these molecules on immune cells are either inhibitory *(indicated by minus symbols)* or stimulatory *(indicated by plus symbols)*. *APC*, antigen-presenting cell; *PD-1*, programmed death 1; *CTLA-4*, cytotoxic T lymphocyte-associated protein 4; *MHC*, major histocompatibility complex; *LAG-3*, lymphocyte-activating gene 3; *TIM-3*, T-cell immunoglobulin and mucin domain-containing 3; *TIGIT*, T-cell Ig and ITIM domain; *PVR*, poliovirus receptor; *GITR*, glucocorticoid-induced TNFR-related protein.

PD-1 is a member of the CD28 receptor family that is typically expressed on activated T and B cells, monocytes, and a subset of thymocytes. The expression of PD-1 on activated T cells enhances the suppressive function of Tregs, and this effect is mediated by the interaction with its ligands, PD-L1 and PD-L2, which are expressed on antigen-presenting cells, endothelial and epithelial cells, and on activated lymphocytes. PD-L1 is the more constitutively expressed of the two and is present on lymphocytes, macrophages, and APCs—in addition to epithelia—including tumor cells. PD-L2 exhibits more restricted expression but elicits similar functional effects upon interaction with PD-1. The interaction between PD-1 and PD-L1 negatively regulates immune responses by decreasing cytokine production and inducing T-cell anergy and apoptosis. PD-1 signaling supports immune tolerance and eventual exhaustion through reduced T-cell receptor signaling, reduced cytokine production, reduced target cell lysis, altered lymphocyte motility, and metabolic reprogramming—in addition to inducing differentiation to a Treg phenotype. PD-1/PD-L1 interaction promotes tolerance by inhibiting effector T-cell mechanisms, including cytokine production, proliferation, and cytotoxicity, and eventually leads to T-cell exhaustion. Upregulation of PD-L1 can occur in HNSCC tumor cells and allows cancer cells to escape from host immune systems by functionally inactivating CTL immune surveillance. The expression of PD-L1 is often high in HNSCC tumors, with positivity being quoted between 50% and 100% across several studies, this wide range likely owing to differences in staining technique, sample preservation, and possibly sampling error. PD-L1 is also upregulated in HNSCC, both in tumor-infiltrating lymphocytes (TILs) and on the tumor surface, and expression of both PD-1 and PD-L1 is preferential for TILs relative to peripheral blood lymphocytes [25].

CTLA4 is constitutively expressed in regulatory T cells but only upregulated in conventional T cells after activation. CTLA-4 expression on the surface of T cells is stimulated by TCR activation. CTLA-4 is the inhibitory counterpart to the stimulatory CD28 signal. CTLA-4 has only transient expression patterns on $CD8^+$ T cells, where it regulates the memory/effector function. CTLA-4, when expressed, interacts with CD80/CD86 on APCs and outcompetes CD28 costimulatory signaling and thereby inhibits T-cell activation. CTLA-4 expression is upregulated in HNSCC tumor cells and preferentially on TILs compared with peripheral lymphocytes [26].

There are other inhibitory immune checkpoints with increasing importance in HNSCC, including TIM3, LAG3, TIGIT, and NKG2A. T-cell immunoglobulin and mucin domain-containing protein 3 (TIM-3) binds with galectin-9, is expressed on both T cells and NK cells, and, when specifically coexpressed with PD-1, signifies an exhausted T-cell phenotype in HNSCC. Blockade of TIM-3 in models of HNSCC produces a antitumor T-cell responses [27]. Lymphocyte-activating gene 3 (LAG-3) is expressed on activated $CD4^+$ and $CD8^+$ T cells, NK cells, B cells, and DCs and functions as a coinhibitory checkpoint molecule in HNSCC. LAG-3 binds with major histocompatibility complex class II (MHC-II) and is highly expressed on Tregs, and its blockade decreases both the inhibitory function of these cells and their overall quantity [28]. T-cell Ig and ITIM domain protein 3 (TIGIT) competes with the stimulatory molecule CD226 for binding with the poliovirus receptor (PVR) and exerts immunosuppressive effects by inhibiting T-cell activation and decreasing DC cytokine production. TIGIT is expressed on Tregs and memory and activated T cells [29]. NKG2A is expressed on both T and NK cells, and its blockade enhances the antitumor immunity mediated by NK and $CD8^+$ T cells in HNSCC [30].

In addition to upregulation of immune inhibitory pathways, perturbation of multiple costimulatory pathways is also seen in HNSCC. The best studied of these molecules are GITR, OX40, and 4-1BB. Glucocorticoid-induced TNFR-related protein (GITR) is a member of the tumor necrosis factor superfamily and is expressed on the surface of Tregs. GITR activation by its ligand reduces Treg recruitment and abrogates their suppressive function [31]. OX40 is expressed on the surface of T cells and promotes proliferation, IFNγ production, and memory formation. Expression of OX40 has been demonstrated on the surface of HNSCC tumor cells and in draining lymph nodes, but expression of its ligand is reduced in the tumor, and thus this pathway is rendered ineffective at generating an antitumor immune response. Another costimulatory molecule expressed on the surface of activated T cells, NK cells, and DCs is 4-1BB, which is downregulated in HNSCC [32].

Recruitment of immunosuppressive cells

The complex interplay of the various immune checkpoint and stimulatory pathways within the HNSCC TME occurs in the context of a diverse cellular milieu, a significant component of which are immunosuppressive cell populations. Outside of direct effector cell suppression, HNSCC regulates and recruits other immune populations capable of modulating T and NK cell responses, including Tregs, myeloid-derived suppressor cells (MDSCs), tumor-associated macrophages (TAMs), and cancer-associated fibroblasts (CAFs), which are discussed in the succeeding text. Immunomodulation enacted by these various cell populations contributes to a tumor-promoting microenvironment.

Tregs promote cancer progression by causing anergy, apoptosis, and cell cycle arrest of T and NK cells via production of IL-10, TGFβ, and direct cell-to-cell contact. In patients with HNSCC, Tregs are increased in peripheral blood and are more potent among T cells infiltrating the tumor, resulting in an immunosuppressed state. Also, Treg numbers are inversely proportional to DC and CTL numbers in HNSCC. In addition, Treg frequency is elevated in patients with HNSCC after treatment, indicating that oncologic treatment increases Treg numbers. Immunosuppressive Tregs have been discussed in several contexts in the preceding text, demonstrating their salient and ubiquitous role in immune evasion in HNSCC. A distinction must be made between natural Tregs, which develop in the thymus and are drawn to the TME subsequently, and inducible Tregs, which are naïve tumor-resident T cells that are driven to regulatory status by the cytokine profile within the TME. Recruitment of Tregs to the TME occurs via binding of CCR4 on the T cell surface by CCL22 produced by the tumor. In normal individuals the Tregs represent only 5%–10% of the population of $CD4^+$ lymphocytes, the remainder being helper T cells. In cancer patients, however, this number may increase to 25%–30%. Together, these various observations suggest that Tregs play an important role in influencing anti-HNSCC immunity [33].

MDSCs are a diverse cellular population of myeloid origin with T-cell suppressive functions. Initial studies in HNSCC found that MDSCs inhibit activated T cells. Also, MDSCs produce nitric oxide and reactive oxygen species that interact to catalyze the nitration of the T-cell receptor, which inhibits T-cell receptor and MHC interaction, signaling, and subsequent activation. The basal levels of MDSCs increase with age and may contribute to increase tumor frequency and growth rate increase with age. Recruitment of MDSCs to the TME occurs through chemotactic stimulus from multiple factors, including IL-6, IL-10, and PGE_2. In vitro HNSCC cell lines, constitutive expression of IL-6 due to upstream genetic mutations, have been shown to directly promote differentiation of peripheral blood mononuclear cells (PBMCs) into MDSCs. Another mechanism of MDSC-mediated immunosuppression involves presentation of neoantigen to experienced T cells, which results in nitration of the TCR and subsequent inability to be stimulated by antigen-presenting cells (APCs). MDSCs also promote activation and expansion of Tregs [34].

TAMs in the tumor microenvironment may be strongly antitumor and possess a so-called M1 phenotype, which is characterized by the production of IFNγ and other type 1 cytokines. Alternately, activated macrophages force a type 2 response, with production of interleukins such as IL-10 that permit tumor growth. TAM-infiltrating tumors correlate with worse clinical outcome and are closely associated with the alternately activated M2 phenotype. These TAMs have been demonstrated to produce IL-6 and IL-10 and have been associated with angiogenesis, local tumor progression, and metastasis. Through these immune/inflammatory cells and mediators, HNSCC induces an immunosuppressed state via multiple potent mechanisms, which is a barrier to effective cancer immunotherapy. Mechanisms of TAM-mediated immunosuppression are similar to that of MDSCs, including depletion of local nutrients required for T-cell function, secretion of immunosuppressive cytokines, expression of PD-L1, and recruitment of Tregs.

Beyond tumor and immune cells, the TME is composed of other supportive stromal cells, including fibroblasts. These TME stromal cells, namely, CAFs, have distinct tumor-supporting properties compared with their normal tissue counterparts, promoting angiogenesis, invasion, and metastasis through cytokine and growth factor production. In HNSCC, CAFs were recently shown to inhibit T-cell proliferation via the PD-1 signaling axis, induce effector T-cell apoptosis, and promote differentiation to a Treg phenotype [35].

Conclusion

In the past 20 years, the ability of cancer cells to evade immune destruction has been proposed as an additional hallmark of cancer [36]. As noted above, a HNSCC tumor is able to do this not only by modulating its own cellular characteristics; but, also, by creating its own tumor microenvironment by recruiting apparently normal immune cells to help shield it from attack by the immune system. Through the production of various cytokines and chemokines, successful HNSCCs and their metastatic derivatives are able to generate an immunosuppressive, protumorigenic, and prometastatic microenvironment by recruiting and "training" immune cells, including Tregs, MDSCs, and TAMs. Alternatively, they may suppress the display of the MHC-I molecules that enable the immune system to detect the presence of the antigen on their surface. In the end the most effective defense mounted by tumor cells may derive from the body's main cellular systems for establishing and maintaining immune tolerance—the mechanisms that normally prevent the development of autoimmune attacks on the body's own tissues [37]. We have only begun to understand the physiology of normal immune function and the possibilities of manipulating it to strengthen anticancer defenses against head and neck cancer [38]. There are likely to be yet other strategies—some still unknown—that allow head and neck cancer cells to evade killing by components of the immune system. This may include emerging fields of exosome biology, tumor microenvironmental metabolism, and host immunogenetics.

Acknowledgments

The authors would like to thank members of the Kim lab, especially Michael Korrer, for critical feedback on this chapter.

References

The authors apologize for being unable to cite the vast amount of excellent literature and research in the field. References are provided to illustrate key points and to link out to further more detailed reading.

[1] Hanahan D, Weinberg RA. The hallmarks of cancer. Cell 2000;100:57–70.
[2] Coley WB. The treatment of malignant tumors by repeated inoculations of erysipelas: with a report of ten original cases. Am J Med Sci (1827–1924) 1893;105:487.
[3] Ehrlich P. Ueber den jetzigen Stand der Karzinomforschun. Ned Tijdschr Geneeskd 1909;5:273–90.
[4] Burnet M. Cancer: a biological approach. Br Med J 1957;1:841–7.

[5] Thomas L, Lawrence HS. Cellular and humoral aspects of the hypersensitive states. New York: Hoeber-Harper; 1959. p. 529–32.
[6] Vesely MD, Kershaw MH, Schreiber RD, Smyth MJ. Natural innate and adaptive immunity to cancer. Annu Rev Immunol 2011;29:235–71.
[7] Martinet L, Smyth MJ. Balancing natural killer cell activation through paired receptors. Nat Rev Immunol 2015;15:243–54.
[8] Coulie PG, Van den Eynde BJ, van der Bruggen P, Boon T. Tumour antigens recognized by T lymphocytes: at the core of cancer immunotherapy. Nat Rev Cancer 2014;14:135–46.
[9] Schreiber RD, Old LJ, Smyth MJ. Cancer immunoediting: integrating immunity's roles in cancer suppression and promotion. Science 2011;331:1565–70.
[10] Dunn GP, Old LJ, Schreiber RD. The immunobiology of cancer immunosurveillance and immunoediting. Immunity 2004;21:137–48.
[11] Meissner M, Reichert TE, Kunkel M, Gooding W, Whiteside TL, Ferrone S, et al. Defects in the human leukocyte antigen class I antigen processing machinery in head and neck squamous cell carcinoma: association with clinical outcome. Clin Cancer Res 2005;11:2552–60.
[12] Tang Q, Zhang J, Qi B, Shen C, Xie W. Downregulation of HLA class I molecules in primary oral squamous cell carcinomas and cell lines. Arch Med Res 2009;40:256–63.
[13] Bandoh N, Ogino T, Katayama A, Takahara M, Katada A, Hayashi T, et al. HLA class I antigen and transporter associated with antigen processing downregulation in metastatic lesions of head and neck squamous cell carcinoma as a marker of poor prognosis. Oncol Rep 2010;23:933–9.
[14] Ogino T, Bandoh N, Hayashi T, Miyokawa N, Harabuchi Y, Ferrone S. Association of tapasin and HLA class I antigen down-regulation in primary maxillary sinus squamous cell carcinoma lesions with reduced survival of patients. Clin Cancer Res 2003;9:4043–51.
[15] Feenstra M, Veltkamp M, van Kuik J, Wiertsema S, Slootweg P, van den Tweel J, et al. HLA class I expression and chromosomal deletions at 6p and 15q in head and neck squamous cell carcinomas. Tissue Antigens 1999;54:235–45.
[16] Grandis JR, Falkner DM, Melhem MF, Gooding WE, Drenning SD, Morel PA. Human leukocyte antigen class I allelic and haplotype loss in squamous cell carcinoma of the head and neck: clinical and immunogenetic consequences. Clin Cancer Res 2000;6:2794–802.
[17] Lu S-H, Reh D, Li AG, Woods J, Corless CL, Kulesz-Martin M, et al. Overexpression of transforming growth factor beta1 in head and neck epithelia results in inflammation, angiogenesis, and epithelial hyperproliferation. Cancer Res 2004;64:4405–10.
[18] Chen W, ten Dijke P. Immunoregulation by members of the TGFβ superfamily. Nat Rev Immunol 2016;16:723–40.
[19] Jebreel A, Mistry D, Loke D, Dunn G, Hough V, Oliver K, et al. Investigation of interleukin 10, 12 and 18 levels in patients with head and neck cancer. J Laryngol Otol 2007;121:246–52.
[20] Bruce JP, Yip K, Bratman SV, Ito E, Liu F-F. Nasopharyngeal cancer: molecular landscape. J Clin Oncol 2015;33:3346–55.
[21] Choudhary MM, France TJ, Teknos TN, Kumar P. Interleukin-6 role in head and neck squamous cell carcinoma progression. World J Otorhinolaryngol Head Neck Surg 2016;2:90–7.
[22] Thomas GR, Chen Z, Leukinova E, Van Waes C, Wen J. Cytokines IL-1 alpha, IL-6, and GM-CSF constitutively secreted by oral squamous carcinoma induce down-regulation of CD80 costimulatory molecule expression: restoration by interferon gamma. Cancer Immunol Immunother 2004;53:33–40.
[23] Tsujikawa T, Yaguchi T, Ohmura G, Ohta S, Kobayashi A, Kawamura N, et al. Autocrine and paracrine loops between cancer cells and macrophages promote lymph node metastasis via CCR4/CCL22 in head and neck squamous cell carcinoma. Int J Cancer 2013;132:2755–66.
[24] Camacho M, León X, Fernández-Figueras M-T, Quer M, Vila L. Prostaglandin E(2) pathway in head and neck squamous cell carcinoma. Head Neck 2008;30:1175–81.
[25] Forster MD, Devlin M-J. Immune checkpoint inhibition in head and neck cancer. Front Oncol 2018;8:310.
[26] Yu G-T, Bu L-L, Zhao Y-Y, Mao L, Deng W-W, Wu T-F, et al. CTLA4 blockade reduces immature myeloid cells in head and neck squamous cell carcinoma. Onco Targets Ther 2016;5:e1151594.
[27] Liu J-F, Wu L, Yang L-L, Deng W-W, Mao L, Wu H, et al. Blockade of TIM3 relieves immunosuppression through reducing regulatory T cells in head and neck cancer. J Exp Clin Cancer Res 2018;37:44.

[28] Mishra AK, Kadoishi T, Wang X, Driver E, Chen Z, Wang X-J, et al. Squamous cell carcinomas escape immune surveillance via inducing chronic activation and exhaustion of CD8+ T cells co-expressing PD-1 and LAG-3 inhibitory receptors. Oncotarget 2016;7:81341–56.
[29] Anderson AC, Joller N, Kuchroo VK. Lag-3, Tim-3, and TIGIT: co-inhibitory receptors with specialized functions in immune regulation. Immunity 2016;44:989–1004.
[30] van Montfoort N, Borst L, Korrer MJ, Sluijter M, Marijt KA, Santegoets SJ, et al. NKG2A blockade potentiates CD8 T cell immunity induced by cancer vaccines. Cell 2018;175:1–28.
[31] Vence L, Bucktrout SL, Fernandez Curbelo I, Blando J, Smith BM, Mahne AE, et al. Characterization and comparison of GITR expression in solid tumors. Clin Cancer Res 2019;25:6501–10.
[32] Liao P, Wang H, Tang Y-L, Tang Y-J, Liang X-H. The common costimulatory and coinhibitory signaling molecules in head and neck squamous cell carcinoma. Front Immunol 2019;10:2457.
[33] Davis RJ, Van Waes C, Allen CT. Overcoming barriers to effective immunotherapy: MDSCs, TAMs, and Tregs as mediators of the immunosuppressive microenvironment in head and neck cancer. Oral Oncol 2016;58:59–70.
[34] Costa NL, Valadares MC, Souza PPC, Mendonca EF, Oliveira JC, Silva TA, et al. Tumor-associated macrophages and the profile of inflammatory cytokines in oral squamous cell carcinoma. Oral Oncol 2013;49:216–23.
[35] Horton JD, Knochelmann HM, Day TA, Paulos CM, Neskey DM. Immune evasion by head and neck cancer: foundations for combination therapy. Trends Cancer 2019;5:208–32.
[36] Hanahan D, Weinberg RA. Hallmarks of cancer: the next generation. Cell 2011;144:646–74.
[37] Chen DS, Mellman I. Elements of cancer immunity and the cancer-immune set point. Nature 2017;541:321–30.
[38] Ferris RL. Immunology and immunotherapy of head and neck cancer. J Clin Oncol 2015;33:3293–304.

CHAPTER

8

Immunotherapeutic approaches in HNSCC

Nicole C. Schmitt[a], *Robert L. Ferris*[b]

[a]Department of Otolaryngology—Head and Neck Surgery and Winship Cancer Institute, Emory University School of Medicine, Atlanta, GA, United States [b]University of Pittsburgh School of Medicine, UPMC Hillman Cancer Center, Pittsburgh, PA, United States

Abstract

Immunotherapy is a rapidly advancing field that has improved survival outcomes for patients with head and neck squamous cell carcinoma (HNSCC). Large phase 3 trials have demonstrated the efficacy of anti-PD-1/PD-L1 monoclonal antibodies in patients with recurrent/metastatic disease, and current investigations will help elucidate the role of these agents in patients with previously untreated, locally advanced disease. Additional immunotherapeutic strategies including antibodies targeting other coinhibitory/costimulatory checkpoints, vaccines, oncolytic viruses, toll-like receptor agonists, adoptive cell therapy, and others are also under active study. As more data emerge regarding the management of immune-related adverse events and potential biomarkers of response, immunotherapeutic approaches will become an increasingly safe and successful therapeutic modality for head and neck cancer.

Abbreviations

ADCC antibody-dependent cell-mediated cytotoxicity
APC antigen-presenting cell
CAR chimeric antigen receptor
CDN cyclic dinucleotide
CPS combined positive score
CTL cytotoxic T lymphocyte
CTLA-4 cytotoxic T lymphocyte antigen 4
DC dendritic cell
EGFR epidermal growth factor receptor
GEP gene expression profile
HLA human leukocyte antigen
HNSCC head and neck squamous cell carcinoma
HPV human papillomavirus
IAP inhibitor of apoptosis protein

Novel Therapies in Head and Neck Cancer: Beyond the Horizon
https://doi.org/10.1016/B978-0-12-820679-9.00008-6

irAE	immune-related adverse event
IDO1	indoleamine 2,3-dioxygenase 1
MDSC	myeloid-derived suppressor cell
MHC	major histocompatibility complex
NK	natural killer
PD-1	programmed cell death 1
PD-L1	programmed cell death ligand 1
RECIST	response evaluation criteria in solid tumors
STING	stimulator of interferon genes
TCR	T-cell receptor
TIGIT	T-cell immunoreceptor with Ig and ITIM domains
TIL	tumor-infiltrating lymphocytes
TLR	Toll-like receptor
TMB	tumor mutational burden
TPS	tumor proportion score
Treg	regulatory T cell

Conflict of interest

NCS has received research funding from Astex Pharmaceuticals. RLF discloses the following: Aduro Biotech, Inc: Consulting; Amgen: Advisory Board; Astra-Zeneca/MedImmune: Advisory Board, Clinical Trial, Research Funding; Bain Capital Life Sciences: Consulting; Bristol-Myers Squibb: Advisory Board, Clinical Trial, Research Funding; EMD Serono: Advisory Board; GlaxoSmithKline: Advisory Board; Iovance Biotherapeutics, Inc.: Consulting; Lilly: Advisory Board; MacroGenics: Advisory Board; Merck: Advisory Board, Clinical Trial; Nanobiotix: Consulting; Numab Therapeutics AG: Advisory Board; Oncorus, Inc.: Advisory Board; Ono Pharmaceutical Co. Ltd: Consulting; Pfizer: Advisory Board; PPD: Advisory Board (Benitec, Immunicum); Regeneron Pharmaceuticals, Inc: Advisory Board; Tesaro: Advisory Board, Research Funding; Torque Therapeutics Inc: Consulting; TTMS: Consulting, Research Funding; VentiRx Pharmaceuticals: Research Funding.

Introduction

Improved understanding of the interactions between tumors and the immune system has ushered in an exciting era in oncology. While surgery, radiation, and cytotoxic chemotherapy have been used for decades to treat head and neck squamous cell carcinoma (HNSCC), survival with these treatment modalities has not dramatically changed in recent decades. Immunotherapy, either used alone or in combination with these more traditional therapies, has improved survival outcomes for HNSCC and other cancers. Despite the improvements in outcomes for recurrent or metastatic HNSCC made possible by immunotherapy [1], several challenges remain. The prognosis for patients with human papillomavirus (HPV)-associated oropharynx is favorable, but these patients often must endure for decades with compromised swallowing function and other permanent adverse effects. For patients with HPV-negative tumors, survival outcomes remain poor. Thus the advent of immunotherapy offers an opportunity to de-escalate therapy for patients with HPV-related tumors and to escalate therapy for higher-risk disease [2].

Basics of tumor immunology and immunotherapy

The antitumor immune response consists of both innate and adaptive immunity (Fig. 1). As part of the innate system, natural killer (NK) cells can detect nonspecific stress or "nonself" signals, followed by release of perforin and granzymes to kill the tumor cells. For a tumor to

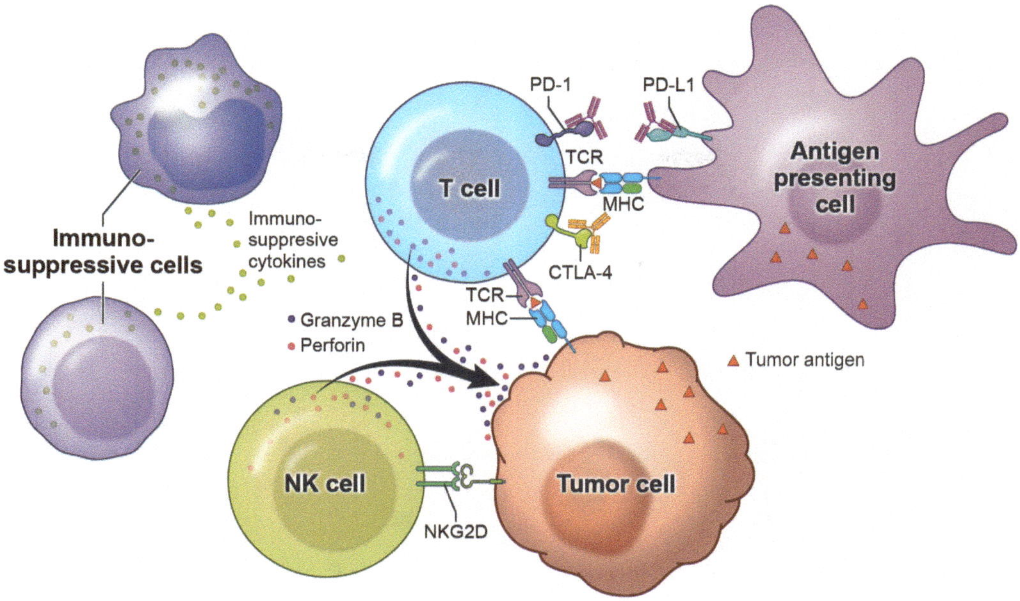

FIG. 1 Overview of tumor immunology. Natural killer cells respond to nonspecific "stress" signals, whereas T cells respond to specific tumor antigens; both secrete perforin and granzymes to induce tumor cell killing. Antigen presenting cells, such as dendritic cells, takes up antigenic peptides and cross-present them to T cells, further enhancing the antigen-specific immune response. The presence of coinhibitory immune checkpoints dampens the antitumor immune response but can be blocked using specific monoclonal antibodies. Immunosuppressive cells, such as myeloid-derived suppressor cells and regulatory T cells, can also dampen the antitumor immune response.

be recognized by the adaptive immune system, it must contain antigens that are recognized as foreign, such as an overexpressed or mutated protein, or a viral oncoprotein. These antigens are then processed and presented on a major histocompatibility complex (MHC) (also called human leukocyte antigen or HLA in humans). The antigen-MHC complex can then be detected by an antigen-specific T-cell receptor (TCR). Antigen-presenting cells can actively take up, process, and present these antigens to naive T cells, which then become activated and expand. The resulting adaptive immune response can be hindered, however, by the presence of inhibitory immune checkpoints such as programmed cell death 1 (PD-1) and cytotoxic T lymphocyte-associated protein 4 (CTLA-4). Further, immunosuppressive cells such as regulatory T cells (Tregs) and myeloid-derived suppressor cells (MDSCs) secrete immunosuppressive factors and cytokines, further dampening the immune response and allowing tumor cells to thrive. A basic overview of tumor immunology is shown in Fig. 1. Tumor immunosurveillance and mechanisms of immune escape in HNSCC are discussed in Chapter 7. This chapter will provide an overview of immunotherapeutic approaches that are approved or under active preclinical and clinical investigations for head and neck cancer.

Immunotherapeutic strategies in head and neck cancer

Tumor antigen-targeting monoclonal antibodies

Cetuximab, a chimeric mouse-human immunoglobulin G1 (IgG1) targeting epidermal growth factor receptor (EGFR), was the first targeted monoclonal antibody to be FDA

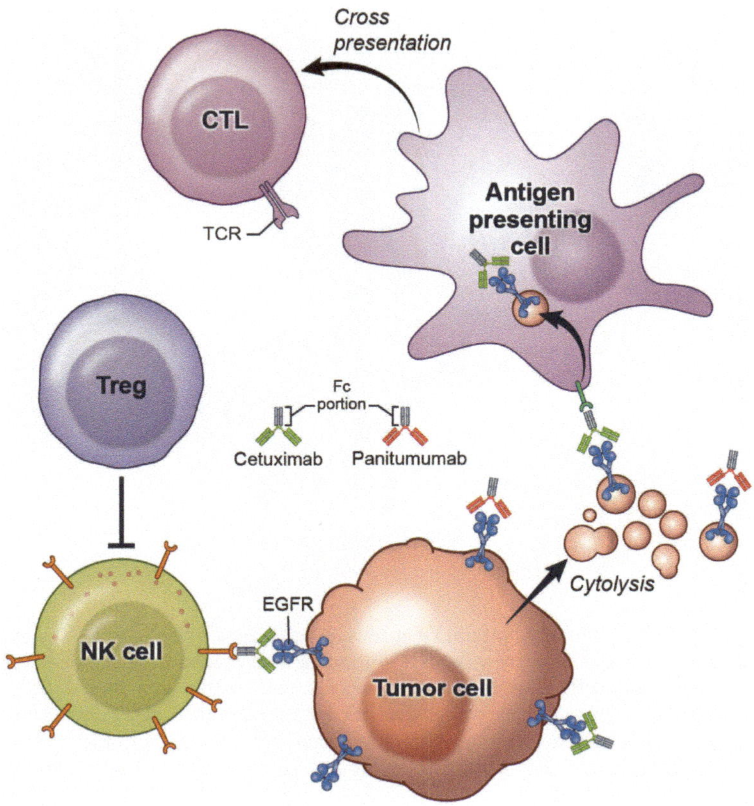

FIG. 2 Cetuximab-mediated antibody-dependent cell-mediated cytotoxicity (ADCC).

approved for the treatment of HNSCC [3,4]. Although EGFR is overexpressed in the vast majority of HNSCC tumors, tumor levels of EGFR do not correlate with response to cetuximab, and only a subset of patients responds [5]. A growing body of literature suggests that responses to cetuximab are mediated primarily by immune mechanisms in HNSCC. Anti-EGFR moAbs induce tumor cell death in part by antibody-dependent cell-mediated cytotoxicity (ADCC), mediated largely by NK cells (Fig. 2), which may be influenced by polymorphisms in the Fcγ receptor IIIa [5,6]. During lysis of tumor cells by ADCC, antibody-coated tumor antigens are released. These antigens can then be engulfed by antigen presenting cells or detected by Fcγ receptors, leading to antigen cross-presentation and activation of tumor antigen-specific cytotoxic T lymphocytes [2,5,7,8]. Indeed, antigen-specific CTLs have been noted in tumor specimens from cetuximab-treated HNSCC patients and appear to correlate with response and clinical outcomes [8,9]. The process of cetuximab-induced ADCC, NK-DC crosstalk, and subsequent adaptive immunity appears to be antibody isotype specific, since panitumumab, a fully humanized IgG2 antibody, fails to elicit these immune responses [10]. Similarly, clinical trials with anti-EGFR tyrosine kinase inhibitors have been disappointing [11].

Cetuximab has also been shown to induce upregulation of HLA class I in HNSCC cells [12]. In a prospective trial of neoadjuvant cetuximab, the variety of unique T-cell receptors increased in the peripheral blood following treatment, suggesting that cetuximab may broaden the antigen-specific immune response [13]. Cetuximab may also have some effects that suppress antitumor immunity. Following cetuximab monotherapy the expression of immune checkpoints including PD-1 and TIM-3 may increase as the T cells become exhausted [14]. Cetuximab may also increase the numbers of immunosuppressive regulatory T cells [15], but it also may favorably alter the ratio of granulocytic to monocytic MDSCs [16]. The immune alterations induced by cetuximab are summarized in Fig. 2. Importantly, these alterations of several components of the tumor immune microenvironment correlate with response to cetuximab [13–16].

Immune checkpoints

The "immunologic synapse" between an antigen presenting cell and cytotoxic T cell contains several different "checkpoints" that serve to dampen or enhance the antitumor immune response (Fig. 3). As discussed in Chapter 7, increased expression of coinhibitory checkpoints and decreased expression/activation of costimulatory checkpoints are important mechanisms of immune escape in head and neck cancer. Counteracting these mechanisms of immune escape comprises the most widely used immunotherapeutic strategy to date for HNSCC and other cancers.

Coinhibitory checkpoints

Programmed cell death 1 pathway

Thus far the most widely studied and efficacious immunotherapeutic agents for HNSCC are monoclonal antibodies that inhibit PD-1. The first trial leading to approval of an anti-PD-1 antibody for HNSCC was KEYNOTE-012. In this phase 1b trial, patients with recurrent and/or metastatic HNSCC who had previously failed treatment with platinum chemotherapy were treated with pembrolizumab every 3 weeks [17]. In the initial cohort, only patients with PD-L1-positive tumors were included; in a later expansion cohort, 132 patients were included regardless of tumor PD-L1 status. The primary endpoint was overall response rate according to response evaluation criteria in solid tumors (RECIST criteria). The overall response rate was 18% for all patients, and this was increased to 22% for patients with PD-L1-positive tumors (vs. 4% for PD-L1-negative tumors). These results led to FDA approval of pembrolizumab for platinum-refractory, recurrent/metastatic HNSCC in 2016. Importantly, long-term follow-up has shown that these responses were long lasting, with 85% of responses lasting more than 6 months and many patients exhibiting responses lasting 30 months or more with few adverse effects [18].

In the first randomized phase 3 trial of anti-PD-1 therapy for HNSCC (CheckMate 141), patients with recurrent/metastatic HNSCC refractory to platinum-based chemotherapy were randomized to receive nivolumab (anti-PD-1) or the investigator's choice of single-agent systemic therapy (docetaxel, methotrexate, or cetuximab). This trial was stopped early due to meeting its endpoint of improved overall survival, the proportion of patients surviving for 1 year more than doubled to 36% with nivolumab versus 16% for standard therapy [19]. Based on these results, nivolumab was also FDA approved in 2016 for patients with platinum-refractory disease. A very similar phase 3 trial of pembrolizumab (KEYNOTE-040) had similar survival and response outcomes [20].

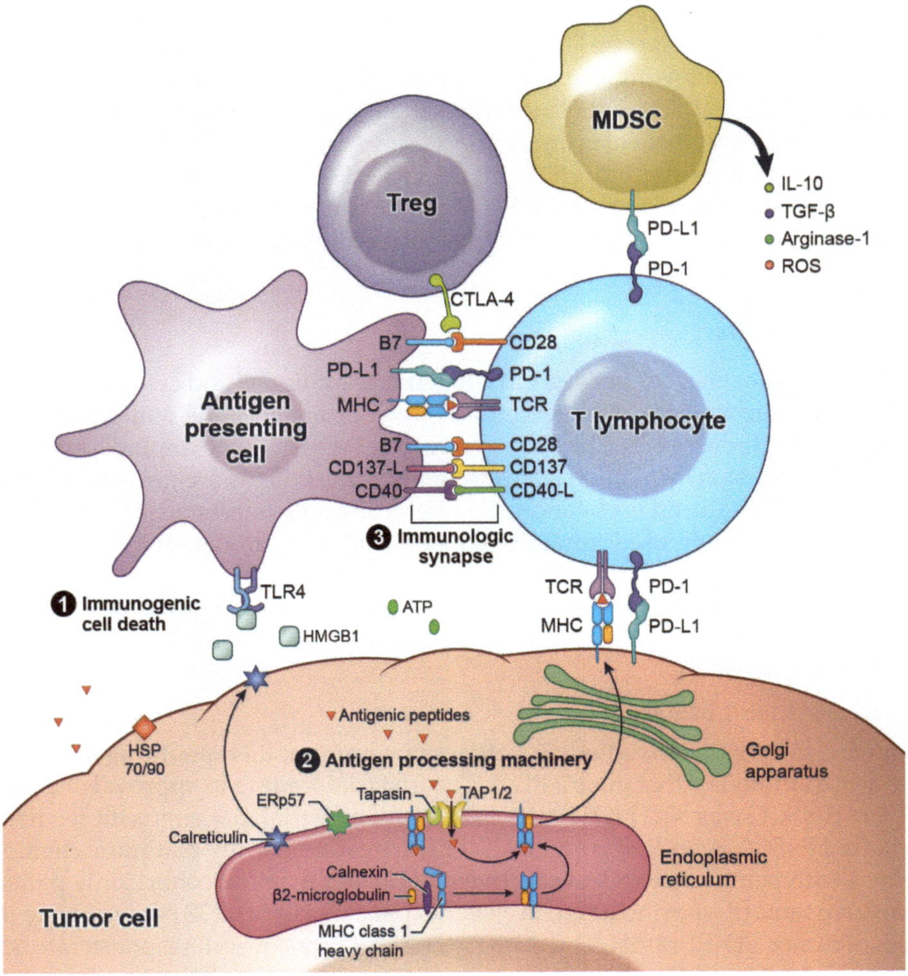

FIG. 3 Overview of immunogenic cell death (1), antigen-processing machinery (2), and coinhibitory and costimulatory checkpoints of the immunological synapse (3). Radiation and certain chemotherapy drugs can induce immunogenic cell death (ICD) (1), leading to release of damage-associated molecular patterns that are detected by toll-like receptors and other receptors on antigen-presenting cells. Tumor cells also process antigenic peptides within the endoplasmic reticulum (2), which involves several chaperone proteins and results in the presentation of the antigenic peptide on MHC class I at the cell surface, where it can then be detected by an antigen-specific T-cell receptor (TCR). The immunologic synapse (3) consists of several signals that occur between antigen-presenting cells and T cells. After an MHC molecule loaded with antigenic peptide is detected by a TCR, "costimulatory" checkpoints and their ligands stimulate the T-cell response, while "coinhibitory" checkpoints and their ligands inhibit the T-cell response.

All of these clinical trials have demonstrated that the PD-L1 status of the tumor impacts responses to anti-PD-1 therapy. As discussed earlier the KEYNOTE-012 trial showed a 22% response rate for tumors that stained positive for PD-L1 expression versus 4% response rate for tumors that were PD-L1 negative [17]. Two-year follow-up of CheckMate 141 data showed that patients with PD-L1-expressing tumors had a 45% reduction in the risk of death with nivolumab versus a 27% reduction in risk of death for PD-L1-negative tumors [21]. The KEYNOTE-040 trial established the combined positive score (CPS), which is determined as the total number of cells (tumor cells, lymphocytes, and macrophages) expressing PD-L1 divided by the total number of tumor cells, then multiplied by 100 [20]. The tumor proportion score (TPS) was defined as the proportion of tumor cells expressing membranous PD-L1 [20]. The best responses to pembrolizumab were noted in patients with a CPS ≥ 1 and TPS $>50\%$ [20].

In addition to the PD-L1 status of the tumor, the HPV status may be important for sensitivity to anti-PD-1 therapy. The KEYNOTE-012 trial showed improved overall response rates for HPV-positive versus HPV-negative tumors [17], whereas the CheckMate 141 trial did not [21]. In the HAWK trial focusing on patients with recurrent/metastatic disease with high PD-L1 expression, the overall response rate to the PD-L1 inhibitor durvalumab was 16% for all patients but increased to 29% in HPV-positive patients [22]. Multiple smaller studies have demonstrated that HPV-positive tumors tend to have increased numbers of immune effector cells and may respond better to anti-PD-1/PD-L1 therapy [23,24].

When reviewing the response rates noted earlier, it is important to note that most patients in these studies had aggressive disease that had already failed prior therapy. The majority of patients in the KEYNOTE-012, KEYNOTE-040, and CheckMate 141 studies had already failed at least two prior lines of therapy for recurrent or metastatic disease [17,19,20]. A subsequent analysis of data from CheckMate 141 showed that the survival benefit of nivolumab versus standard therapy was better for patients without prior cetuximab exposure [25]. In KEYNOTE-055, a single-arm study of patients who had already failed both platinum-based chemotherapy and cetuximab, the overall response rate was 16% [26]. The potential mechanisms for overlapping resistance to cetuximab and anti-PD-1 therapy are unclear.

Another important aspect of ant-PD-1 therapy is the low rate of adverse effects compared with cytotoxic chemotherapy. These so-called immune-related adverse events (irAEs) are common, occurring in the majority of patients [26]. However, the proportion of patients experiencing $>$ grade 3 AEs is much lower, and for most patients, anti-PD-1 therapy is well tolerated [1,17,19,20]. The most common adverse effects include nausea, hypothyroidism, and fatigue [17,19,20,22].

Once established that the anti-PD-1/PD-L1 therapy is effective for platinum-refractory, recurrent/metastatic disease, the role of these agents as first-line treatment in recurrent/metastatic disease was investigated. The EXTREME trial has previously established cisplatin, cetuximab, and fluorouracil as the standard first-line therapy for recurrent/metastatic HNSCC [27]. In the KEYNOTE-048 trial, patients were randomized to receive the EXTREME regimen versus first-line, single-agent pembrolizumab versus pembrolizumab (instead of cetuximab) in combination with the platinum/fluorouracil doublet. The second interim analysis showed improved overall survival with first-line pembrolizumab alone in patients with a combined positive score (CPS) ≥ 1, leading to FDA approval for pembrolizumab in this situation. On final analysis, pembrolizumab/chemotherapy was associated with better response rates and improved survival versus cetuximab/chemotherapy (EXTREME), regardless of PD-L1 status [28]. This regimen has thus supplanted the EXTREME regimen as the first-line standard of care for patients with recurrent/metastatic disease and a CPS <1.

Based on the promising responses in heavily pretreated patients and favorable toxicity profiles, anti-PD-1/PD-L1 therapy is now under study as a way of escalating therapy in patients with previously untreated, locally advanced disease with high risk of recurrence, that is, in patients with HPV-negative disease and/or smoking history. Studies of checkpoint blockade in the neoadjuvant setting suggest that the responses are much higher than in the recurrent/metastatic setting. For example, preliminary results from a trial of neoadjuvant pembrolizumab in high-risk HNSCC showed a 43% rate of pathologic response in surgical specimens, with some patients showing a complete pathologic response [29]. Multiple similar studies with various checkpoint blocking drugs are currently underway. The role of anti-PD-1 in patients treated with definitive or adjuvant radiation therapy is discussed in the succeeding text (combination therapies).

In summary, patients with recurrent/metastatic HNSCC demonstrate reasonable response rates to anti-PD-1/PD-L1 therapy. Responses are higher in patients with PD-L1-positive tumors and lower in patients who have failed prior therapy including platinum chemotherapy and cetuximab. The incorporation of anti-PD-1 therapy into regimens for previously untreated, locally advanced HNSCC is under active study in multiple clinical trials.

Cytotoxic T lymphocyte-associated protein 4 (CTLA-4)

Though well established as effective therapy in melanoma [30], CTLA-4 inhibitors have been less studied for head and neck squamous cell carcinoma. Clinical trials for HNSCC thus far have consisted of therapeutic combinations including monoclonal antibodies against CTLA-4 and other forms of immunotherapy and/or standard therapy; these combinations are further discussed later in this chapter. Most notably the CONDOR trial was a phase 2 study that randomized HNSCC patients with recurrent/metastatic, platinum-refractory, PD-L1-low tumors to treatment with durvalumab (anti-PD-L1), tremelimumab (anti-CTLA-4), or the combination [31]. The rationale was that patients with PD-L1-low tumors might benefit from adding an anti-CTLA-4 antibody. Very few patients (1.6%) responded to tremelimumab alone. The overall response rate to durvalumab in these PD-L1-low patients was 9.2%. Adding tremelimumab resulted in a higher incidence of grade 3/4 immune-related adverse events without improving the overall response rate (7.8%), suggesting that adding an anti-CTLA-4 antibody may not be an effective strategy for improving responses to anti-PD-1/PD-L1 therapy for PD-L1-low tumors [31].

Other coinhibitory checkpoints

One mechanism for resistance to anti-PD-1 therapy seems to be the upregulation of other coinhibitory checkpoints, including Tim-3, LAG-3, TIGIT, and IDO1 [32–36]. Interestingly, HPV-related tumors tend to have higher expression of these T-cell exhaustion markers [32, 37]. IDO1 gene expression was found to be elevated in HNSCC specimens that also expressed PD-L1 [32]. One study found elevated levels of TIGIT on CD4+ and CD8+ T cells from HNSCC patients, and anti-TIGIT therapy effectively delayed tumor growth in a murine model, especially when anti-PD-1 therapy was added [33]. Similarly, tumor specimens from HNSCC patients are enriched in PD-1+ and TIM-3+ T cells [38,39]. Since TIM-3 is expressed on both exhausted CD8+ T cells and immunosuppressive Tregs [34,35], TIM-3 blockade is a particularly attractive strategy; indeed, TIM-3 blockade has shown efficacy in multiple preclinical studies [35,38]. Targeting LAG3 in addition to PD-1 has also proven efficacious in preclinical/murine studies [36].

Costimulatory checkpoints

Another strategy is to use antibodies that activate costimulatory receptors, such as OX40, CD137 (also known as 4-1BB), CD40, or their ligands to stimulate T-cell function. High levels of intratumoral OX40-positive T cells have been noted in HNSCC patients [40], and preclinical studies have shown enhanced antitumor immunity with OX40 agonists [40]. In contrast, another study showed decreased levels of OX40 and CD137/4-1BB on peripheral blood T cells from HNSCC patients compared with healthy control patients, with the lower levels associated with advanced disease [41]. A study of patients with recurrent/metastatic HNSCC found that high OX40 ligand expression correlated with poor outcomes [42]. Despite these mixed data the use of costimulatory agonists to tip the balance in favor of an immunostimulatory response is a rational strategy, especially in combination with coinhibitory checkpoint inhibitors. A phase 1 trial of neoadjuvant anti-OX40 antibody for HNSCC is underway (NCT02274155). A preclinical study showed enhanced tumor growth delay with an agonist of CD137 added to chemoradiation in a mouse model of HPV-positive HNSCC [43], and two CD137-targeting antibodies are in phase 1 trials for HNSCC [43]. Combinations of these agents with coinhibitory checkpoint inhibitors are under study in clinical trials for multiple solid tumor types [40].

Vaccines for HNSCC

Vaccines used as immunotherapy for cancer may consist of peptides, DNA, whole tumor cells, or dendritic cells loaded ex vivo with antigenic material [44]. In a phase 1 study of autologous dendritic cells loaded with HLA class I p53 peptides, vaccinated patients showed modest p53-specific immunity, decrease in the number of Tregs, and favorable outcomes [45]. Transfection of dendritic cells with lysyl oxidase-like 4 (LOXL4) for use as a vaccine is also under investigation for HNSCC [46].

Tumor vaccines can also be delivered by viral or bacterial vectors. Oncolytic viruses are natural or engineered viruses that preferentially infect and lyse tumor cells more than normal cells [47]. In addition to releasing antigenic material from killed tumor cells to the tumor microenvironment, these viruses can activate interferon responses, toll-like receptors, and other pathways that activate antitumor immunity [47]. The most successful oncolytic virus thus far is talimogene laherparepvec (T-VEC), an attenuated herpes simplex 1 (HSV-1) virus that also encodes for granulocyte-macrophage colony-stimulating factor (GM-CSF). T-VEC has shown efficacy alone and with anti-PD-1 checkpoint blockade, now an FDA-approved regimen for melanoma [48]. An oncolytic virus similar to an early version of T-VEC (HSV-1, encoding GM-CSF) was delivered in a phase 1 study to patients with previously untreated, locally advanced HNSCC, in addition to standard chemoradiation. On posttreatment neck dissection, most patients demonstrated a complete response. The vaccine was well tolerated, and an optimal dose was determined for future study [49]. More recently the oncolytic vaccinia virus GL-ONC1 was successfully used in combination with chemoradiation for oropharyngeal carcinoma, with favorable outcomes in a phase 1 study [50].

HPV-specific therapies

Although patients with HPV-related HNSCC overall tend to have a favorable prognosis, there is a need to provide novel therapies for the minority of patients who do not fully respond to standard therapies. There is also a desire to replace standard therapies with less toxic alternatives for these patients, who often must live for decades with dysphagia and other adverse effects. The presence of antigenic oncoproteins in HPV-related disease provides a unique opportunity to pursue multiple avenues for enhancing antitumor immunity.

Preventive HPV vaccines

Most individuals are exposed to HPV strains and will mount an antibody response to clear the virus. The natural antibody response to the virus is a slow process that occurs in response to the L1 capsid protein [51,52]. Preventive vaccines were initially developed to eradicate cervical cancer and other anogenital malignancies but are likely to reduce the incidence of HPV-related oropharyngeal carcinoma as well. Both commercially available vaccines include virus-like particles (VLPs) resembling the L1 capsid protein, and unlike natural infection, vaccination introduces these VLPs to the lymphatics [52]. Two preventive vaccines are currently available: a bivalent vaccine that protects against high-risk subtypes 16 and 18 (Gardasil, Merck and Company, Whitehouse Station, New Jersey) and a quadrivalent vaccine that also protects against the low-risk subtypes 6 and 11 (Cervarix, GlaxoSmithKline Biologicals, Rixensart, Belgium). Both vaccines are given in a series of three injections, at 0, 1 or 2, and 6 months [51,52]. Multiple phase 3, randomized and double-blinded, controlled trials involving thousands of young women have shown these vaccines to be 90%–100% effective, leading to public health initiatives to promote worldwide vaccination of preadolescent girls [53].

More recently, studies have examined how these two vaccines affect HPV infection in the oral cavity and in men. A randomized, double-blind controlled trial of over 4000 young men showed Gardasil to be over 90% effective in preventing HPV-related anogenital lesions [53,54]. The Centers for Disease Control and Prevention currently recommend vaccination for both preadolescent boys and girls, which is likely to reduce the future incidence of oropharyngeal carcinoma. The effects of these vaccines on oral HPV prevalence and oropharyngeal malignancy are not yet well understood but are currently under investigation [55,56].

Therapeutic HPV vaccines

Once infection with HPV occurs and promotes oncogenesis, prophylactic vaccines are no longer of benefit. Therapeutic vaccines aim to leverage the immunogenic properties of HPV oncoproteins to enhance responses of the immune system to HPV-infected and transformed tumor cells. Several types of therapeutic vaccines are under investigation for HPV-related malignancies, many of them for cervical cancer [57]. Several candidate vaccines have used recombinant HPV proteins or synthetic peptides [57]. For HPV-related HNSCC a few clinical and preclinical vaccine studies have shown encouraging results. Using TC-1 a syngeneic mouse carcinoma model engineered to express HPV E6 and E7, intratumoral injection of an HPV-16 E7 peptide vaccine with poly(I:C) adjuvant induced a robust antitumor immune response [58]. In another preclinical study, vaccination with synthetic HPV-16 peptides induced synergistic tumor cell killing when given with cisplatin chemotherapy [59]. A peptide vaccine consisting of HLA class I and HLA class II restricted melanoma antigen E (MAGE)-A3 or HPV-16 peptides was designed with a "Trojan" approach to prevent degradation of the antigenic peptides and deliver them directly to the endoplasmic reticulum/Golgi apparatus [60]. In four of the five vaccinated HNSCC patients, antigen-specific peripheral blood responses were noted, and the vaccine was well tolerated. However, none of the patients had a clinical response by RECIST criteria [44,60]. A phase 1 trial of a $p16^{INK4A}$ peptide vaccine, including patients with HPV-related anogenital and head and neck cancer, showed similar results: antigen-specific immune responses were noted and several patients demonstrated stable disease, but there were no objective clinical responses [61]. A DNA vaccine comprising three plasmids expressing HPV E6 and E7 proteins along with adjuvant IL-12 (MEDI0457/INO-3112) was well tolerated in a pilot study of HNSCC patients with advanced HPV-related disease, and E6/E7-specific immune responses were noted [62].

Vaccines employing viral or bacterial vectors for tumor antigen delivery are also under study. Human dendritic cells transduced with adenovirus expressing an E6/E7/p16 fusion protein were shown to induce stimulation and antitumor activity of autologous T cells in vitro, and this was further enhanced by transducing an adenovirus expressing anti-PD-1 [63]. In a preclinical tumor model, a *Listeria*-based HPV-16 E7 vaccine delayed tumor growth and increased levels of antigen-specific CD8+ T cells in the peripheral blood, spleen, and tumor microenvironment [64]. A phase 1 trial with the *Listeria*-based HPV-16 antigen vaccine ADXS11-001 was opened for oropharyngeal cancer patients (NCT01598792) but was terminated after one patient developed systemic listeriosis [44,65]. A phase 2 window-of-opportunity trial using ADXS11-001 vaccine prior to transoral robotic resection for HPV-related HNSCC is currently recruiting patients, with change in the E6/E7-specific T cells as the primary endpoint (NCT02002182). Another trial is comparing ADXS11-001 with the PD-L1 inhibitor MEDI4736 alone or in combination for patients with recurrent or metastatic HPV-related malignancies (NCT02291055).

Adoptive T cell therapy

Adoptive T-cell therapy involves the infusion of autologous T cells that are high in number and equipped to kill tumor cells expressing specific antigens. Much of the work thus far on T-cell transfer in HNSCC has focused on HPV-related disease. While adoptive T-cell therapy is technically challenging and costly, effective methods for harvesting T cells from patients with HPV-related HNSCC and generating HPV E6/E7-specific T cells on a large scale have been developed [66]. Clinical trials of HPV E6/E7-specific T cells for patients with HPV-related malignancies, including HNSCC, are enrolling at multiple centers (NCT01585428, NCT02280811, and NCT02379520). In a phase 2 study including HPV-related cervical and oropharyngeal cancers, infusion of HPV-reactive tumor-infiltrating lymphocytes induced some durable responses [67]. To enhance the efficacy of infused T cells, another strategy used at the National Cancer Institute involves engineering the T-cell receptors (TCRs) to specifically recognize HPV-16 E6 or E7 peptides. These T cells usually must be restricted to a specific human leukocyte antigen (HLA) subtype. Infusion of T cells engineered to recognize E6 provided some durable responses. Patients who did not respond demonstrated defects in interferon responses or antigen presentation [68]. Trials involving the infusion of HLA-A2:01-restricted E7 TCR T cells are enrolling for patients with recurrent/metastatic disease

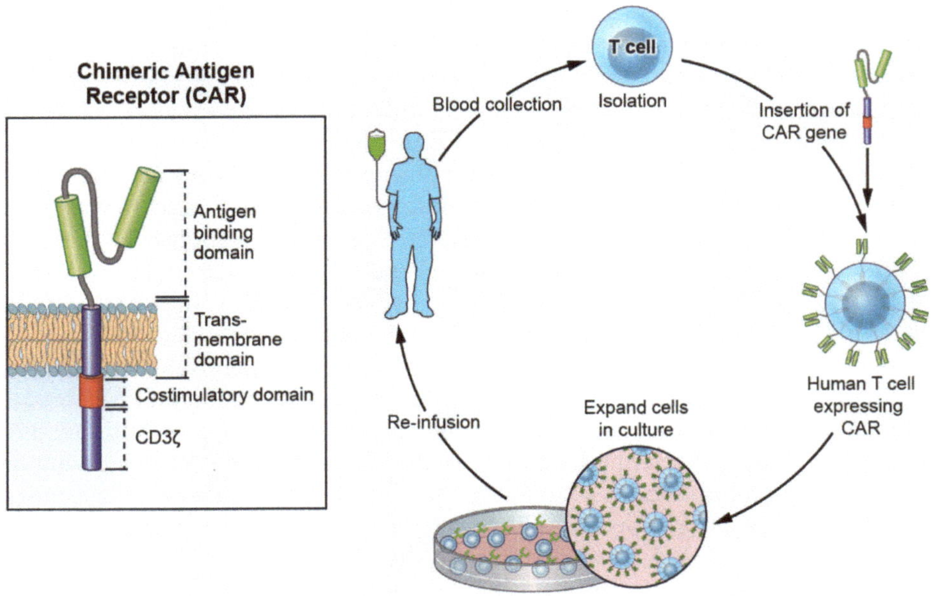

FIG. 4 Structure of a chimeric antigen receptor (CAR) and schematic of CAR T-cell therapy.

(NCT02858310) or previously untreated, locally advanced disease (NCT04015336 and NCT04044950).

Chimeric antigen receptor (CAR) T cells (Fig. 4) have also been engineered for use in HNSCC. Unlike traditional T cells, the recognition of tumor antigens by CAR T cells does not require processing and presentation of the antigen on MHC/HLA. Newer generation CARs also express costimulatory molecules, such as CD137/41BB. In similar fashion to other adoptive T-cell strategies, autologous T cells can be harvested, engineered to express the CAR, then expanded and reinfused (Fig. 4). T4 immunotherapy consists of the transfer of autologous T cells engineered to express chimeric receptors for ErbB dimers (upregulated in HNSCC) and the 4ab receptor that converts IL-4 into a signal for enrichment and expansion of T cells [69]. Following promising preclinical results [69], T4 CAR-T immunotherapy is currently under investigation in a phase 1 clinical trial (NCT01818323).

Toll-like receptor agonists

Toll-like receptors are an important part of the immune response to viruses, and they play an integral role in immunogenic cell death (Fig. 2), inflammation, and innate immunity [70]. TLR levels may differ between HPV-positive and HPV-negative HNSCC, and these varying TLR levels may correlate with the degree of T-cell infiltration and with prognosis [70,71]. An agonist of toll-like receptor 8 (TLR8), VTX-2337 (motolimod), was found in preclinical studies to enhance the killing of tumor cells by NK cells in combination with cetuximab [72,73]. Another study showed excellent antitumor activity of TLR7/9 agonists in multiple murine models, which was further enhanced by anti-PD-1 therapy [74]. A phase 1b trial of motolimod with cetuximab for recurrent/metastatic HNSCC showed some partial responses; about half of the patients demonstrated stable disease, and NK cells were more activated than at baseline [75]. A randomized, controlled clinical trial comparing motolimod with placebo in combination with the EXTREME regimen (platinum, fluorouracil, and cetuximab) for recurrent or metastatic HNSCC showed no difference in overall or progression-free survival [76]. However, improved outcomes were noted for patients with HPV-related disease or patients who exhibited a reaction at the injection site [76].

STING agonists

Cyclic dinucleotides (CDN) activate the STING receptor, which in turn leads to the production of interferons and proinflammatory cytokines. Natural and synthetic CDN have shown encouraging results in preclinical and clinical studies [77]. In preclinical murine models of HNSCC, intratumoral injection of CDN induced tumor regression and enhanced antitumor immune responses, especially when combined with anti-PD-1 therapy [78,79]. A phase 1 clinical trial of intratumoral injection of the STING agonist MK-1454 alone or with systemic pembrolizumab is underway for lymphomas and solid tumors, including HSNCC (NCT03010176). Preliminary results from that trial showed encouraging safety and evidence of immune responses in both arms, but measurable responses were only seen in the dual-treatment arm [80]. Another clinical trial combining the STING agonist ADU-S100 (also known as MIW815) with pembrolizumab (anti-PD-1) is currently enrolling patients with PD-L1-positive, recurrent/metastatic HNSCC (NCT03937141). Combinations with anti-CTLA-4 antibodies are also under investigation, including a combination of ADU-S100 plus ipilimumab for solid tumors and lymphomas (NCT02675439).

Inhibition of myeloid derived suppressor cells

One of the many causes of resistance to immunotherapy involving T cells or NK cells is the presence of immunosuppressive cells, such as regulatory T cells or myeloid-derived suppressor cells (MDSCs). Murine studies have shown that responses to immunotherapy can be enhanced by depleting immunosuppressive myeloid cells or by inhibiting the trafficking of these cells to the tumor microenvironment [81–83]. A monoclonal antibody inhibiting semaphorin4D improved responses to immune checkpoint blockade in mouse models by inhibiting MDSC recruitment and function [84], and a window-of-opportunity trial is underway with this agent in combination with immune checkpoint blockade for previously untreated, surgically resectable HNSCC (NCT03690986).

Combination therapies

Given the high rates of resistance to single-agent immune checkpoint blockade, there are considerable efforts underway to determine how best to combine immunotherapies with other forms of immunotherapy, with standard therapies, and with novel targeted therapies.

While an exhaustive list of all possible therapeutic combinations including immunotherapy is beyond the scope of this chapter, some of the key concepts and studies are highlighted here.

Combinations of immunotherapy

As mentioned earlier in this chapter, cetuximab enhances innate and adaptive antitumor immunity in HNSCC, but response may be limited by the presence of CTLA-4-expressing Tregs and other immunosuppressive cells [6–8,15,16]. It follows that combining cetuximab with other forms of immunotherapy is an attractive strategy. In a trial of cetuximab and the anti-CTLA-4 moAb ipilimumab with radiation for patients with previously untreated, locally advanced, HPV-negative or high-risk HPV-positive HNSCC (NCT01935921), 18 patients were enrolled, and the combination therapy was well tolerated with encouraging response and survival outcomes at 2 years posttreatment [85]. In a phase 1b trial (NCT02110082), the combination of cetuximab and the CD137 agonist urelumab was well tolerated and increased antigen-processing machinery (HLA-DR) in peripheral blood dendritic cells [86].

Other combinations of multiple forms of immunotherapy include the use of multiple coinhibitory checkpoint inhibitors, or a coinhibitory checkpoint inhibitor in combination with a costimulatory checkpoint agonist such as the CD137 agonist, urelumab. As discussed earlier the phase 2 CONDOR trial showed no added benefit from adding anti-CTLA-4 therapy (tremelimumab) to anti-PD-L1 therapy (durvalumab) in patients with PD-L1-low tumors [22]. Similar results were seen in the phase 3 EAGLE study [87]. Additional trials are underway using the IDO1 inhibitor epacadostat in combination with pembrolizumab (ECHO-202/KEYNOTE-037) or nivolumab (ECHO-204); both combinations have been well tolerated with encouraging response rates [88,89]. A phase 1 trial combining nivolumab (anti-PD-1) and urelumab (CD137 agonist) in multiple tumor types including HNSCC (NCT02253992) was recently completed.

Checkpoint inhibitors may also be combined with vaccines or oncolytic viruses. Multiple clinical trials combining HPV-specific vaccines with checkpoint inhibitors are open or under development for HPV-related malignancies, including HNSCC [2]. A multicenter phase 1b study combining pembrolizumab (anti-PD-1) with the HSV-1-based oncolytic virus, T-VEC, enrolled 36 patients with recurrent/metastatic HNSCC. The overall response rate was 16.7%, and the disease control rate was 38.9% [90], indicating that this particular combination was not more effective than anti-PD-1 antibodies alone. Other oncolytic viruses in preclinical development are designed to actually express checkpoint-blocking antibodies, which will allow delivery of the checkpoint-blocking antibody to the tumor while limiting off-target immune effects [91].

Lastly, TLR and STING agonists can be combined with checkpoint inhibitors. As discussed earlier a preclinical study using multiple murine models of HNSCC showed additive activity with TLR7/9 inhibitors and anti-PD-1 [74]. A phase 1b/2 study is combining pembrolizumab with intratumoral injection of SD-101, a synthetic agonist of TLR9, for recurrent/metastatic HNSCC. Thus far, this combination has been well tolerated with a disease control rate of 48% [92]. As discussed earlier, STING agonists alone enhance intratumoral immune responses, which may result in partial or complete clinical responses when combined with checkpoint inhibitors [79,80]. A preclinical study also showed that STING activation can enhance cetuximab-induced activation of NK cells and dendritic cell maturation [93], suggesting a rationale for combining STING agonists with cetuximab.

Immunotherapy combined with standard therapy

While immune checkpoint blockade and immunotherapeutic combinations hold promise for challenging cases of HNSCC, many patients remain resistant. Immune checkpoint blockade works in large part by enhancing preexisting immune responses in immunologically "hot" tumors; these drugs do not work well in tumors that have no underlying immune response, or so-called "cold" tumors. Thus there is a major, ongoing research effort to find ways to induce new immune responses in these "noninflamed" tumors, which can then be further enhanced with immune checkpoint-blocking antibodies and other immune therapies.

A growing body of evidence suggests that standard therapies such as radiation and cytotoxic chemotherapy may stimulate antitumor immunity. In addition to the release of antigen from dying cells following treatment, radiation and cytotoxic drugs may actually induce or promote immunogenic cell death, antigen processing/presentation, and adaptive immunity [94–96]. These processes are illustrated in Fig. 3. In a syngeneic mouse model of HPV-positive HNSCC, Rag1 mice lacking functional B and T cells did not respond to cisplatin or radiation, suggesting that these standard therapies have mechanisms involving the adaptive immune system [97]. Experiments in preclinical models of HNSCC suggest that cisplatin chemotherapy may increase tumor cell expression of APM components and of PD-L1; in a murine model of HNSCC, synergistic antitumor activity was seen when cisplatin was paired with anti-PD-1/PD-L1 [98]. Cisplatin and other platinum chemotherapy drugs can also induce immunogenic cell death, leading to activation of dendritic cells and enhanced adaptive immune responses [99–102].

Multiple preclinical studies combining checkpoint inhibitors with radiation have shown additive or synergistic effects, resulting in an enhanced "abscopal" effect of radiation-induced cell death outside of the radiation field [2,103,104]. One clinical study showed increases in CD8+ T cells in the peripheral blood following radiotherapy in patients with HNSCC [105], supporting the notion that checkpoint inhibitors may enhance the effects of radiotherapy in HNSCC patients.

Multiple clinical trials are currently investigating combinations of immunotherapy with standard therapies for HNSCC. One of the most exciting avenues of study is the use of neoadjuvant immune checkpoint blockade in patients for whom surgical resection is planned. These "window-of-opportunity" trials enable comparison of immune cell subsets in tumor resection specimens to baseline biopsy specimens, offering insight into potential mechanisms and efficacy of novel immune therapies [106]. In an era where hundreds of novel therapies and combinations are under study, window-of-opportunity trials are providing valuable information for the design of future larger studies [106]. Preclinical studies suggest that immunotherapy may be most efficacious when delivered prior to surgery, allowing more robust antigen-specific responses [106,107]. Indeed, response rates thus far in trials of neoadjuvant immune checkpoint blockade are much higher than those seen in heavily pretreated patients with recurrent/metastatic disease [29,108]. In two phase 2 studies where pembrolizumab was administered preoperatively and along with postoperative chemoradiation in high-risk patients, pathologic responses were >40%; some patients demonstrated a complete pathologic response [29,109]. Similar results were obtained in a trial of neoadjuvant nivolumab [108]. It is important to note that some patients failed to respond to neoadjuvant PD-1 blockade, demonstrating progressive disease while awaiting surgery; however, the PD-1 blockade was well tolerated and did not usually delay the timing of surgery [29,108]. A phase 3 study investing neoadjuvant and adjuvant delivery of pembrolizumab to patients with high-risk, resectable HNSCC (KEYNOTE-689) is currently underway [110].

Another strategy recently explored in a window-of-opportunity trial is the combination of motolimod (TLR8 agonist) and cetuximab. Preclinical in vitro studies suggested that this combination shifted macrophages toward an antitumor M1 phenotype and decreased the ability of MDSCs to inhibit T-cell proliferation [111]. In a phase 1b neoadjuvant study of 14 patients, motolimod plus cetuximab resulted in fewer MDSCs, increased M1 macrophages and CD8+ T cells, and decreased expression of coinhibitory checkpoint receptors [111].

In addition to neoadjuvant studies, other trials are exploring the use of immune checkpoint blockade in the adjuvant setting for high-risk or intermediate-risk disease. A phase 1 study has been conducted to determine the safe dose of pembrolizumab when given with adjuvant cisplatin chemoradiation for high-risk disease [112]. Another phase 1 study will use durvalumab (anti-PD-L1) and tremelimumab (anti-CTLA-4) instead of cisplatin as the systemic adjuvant therapy for intermediate-risk disease, defined as having a T3 or T4 primary tumor, the presence of perineural or lymphovascular invasion, or N+ neck disease without extracapsular extension [113]. Based on results of all of these studies, it is likely that neoadjuvant and adjuvant immune checkpoint blockade may become a standard therapeutic modality in patients with intermediate- or high-risk, surgically resectable HNSCC.

Immune checkpoint blockade has also been used successfully with definitive radiation and chemoradiation. Given the renal toxicity and hearing loss associated with cisplatin chemotherapy, many patients are ineligible for standard cisplatin-based chemoradiation. For these cisplatin-ineligible patients, durvalumab and pembrolizumab have been administered concurrently with radiation, with favorable safety and feasibility [114,115]. Another study used a combination of anti-PD-1/CTLA-4 therapy (nivolumab/ipilimumab) with concurrent radiation, also showing reasonable toxicity profiles [116]. Larger studies and long-term follow-up are needed to compare the efficacy of these regimens with standard cisplatin-based chemoradiation.

For patients with high-risk HNSCC, immune checkpoint blockade has been added to definitive chemoradiation regimens, with the hope of improving survival outcomes. One study adding pembrolizumab to radiation and weekly cisplatin showed reasonable safety, with all patients receiving the intended 70Gy dose of radiation and 85% receiving the goal cisplatin dose of $\geq$200 mg/m^2 [117]. In RTOG 3504, nivolumab was added to four different regimens for patients with intermediate- or high-risk disease: (1) weekly cisplatin with radiation, (2) high-dose cisplatin given every 3 weeks with radiation, (3) cetuximab and radiation, and (4) radiation alone. Concomitant nivolumab was delivered safely with all regimens, but additional/adjuvant nivolumab was not feasible after high-dose cisplatin or in cisplatin-ineligible patients [118]. Two large randomized, placebo-controlled studies adding anti-PD-1/anti-PD-L1 therapy to cisplatin-based chemoradiation are currently underway [119,120].

Immunotherapy combined with targeted therapy

Another treatment option under study for HNSCC is the addition of immunotherapy to targeted therapies. Studies using multiple mouse models of HNSCC have shown that drugs antagonizing inhibitor of apoptosis proteins (IAPs) have beneficial effects on antitumor immunity, especially when paired with radiation or immune checkpoint blockade [121]. In support of this idea, a window-of-opportunity study using the IAP antagonist Debio 1143 prior to surgical resection for HNSCC showed increased PD-1/PD-L1 staining and increased numbers of CD8 + T cells [122]. Inhibitors of WEE1 kinase, which have shown activity against HNSCC [123], have been shown in preclinical studies to enhance the antitumor activity of NK cells and T cells. Similarly to IAP antagonists the WEE1 kinase inhibitor AZD1775 enhances antitumor immunity when paired with radiation and/or immune checkpoint blockade [124,125]. Another preclinical study showed that a MEK inhibitor enhanced response to anti-PD-1 therapy in a syngeneic model of HNSCC, in part via increased MHC class I and PD-L1 expression [126].

Biomarkers of response to immunotherapy in HNSCC

Although immune checkpoint blockade and the other immunotherapeutic strategies described earlier are promising, not all patients will respond to these drugs. Biomarkers are sorely needed to predict which patients are most likely to respond. One concept that is critical to the understanding of immunotherapy is that most of these agents, especially anti-PD-1/PD-L1 therapy, simply serve to improve the efficacy of the underlying immune response. For example, if T cells are presented but "exhausted" as demonstrated by high PD-1 expression, then anti-PD-1 therapy may work quite well. For tumors that are "immunologically cold" or have adopted other mechanisms of immune escape as described in Chapter 7, responses to immunotherapeutic drugs tend to be less impressive. In general, virally mediated tumors, that is, HPV-related HNSCC or Epstein-Barr virus (EBV)-related nasopharyngeal carcinoma, tend to have higher numbers of tumor-infiltrating lymphocytes [24]. In addition, tumors with a high number of mutations generally also respond better to immune checkpoint blockade. For this reason, tumors with a high "tumor mutational burden" (TMB) such as melanoma, lung cancer, and smoking-related HNSCC respond relatively well to immunotherapy [127,128]. Indeed, smoking history correlates with improved responses to immune checkpoint blockade in HNSCC [128,129]. In addition to overall mutational burden, mutations in *NOTCH1*, *SMARCA4*, and frameshift mutations in tumor suppressor genes also correlate with responses to anti-PD-1/PD-L1 therapy in HPV-negative HNSCC [128].

As described earlier, patients with PD-L1-positive tumors tend to have better responses to anti-PD-1/PD-L1 therapy. The degree of PD-1 expression on tumor-infiltrating lymphocytes (TIL) may also be important. Kansy and colleagues evaluated PD-1 expression on more than 40 HNSCC tumor specimens and found that although PD-1 + TIL numbers were higher in HPV-related tumors, TIL with very high surface levels of PD-1 were found more frequently in HPV-negative tumors and predicted poor prognosis [24]. The PD-1-high TIL were less functional, as measured by IFN-γ production, and were depleted in a mouse model of HNSCC upon treatment with anti-PD-1 antibody [24]. Taken together, these findings suggest that PD-L1 expression correlates with improved responses to anti-PD-1 therapy, whereas high levels of PD-1 on TIL may have the opposite effect.

In addition to PD-1/PD-L1 the expression of other costimulatory and coinhibitory checkpoints may also predict immunotherapeutic responses. In a 3-year observational cohort of HNSCC patients treated with anti-PD-1/PD-L1 for whom sequencing and flow data were available, Hanna and colleagues noted that TIM-3/LAG-3 co-expression with PD-1 was higher on TIL from nonresponders [128]. Similarly, analysis of potential biomarkers from the CheckMate 141 trial showed that responders to nivolumab demonstrated decreases in CTLA-4+ T cells, in addition to lower numbers of PD-1+ CD8+ T cells and PD-1+ Tregs at baseline [130].

Given that PD-L1 expression is tightly linked to IFN-γ, which is also a critical component of the antitumor immune response, IFN-related genes have also been investigated as potential markers of response. Prat et al. profiled the expression of gene expression signatures associated with CD8/CD4+ T cells activation, natural killer cell activity, and IFN activation in tumor samples from patients with HNSCC and other tumors treated with anti-PD-1 therapy, finding that these gene signatures were significantly associated with nonprogressive disease and progression-free survival [131]. In a similar study, baseline mRNA samples from tumors of pembrolizumab-treated patients were used to create a robust T cell-inflamed gene expression profile (GEP). The GEP contains IFN-γ-related genes important for antigen presentation, adaptive immunity, and chemokine expression [132]. While most patients demonstrating a response or stable disease after pembrolizumab had a high IFN-γ signature score, a subset of patients with a high score had progressive disease, suggesting that the expression of IFN-γ-related genes is necessary but not sufficient for a good response to pembrolizumab [132].

Immune-related adverse events

The topic of immune-related adverse events (irAEs) deserves further discussion. While boosting the antitumor immune response, immune checkpoint-blocking antibodies and other forms of immunotherapy can also activate the immune system to cause undesirable inflammatory or autoimmune side effects. As discussed earlier, these irAEs are generally mild. Common irAEs include rash, vitiligo, hypothyroidism, and diarrhea. More serious irAEs include colitis, hepatitis, pneumonitis, Stevens-Johnson syndrome or other severe skin toxicities, myocarditis, and neurologic irAEs [133]. Most irAEs can be successfully managed so that immune checkpoint blockade can be continued [134].

While a history of autoimmune disease is not an absolute contraindication to immune checkpoint blockade, these patients often will exhibit a flare of their preexisting disease [133], and thus these patients are often excluded from clinical trials of novel immunotherapies. Patients with preexisting autoimmune disease do not appear to be more likely to experience other irAEs [133].

Interestingly, irAEs may correlate with responses to immunotherapy. A cetuximab-induced rash, thought to be of immune etiology, correlates with improved response [135]. In melanoma, immunotherapy-related vitiligo is associated with improved survival [136].

The study of irAEs is in its infancy, and further studies are needed to fully characterize these adverse events and how they may correlate with responses or resistance to immunotherapy. The Society for the Immunotherapy of Cancer has created consensus guidelines for managing these toxicities [134].

Conclusions

The field of head and neck cancer treatment has entered an exciting era, in which immunotherapeutic approaches offer novel, promising opportunities for improving survival outcomes for patients with aggressive or recurrent disease and decreasing toxicities associated with other treatment modalities in patients with HPV-related disease. However, several remaining challenges must be addressed to harness the full potential of immunotherapies. First, we must elucidate which of the many possible therapeutic combinations are most deserving of study in large phase 3 clinical trials. For patients with surgically resectable disease, "window-of-opportunity" studies can be particularly helpful in this regard [106]. Secondly, expertise in the management of irAEs is emerging. Lastly since not all patients respond to immunotherapy, biomarkers to predict which patients are most likely to benefit are still needed.

Acknowledgments

The authors thank Mr. Alan Hoofring of the NIH Medical Arts team for his expert illustrations. This work was funded in part by NIH, NIDCD, intramural project ZIA-DC000090.

References

[1] EEW C, Bell RB, Bifulco CB, Burtness B, Gillison ML, Harrington KJ, et al. The Society for Immunotherapy of Cancer consensus statement on immunotherapy for the treatment of squamous cell carcinoma of the head and neck (HNSCC). J Immunother Cancer 2019;7(1):184.

[2] Ferris RL. Immunology and immunotherapy of head and neck cancer. J Clin Oncol 2015;33(29):3293–304.

[3] Vermorken JB, Psyrri A, Mesia R, Peyrade F, Beier F, de Blas B, et al. Impact of tumor HPV status on outcome in patients with recurrent and/or metastatic squamous cell carcinoma of the head and neck receiving chemotherapy with or without cetuximab: retrospective analysis of the phase III EXTREME trial. Ann Oncol 2014; 25(4):801–7.

[4] Bonner JA, Harari PM, Giralt J, Azarnia N, Shin DM, Cohen RB, et al. Radiotherapy plus cetuximab for squamous-cell carcinoma of the head and neck. N Engl J Med 2006;354(6):567–78.

[5] Ferris RL, Jaffee EM, Ferrone S. Tumor antigen-targeted, monoclonal antibody-based immunotherapy: clinical response, cellular immunity, and immunoescape. J Clin Oncol 2010;28(28):4390–9.

[6] Lopez-Albaitero A, Lee SC, Morgan S, Grandis JR, Gooding WE, Ferrone S, et al. Role of polymorphic Fc gamma receptor IIIa and EGFR expression level in cetuximab mediated, NK cell dependent in vitro cytotoxicity of head and neck squamous cell carcinoma cells. Cancer Immunol Immunother 2009;58(11):1853–64.

[7] Lee SC, Srivastava RM, Lopez-Albaitero A, Ferrone S, Ferris RL. Natural killer (NK): dendritic cell (DC) cross talk induced by therapeutic monoclonal antibody triggers tumor antigen-specific T cell immunity. Immunol Res 2011;50(2–3):248–54.

[8] Srivastava RM, Lee SC, Andrade Filho PA, Lord CA, Jie HB, Davidson HC, et al. Cetuximab-activated natural killer and dendritic cells collaborate to trigger tumor antigen-specific T-cell immunity in head and neck cancer patients. Clin Cancer Res 2013;19(7):1858–72.

[9] Ferris RL, Kim S, Trivedi S, Srivastava RM, Concha-Benavente F, Heron DE, et al. Correlation of anti-tumor adaptive immunity with clinical response in phase II "window" trial of neoadjuvant cetuximab in patients with resectable stage III-IV head and neck squamous cell carcinoma (HNSCC). J Clin Oncol 2016;34: suppl; abstr 6060.

[10] Trivedi S, Srivastava RM, Concha-Benavente F, Ferrone S, Garcia-Bates TM, Li J, et al. Anti-EGFR targeted monoclonal antibody isotype influences anti-tumor cellular immunity in head and neck cancer patients. Clin Cancer Res 2016;22:5229–37.

[11] Martins RG, Parvathaneni U, Bauman JE, Sharma AK, Raez LE, Papagikos MA, et al. Cisplatin and radiotherapy with or without erlotinib in locally advanced squamous cell carcinoma of the head and neck: a randomized phase II trial. J Clin Oncol 2013;31(11):1415–21.

[12] Srivastava RM, Trivedi S, Concha-Benavente F, Hyun-Bae J, Wang L, Seethala RR, et al. STAT1-induced HLA class I upregulation enhances immunogenicity and clinical response to anti-EGFR mAb cetuximab therapy in HNC patients. Cancer Immunol Res 2015;3(8):936–45.

[13] Kansy BA, Shayan G, Jie HB, Gibson SP, Lei YL, Brandau S, et al. T cell receptor richness in peripheral blood increases after cetuximab therapy and correlates with therapeutic response. Oncoimmunology 2018;7(11): e1494112.

[14] Jie HB, Srivastava RM, Argiris A, Bauman JE, Kane LP, Ferris RL. Increased PD-1(+) and TIM-3(+) TILs during cetuximab therapy inversely correlate with response in head and neck cancer patients. Cancer Immunol Res 2017;5(5):408–16.

[15] Jie HB, Schuler PJ, Lee SC, Srivastava RM, Argiris A, Ferrone S, et al. CTLA-4(+) regulatory T cells increased in cetuximab-treated head and neck cancer patients suppress NK cell cytotoxicity and correlate with poor prognosis. Cancer Res 2015;75(11):2200–10.

[16] Li J, Srivastava RM, Ettyreddy A, Ferris RL. Cetuximab ameliorates suppressive phenotypes of myeloid antigen presenting cells in head and neck cancer patients. J Immunother Cancer 2015;3:54.

[17] Chow LQM, Haddad R, Gupta S, Mahipal A, Mehra R, Tahara M, et al. Antitumor activity of pembrolizumab in biomarker-unselected patients with recurrent and/or metastatic head and neck squamous cell carcinoma: results from the phase Ib KEYNOTE-012 expansion cohort. J Clin Oncol 2016;34(32):3838–45.

[18] Mehra R, Seiwert TY, Gupta S, Weiss J, Gluck I, Eder JP, et al. Efficacy and safety of pembrolizumab in recurrent/metastatic head and neck squamous cell carcinoma: pooled analyses after long-term follow-up in KEYNOTE-012. Br J Cancer 2018;119(2):153–9.

[19] Ferris RL, Blumenschein GJ, Fayette J, Guigay J, Colevas AD, Licitra LF, et al. Nivolumab for recurrent squamous cell carcinoma of the head and neck. N Engl J Med 2016;375:1856–67.

[20] Cohen EE, Machiels J-PH, Harrington K, Burtness B, Shin SW, Gause CK, et al. KEYNOTE-040: a phase III randomized trial of pembrolizumab (MK-3475) versus standard treatment in patients with recurrent or metastatic head and neck cancer. J Clin Oncol 2015;33. https://doi.org/10.1200/jco.2015.33.15_suppl.tps6084. suppl; abstr TPS6084.

[21] Ferris RL, Blumenschein Jr G, Fayette J, Guigay J, Colevas AD, Licitra L, et al. Nivolumab vs investigator's choice in recurrent or metastatic squamous cell carcinoma of the head and neck: 2-year long-term survival update of CheckMate 141 with analyses by tumor PD-L1 expression. Oral Oncol 2018;81:45–51.

[22] Zandberg DP, Algazi AP, Jimeno A, Good JS, Fayette J, Bouganim N, et al. Durvalumab for recurrent or metastatic head and neck squamous cell carcinoma: results from a single-arm, phase II study in patients with ≥25% tumour cell PD-L1 expression who have progressed on platinum-based chemotherapy. Eur J Cancer 2019;107:142–52.

[23] Wang J, Sun H, Zeng Q, Guo XJ, Wang H, Liu HH, et al. HPV-positive status associated with inflamed immune microenvironment and improved response to anti-PD-1 therapy in head and neck squamous cell carcinoma. Sci Rep 2019;9(1):13404.

[24] Kansy BA, Concha-Benavente F, Srivastava RM, Jie HB, Shayan G, Lei Y, et al. PD-1 status in CD8(+) T cells associates with survival and anti-PD-1 therapeutic outcomes in head and neck cancer. Cancer Res 2017; 77(22):6353–64.

[25] Ferris RL, Licitra L, Fayette J, Even C, Blumenschein Jr G, Harrington KJ, et al. Nivolumab in patients with recurrent or metastatic squamous cell carcinoma of the head and neck: efficacy and safety in CheckMate 141 by prior cetuximab use. Clin Cancer Res 2019;25(17):5221–30.

[26] Bauml J, Seiwert TY, Pfister DG, Worden F, Liu SV, Gilbert J, et al. Pembrolizumab for platinum- and cetuximab-refractory head and neck cancer: results from a single-arm, phase II study. J Clin Oncol 2017; 35(14):1542–9.

[27] Vermorken JB, Mesia R, Rivera F, Remenar E, Kawecki A, Rottey S, et al. Platinum-based chemotherapy plus cetuximab in head and neck cancer. N Engl J Med 2008;359(11):1116–27.

[28] Rischin D, Harrington K, Greil R, Soulieres D, Tahara M, de Castro G, et al. Protocol-specified final analysis of the phase 3 KEYNOTE-048 trial of pembrolizumab (Pembro) as first-line therapy for recurrent/metastatic head and neck squamous cell carcinoma (R/M HNSCC). J Clin Oncol 2019;37. https://doi.org/10.1200/JCO.2019.37.15_suppl.6000. suppl; abstr 6000.

[29] Uppaluri R, Zolkind P, Lin T, Nussenbaum B, Jackson RS, Rich J, et al. Neoadjuvant pembrolizumab in surgically resectable, HPV negative, locally advanced head and neck squamous cell carcinoma (HNSCC). J Clin Oncol 2017;35. https://doi.org/10.1200/JCO.2017.35.15_suppl.6012. Suppl; abst 6012.

[30] Larkin J, Chiarion-Sileni V, Gonzalez R, Grob JJ, Rutkowski P, Lao CD, et al. Five-year survival with combined nivolumab and ipilimumab in advanced melanoma. N Engl J Med 2019;381:1535–46.
[31] Siu LL, Even C, Mesia R, Remenar E, Daste A, Delord JP, et al. Safety and efficacy of durvalumab with or without tremelimumab in patients with PD-L1-low/negative recurrent or metastatic HNSCC: the phase 2 CONDOR randomized clinical trial. JAMA Oncol 2019;5(2):195–203.
[32] Wirth L, Burtness B, Mehra R, Bauman JR, Lee J, Stauffer Smith NM, et al. IDO1 as a mechanism of adaptive immune resistance to anti-PD-1 monotherapy in HNSCC. J Clin Oncol 2017;35. https://doi.org/10.1200/JCO.2017.35.15_suppl.6053. suppl; abstr 6053.
[33] Wu L, Mao L, Liu JF, Chen L, Yu GT, Yang LL, et al. Blockade of TIGIT/CD155 signaling reverses T-cell exhaustion and enhances antitumor capability in head and neck squamous cell carcinoma. Cancer Immunol Res 2019;7:1700–13. https://doi.org/10.1158/2326-6066.CIR-18-0725.
[34] Liu Z, McMichael EL, Shayan G, Li J, Chen K, Srivastava R, et al. Novel effector phenotype of Tim-3(+) regulatory T cells leads to enhanced suppressive function in head and neck Cancer patients. Clin Cancer Res 2018; 24(18):4529–38.
[35] Oweida A, Hararah MK, Phan A, Binder D, Bhatia S, Lennon S, et al. Resistance to radiotherapy and PD-L1 blockade is mediated by TIM-3 upregulation and regulatory T-cell infiltration. Clin Cancer Res 2018; 24(21):5368–80.
[36] Woo SR, Turnis ME, Goldberg MV, Bankoti J, Selby M, Nirschl CJ, et al. Immune inhibitory molecules LAG-3 and PD-1 synergistically regulate T-cell function to promote tumoral immune escape. Cancer Res 2012; 72(4):917–27.
[37] Gameiro SF, Ghasemi F, Barrett JW, Koropatnick J, Nichols AC, Mymryk JS, et al. Treatment-naive HPV+ head and neck cancers display a T-cell-inflamed phenotype distinct from their HPV- counterparts that has implications for immunotherapy. Oncoimmunology 2018;7(10):e1498439.
[38] Pfannenstiel LW, Diaz-Montero CM, Tian YF, Scharpf J, Ko JS, Gastman BR. Immune-checkpoint blockade opposes CD8(+) T-cell suppression in human and murine cancer. Cancer Immunol Res 2019;7(3):510–25.
[39] Shayan G, Srivastava R, Li J, Schmitt N, Kane LP, Ferris RL. Adaptive resistance to anti-PD1 therapy by Tim-3 upregulation is mediated by the PI3K-Akt pathway in head and neck cancer. Oncoimmunology 2017;6(1): e1261779.
[40] Bell RB, Leidner RS, Crittenden MR, Curti BD, Feng Z, Montler R, et al. OX40 signaling in head and neck squamous cell carcinoma: overcoming immunosuppression in the tumor microenvironment. Oral Oncol 2016;52:1–10.
[41] Baruah P, Lee M, Odutoye T, Williamson P, Hyde N, Kaski JC, et al. Decreased levels of alternative co-stimulatory receptors OX40 and 4-1BB characterise T cells from head and neck cancer patients. Immunobiology 2012;217(7):669–75.
[42] Lecerf C, Kamal M, Vacher S, Chemlali W, Schnitzler A, Morel C, et al. Immune gene expression in head and neck squamous cell carcinoma patients. Eur J Cancer 2019;121:210–23.
[43] Lucido CT, Vermeer PD, Wieking BG, Vermeer DW, Lee JH. CD137 enhancement of HPV positive head and neck squamous cell carcinoma tumor clearance. Vaccines (Basel) 2014;2(4):841–53.
[44] Gildener-Leapman N, Lee J, Ferris RL. Tailored immunotherapy for HPV positive head and neck squamous cell cancer. Oral Oncol 2014;50(9):780–4.
[45] Schuler PJ, Harasymczuk M, Visus C, Deleo A, Trivedi S, Lei Y, et al. Phase I dendritic cell p53 peptide vaccine for head and neck cancer. Clin Cancer Res 2014;20(9):2433–44.
[46] Weise JB, Csiszar K, Gottschlich S, Hoffmann M, Schmidt A, Weingartz U, et al. Vaccination strategy to target lysyl oxidase-like 4 in dendritic cell based immunotherapy for head and neck cancer. Int J Oncol 2008; 32(2):317–22.
[47] Hennessy ML, Bommareddy PK, Boland G, Kaufman HL. Oncolytic immunotherapy. Surg Oncol Clin N Am 2019;28(3):419–30.
[48] Ribas A, Dummer R, Puzanov I, VanderWalde A, Andtbacka RHI, Michielin O, et al. Oncolytic virotherapy promotes intratumoral T cell infiltration and improves anti-PD-1 immunotherapy. Cell 2018;174(4):1031–2.
[49] Harrington KJ, Hingorani M, Tanay MA, Hickey J, Bhide SA, Clarke PM, et al. Phase I/II study of oncolytic HSV GM-CSF in combination with radiotherapy and cisplatin in untreated stage III/IV squamous cell cancer of the head and neck. Clin Cancer Res 2010;16(15):4005–15.

[50] Mell LK, Brumund KT, Daniels GA, Advani SJ, Zakeri K, Wright ME, et al. Phase I trial of intravenous oncolytic vaccinia virus (GL-ONC1) with cisplatin and radiotherapy in patients with locoregionally advanced head and neck carcinoma. Clin Cancer Res 2017;23(19):5696–702.
[51] Frazer IH. Interaction of human papillomaviruses with the host immune system: a well evolved relationship. Virology 2009;384(2):410–4.
[52] Stanley M, Pinto LA, Trimble C. Human papillomavirus vaccines—immune responses. Vaccine 2012;30(Suppl 5):F83–7.
[53] Schiller JT, Castellsague X, Garland SM. A review of clinical trials of human papillomavirus prophylactic vaccines. Vaccine 2012;30(Suppl 5):F123–38.
[54] Giuliano AR, Palefsky JM, Goldstone S, Moreira Jr ED, Penny ME, Aranda C, et al. Efficacy of quadrivalent HPV vaccine against HPV infection and disease in males. N Engl J Med 2011;364(5):401–11.
[55] Gildener-Leapman N, Ferris RL, Bauman JE. Promising systemic immunotherapies in head and neck squamous cell carcinoma. Oral Oncol 2013;49(12):1089–96.
[56] Guo T, Eisele DW, Fakhry C. The potential impact of prophylactic human papillomavirus vaccination on oropharyngeal cancer. Cancer 2016;122:2313–23.
[57] Vonsky MS, Runov AL, Gordeychuk IV, Isaguliants MG. Therapeutic vaccines against human papilloma viruses: achievements and prospects. Biochemistry (Mosc) 2019;84(7):800–16.
[58] Ishida E, Lee J, Campbell JS, Chakravarty PD, Katori Y, Ogawa T, et al. Intratumoral delivery of an HPV vaccine elicits a broad anti-tumor immune response that translates into a potent anti-tumor effect in a preclinical murine HPV model. Cancer Immunol Immunother 2019;68(8):1273–86.
[59] van der Sluis TC, van Duikeren S, Huppelschoten S, Jordanova ES, Beyranvand Nejad E, Sloots A, et al. Vaccine-induced tumor necrosis factor-producing T cells synergize with cisplatin to promote tumor cell death. Clin Cancer Res 2015;21(4):781–94.
[60] Voskens CJ, Sewell D, Hertzano R, DeSanto J, Rollins S, Lee M, et al. Induction of MAGE-A3 and HPV-16 immunity by Trojan vaccines in patients with head and neck carcinoma. Head Neck 2012;34(12):1734–46.
[61] Reuschenbach M, Pauligk C, Karbach J, Rafiyan MR, Kloor M, Prigge ES, et al. A phase 1/2a study to test the safety and immunogenicity of a p16(INK4a) peptide vaccine in patients with advanced human papillomavirus-associated cancers. Cancer 2016;122(9):1425–33.
[62] Aggarwal C, Halmos B, Porosnicu M, Saba NF, Sukari A, Grethlein SJ, et al. A phase 1b/2a, multi-center, open-label study to evaluate the safety and efficacy of combination treatment with MEDI0457 (INO-3112) and durvalumab (MEDI4736) in patients with recurrent/metastatic human papilloma virus–associated head and neck squamous cell cancer. J Clin Oncol 2018;36. https://doi.org/10.1200/JCO.2018.36.15_suppl.TPS6093. suppl; abstr TPS6093.
[63] Garcia-Bates TM, Kim E, Concha-Benavente F, Trivedi S, Mailliard RB, Gambotto A, et al. Enhanced cytotoxic CD8 T cell priming using dendritic cell-expressing human papillomavirus-16 E6/E7-p16INK4 fusion protein with sequenced anti-programmed death-1. J Immunol 2016;196(6):2870–8.
[64] Sewell DA, Pan ZK, Paterson Y. Listeria-based HPV-16 E7 vaccines limit autochthonous tumor growth in a transgenic mouse model for HPV-16 transformed tumors. Vaccine 2008;26(41):5315–20.
[65] Sacco JJ, Evans M, Harrington KJ, Man S, Powell N, Shaw RJ, et al. Systemic listeriosis following vaccination with the attenuated listeria monocytogenes therapeutic vaccine, ADXS11-001. Hum Vaccin Immunother 2016;12(4):1085–6.
[66] Ramos CA, Narala N, Vyas GM, Leen AM, Gerdemann U, Sturgis EM, et al. Human papillomavirus type 16 E6/E7-specific cytotoxic T lymphocytes for adoptive immunotherapy of HPV-associated malignancies. J Immunother 2013;36(1):66–76.
[67] Stevanovic S, Helman SR, Wunderlich JR, Langhan MM, Doran SL, Kwong MLM, et al. A phase II study of tumor-infiltrating lymphocyte therapy for human papillomavirus-associated epithelial cancers. Clin Cancer Res 2019;25(5):1486–93.
[68] Doran SL, Stevanovic S, Adhikary S, Gartner JJ, Jia L, Kwong ML, et al. T-cell receptor gene therapy for human papillomavirus-associated epithelial cancers: a first-in-human, phase I/II study. J Clin Oncol 2019;37(30):2759–69.
[69] van Schalkwyk MC, Papa SE, Jeannon JP, Guerrero Urbano T, Spicer JF, Maher J. Design of a phase I clinical trial to evaluate intratumoral delivery of ErbB-targeted chimeric antigen receptor T-cells in locally advanced or recurrent head and neck cancer. Hum Gene Ther Clin Dev 2013;24(3):134–42.

[70] Jouhi L, Mohamed H, Makitie A, Remes SM, Haglund C, Atula T, et al. Toll-like receptor 5 and 7 expression may impact prognosis of HPV-positive oropharyngeal squamous cell carcinoma patients. Cancer Immunol Immunother 2017;66(12):1619–29.

[71] Haeggblom L, Nasman A, Ramqvist T, Haglund C, Hagstrom J, Makitie A, et al. TLR5 and TLR7 are differentially expressed in human papillomavirus-positive and negative base of tongue squamous cell carcinoma, and TLR7 may have an independent prognostic influence. Acta Otolaryngol 2019;139(2):206–10.

[72] Stephenson RM, Lim CM, Matthews M, Dietsch G, Hershberg R, Ferris RL. TLR8 stimulation enhances cetuximab-mediated natural killer cell lysis of head and neck cancer cells and dendritic cell cross-priming of EGFR-specific CD8+ T cells. Cancer Immunol Immunother 2013;62(8):1347–57.

[73] Dietsch GN, Lu H, Yang Y, Morishima C, Chow LQ, Disis ML, et al. Coordinated activation of Toll-like receptor8 (TLR8) and NLRP3 by the TLR8 agonist, VTX-2337, ignites tumoricidal natural killer cell activity. PLoS One 2016;11(2):e0148764.

[74] Sato-Kaneko F, Yao S, Ahmadi A, Zhang SS, Hosoya T, Kaneda MM, et al. Combination immunotherapy with TLR agonists and checkpoint inhibitors suppresses head and neck cancer. JCI Insight 2017;2(18). https://doi.org/10.1172/jci.insight.93397.

[75] Chow LQM, Morishima C, Eaton KD, Baik CS, Goulart BH, Anderson LN, et al. Phase Ib trial of the Toll-like receptor 8 agonist, motolimod (VTX-2337), combined with cetuximab in patients with recurrent or metastatic SCCHN. Clin Cancer Res 2017;23(10):2442–50.

[76] Ferris RL, Saba NF, Gitlitz BJ, Haddad R, Sukari A, Neupane P, et al. Effect of adding motolimod to standard combination chemotherapy and cetuximab treatment of patients with squamous cell carcinoma of the head and neck: the Active8 randomized clinical trial. JAMA Oncol 2018;4(11):1583–8.

[77] Fu J, Kanne DB, Leong M, Glickman LH, McWhirter SM, Lemmens E, et al. STING agonist formulated cancer vaccines can cure established tumors resistant to PD-1 blockade. Sci Transl Med 2015;7(283).

[78] Gadkaree SK, Fu J, Sen R, Korrer MJ, Allen C, Kim YJ. Induction of tumor regression by intratumoral STING agonists combined with anti-programmed death-L1 blocking antibody in a preclinical squamous cell carcinoma model. Head Neck 2017;39(6):1086–94.

[79] Moore E, Clavijo PE, Davis R, Cash H, Van Waes C, Kim Y, et al. Established T cell-inflamed tumors rejected after adaptive resistance was reversed by combination STING activation and PD-1 pathway blockade. Cancer Immunol Res 2016;4(12):1061–71.

[80] Harrington KJ, Brody J, Ingham M, Strauss J, Cemerski S, Wang M, et al. LBA15Preliminary results of the first-in-human (FIH) study of MK-1454, an agonist of stimulator of interferon genes (STING), as monotherapy or in combination with pembrolizumab (pembro) in patients with advanced solid tumors or lymphomas. Ann Oncol 2018;29(Suppl_8). viii712.

[81] Clavijo PE, Moore EC, Chen J, Davis RJ, Friedman J, Kim Y, et al. Resistance to CTLA-4 checkpoint inhibition reversed through selective elimination of granulocytic myeloid cells. Oncotarget 2017;8(34):55804–20.

[82] Davis RJ, Moore EC, Clavijo PE, Friedman J, Cash H, Chen Z, et al. Anti-PD-L1 efficacy can be enhanced by inhibition of myeloid-derived suppressor cells with a selective inhibitor of PI3Kdelta/gamma. Cancer Res 2017;77(10):2607–19.

[83] Sun L, Clavijo PE, Robbins Y, Patel P, Friedman J, Greene S, et al. Inhibiting myeloid-derived suppressor cell trafficking enhances T cell immunotherapy. JCI Insight 2019;4(7). https://doi.org/10.1172/jci.insight.126853.

[84] Clavijo PE, Friedman J, Robbins Y, Moore EC, Smith E, Zauderer M, et al. Semaphorin4D inhibition improves response to immune-checkpoint blockade via attenuation of MDSC recruitment and function. Cancer Immunol Res 2019;7(2):282–91.

[85] Theodoraki MN, Yerneni S, Gooding WE, Ohr J, Clump DA, Bauman JE, et al. Circulating exosomes measure responses to therapy in head and neck cancer patients treated with cetuximab, ipilimumab, and IMRT. Oncoimmunology 2019;8(7):1593805.

[86] Srivastava RM, Trivedi S, Concha-Benavente F, Gibson SP, Reeder C, Ferrone S, et al. CD137 stimulation enhances cetuximab-induced natural killer: dendritic cell priming of antitumor T-cell immunity in patients with head and neck cancer. Clin Cancer Res 2017;23(3):707–16.

[87] Licitra L, Haddad R, Even C, Tahara M, Dvorkin M, Ciuleanu TE, et al. EAGLE: a phase 3, randomized, open-label study of durvalumab (D) with or without tremelimumab (T) in patients (pts) with recurrent or metastatic head and neck squamous cell carcinoma (R/M HNSCC). J Clin Oncol 2019;37. https://doi.org/10.1200/JCO.2019.37.15_suppl.6012. suppl; abstr 6012.

[88] Daud A, Saleh MN, Hu J, Bleeker JS, Riese MJ, Meier R, et al. Epacadostat plus nivolumab for advanced melanoma: updated phase 2 results of the ECHO-204 study. J CLin Oncol 2018;36. https://doi.org/10.1200/JCO.2018.36.15_suppl.9511. suppl; abstr 9511.
[89] Hamid O, Bauer TM, Spira AI, Olszanski AJ, Patel SP, Wasser JS, et al. Epacadostat plus pembrolizumab in patients with SCCHN: preliminary phase 1/2 results from ECHO-202/KEYNOTE-037. J Clin Oncol 2017;35. https://doi.org/10.1200/JCO.2017.35.15_suppl.6010. suppl; abstr 6010.
[90] Harrington KJ, King AH, Mach N, Rordorf T, Jaime JC, Espeli V, et al. Safety and preliminary efficacy of talimogene laherparepvec (T-VEC) in combination (combo) with pembrolizumab (Pembro) in patients (pts) with recurrent or metastatic squamous cell carcinoma of the head and neck (R/M HNSCC): a multicenter, phase 1b study (MASTERKEY-232). J Clin Oncol 2018;36. https://doi.org/10.1200/JCO.2018.36.15_suppl.6036. suppl; abstr 6036.
[91] Thomas S, Kuncheria L, Roulstone V, Kyula JN, Mansfield D, Bommareddy PK, et al. Development of a new fusion-enhanced oncolytic immunotherapy platform based on herpes simplex virus type 1. J Immunother Cancer 2019;7(1):214.
[92] Cohen EE, Nabell L, Wong DJL, Day TA, Daniels GA, Milhem MM, et al. Phase 1b/2, open label, multicenter study of intratumoral SD-101 in combination with pembrolizumab in anti-PD-1 treatment naïve patients with recurrent or metastatic head and neck squamous cell carcinoma (HNSCC). J Clin Oncol 2019;37. https://doi.org/10.1200/JCO.2019.37.15_suppl.6039. suppl; abstr 6039.
[93] Lu S, Concha-Benavente F, Shayan G, Srivastava RM, Gibson SP, Wang L, et al. STING activation enhances cetuximab-mediated NK cell activation and DC maturation and correlates with HPV(+) status in head and neck cancer. Oral Oncol 2018;78:186–93.
[94] de Biasi AR, Villena-Vargas J, Adusumilli PS. Cisplatin-induced antitumor immunomodulation: a review of preclinical and clinical evidence. Clin Cancer Res 2014;20(21):5384–91.
[95] Dovedi SJ, Illidge TM. The antitumor immune response generated by fractionated radiation therapy may be limited by tumor cell adaptive resistance and can be circumvented by PD-L1 blockade. Oncoimmunology 2015;4(7).
[96] Hato SV, Khong A, de Vries IJ, Lesterhuis WJ. Molecular pathways: the immunogenic effects of platinum-based chemotherapeutics. Clin Cancer Res 2014;20(11):2831–7.
[97] Spanos WC, Nowicki P, Lee DW, Hoover A, Hostager B, Gupta A, et al. Immune response during therapy with cisplatin or radiation for human papillomavirus-related head and neck cancer. Arch Otolaryngol Head Neck Surg 2009;135(11):1137–46.
[98] Tran L, Allen CT, Xiao R, Moore E, Davis R, Park SJ, et al. Cisplatin alters antitumor immunity and synergizes with PD-1/PD-L1 inhibition in head and neck squamous cell carcinoma. Cancer Immunol Res 2017; 5(12):1141–51.
[99] Park SJ, Ye W, Xiao R, Silvin C, Padget M, Hodge JW, et al. Cisplatin and oxaliplatin induce similar immunogenic changes in preclinical models of head and neck cancer. Oral Oncol 2019;95:127–35.
[100] Bezu L, Gomes-de-Silva LC, Dewitte H, Breckpot K, Fucikova J, Spisek R, et al. Combinatorial strategies for the induction of immunogenic cell death. Front Immunol 2015;6:187.
[101] Kepp O, Galluzzi L, Martins I, Schlemmer F, Adjemian S, Michaud M, et al. Molecular determinants of immunogenic cell death elicited by anticancer chemotherapy. Cancer Metastasis Rev 2011;30(1):61–9.
[102] Martins I, Kepp O, Schlemmer F, Adjemian S, Tailler M, Shen S, et al. Restoration of the immunogenicity of cisplatin-induced cancer cell death by endoplasmic reticulum stress. Oncogene 2011;30(10):1147–58.
[103] Schoppy DW, Sunwoo JB. Immunotherapy for head and neck squamous cell carcinoma. Hematol Oncol Clin North Am 2015;29(6):1033–43.
[104] Morisada M, Clavijo PE, Moore E, Sun L, Chamberlin M, Van Waes C, et al. PD-1 blockade reverses adaptive immune resistance induced by high-dose hypofractionated but not low-dose daily fractionated radiation. Oncoimmunology 2018;7(3):e1395996.
[105] Sridharan V, Margalit DN, Curreri SA, Severgnini M, Hodi FS, Haddad RI, et al. Systemic immunologic effects of definitive radiation in head and neck cancer. Int J Radiat Oncol Biol Phys 2016;94(4):864.
[106] Zandberg DP, Ferris RL. Window studies in squamous cell carcinoma of the head and neck: values and limits. Curr Treat Options Oncol 2018;19(12):68.
[107] Friedman J, Moore E, Zolkind P, Robbins Y, Clavijo PE, Sun L, et al. Neoadjuvant PD-1 immune checkpoint blockade reverses functional immunodominance among tumor-antigen specific T cells. Clin Cancer Res 2019. https://doi.org/10.1158/1078-0432.CCR-19-2209 [Epub ahead of print].

[108] Horton JD, Knochelmann H, Armeson K, Kaczmar JM, Paulos C, Neskey D. Neoadjuvant presurgical PD-1 inhibition in oral cavity squamous cell carcinoma. J Clin Oncol 2019;37. https://doi.org/10.1200/JCO.2019.37.15_suppl.2574. suppl; abstr 2574.

[109] Wise-Draper TM, Old MO, Worden F, O'Brien PM, Cohen EE, Dunlap N, et al. Phase II multi-site investigation of neoadjuvant pembrolizumab and adjuvant concurrent radiation and pembrolizumab with or without cisplatin in resected head and neck squamous cell carcinoma. J Clin Oncol 2018;36. https://doi.org/10.1200/JCO.2018.36.15_suppl.6017. suppl; abstr 6017.

[110] Uppaluri R, Lee NY, Westra W, Cohen EE, Haddad R, Temam S, et al. KEYNOTE-689: phase 3 study of adjuvant and neoadjuvant pembrolizumab combined with standard of care (SOC) in patients with resectable, locally advanced head and neck squamous cell carcinoma. J Clin Oncol 2019;37. https://doi.org/10.1200/JCO.2019.37.15_suppl.TPS6090. suppl; abstr TPS6090.

[111] Shayan G, Kansy BA, Gibson SP, Srivastava RM, Bryan JK, Bauman JE, et al. Phase Ib study of immune biomarker modulation with neoadjuvant CETUXIMAB and TLR8 stimulation in head and neck cancer to overcome suppressive myeloid signals. Clin Cancer Res 2018;24(1):62–72.

[112] Bauman JE, Harris J, Uppaluri R, Yao M, Ferris RL, Chen J, et al. NRG-HN003: phase I and expansion cohort study of adjuvant cisplatin, intensity-modulated radiation therapy (IMRT), and MK-3475 (Pembrolizumab) in high risk head and neck squamous cell carcinoma (HNSCC). J Clin Oncol 2019;37. https://doi.org/10.1200/JCO.2019.37.15_suppl.6023. suppl; abstr 6023.

[113] Sheth S, Chera BS, Vincent BG, Shen CJ, Weissler MC, Blumberg J, et al. Phase I study of durvalumab and tremelimumab together with radiotherapy for the adjuvant treatment of intermediate-risk head and neck squamous cell carcinoma. J Clin Oncol 2019;37. https://doi.org/10.1200/JCO.2019.37.8_suppl.TPS73. suppl 8; abstr TPS73.

[114] Mell LK, Torres-Saavedra PA, Wong SJ, Chang S, KIsh JA, Minn A, et al. Safety of radiotherapy with concurrent and adjuvant MEDI4736 (durvalumab) in patients with locoregionally advanced head and neck cancer with a contraindication to cisplatin: NRG-HN004. J Clin Oncol 2019;37. https://doi.org/10.1200/JCO.2019.37.15_suppl.6065. suppl; abstr 6065.

[115] Weiss J, Bauman JR, Deal AM, Sheth S, Chera BS, Shen CJ, et al. Preliminary toxicity data from the combination of pembrolizumab and definitive-dose radiotherapy for locally advanced head and neck cancer with contraindication to cisplatin therapy. J Clin Oncol 2018;36. https://doi.org/10.1200/JCO.2018.36.15_suppl.6069. suppl; abstr 6069.

[116] Johnson JM, Ad VB, Lorber E, Poller D, Luginbuhl A, Curry JM, et al. Safety of nivolumab and ipilimumab in combination with radiotherapy in patients with locally advanced squamous cell carcinoma of the head and neck (LA SCCHN). J Clin Oncol 2019;37. https://doi.org/10.1200/JCO.2019.37.15_suppl.6070. suppl; abstr 6070.

[117] Powell SF, Gitau MM, Sumey CJ, Reynolds JT, Lohr M, McGraw S, et al. Safety of pembrolizumab with chemoradiation (CRT) in locally advanced squamous cell carcinoma of the head and neck (LA-SCCHN). J Clin Oncol 2017;35. https://doi.org/10.1200/JCO.2017.35.15_suppl.6011. suppl; abstr 6011.

[118] Gillison ML, Ferris RL, Harris J, Colevas AD, Mell LK, Kong C, et al. Safety and disease control achieved with the addition of nivolumab (Nivo) to chemoradiotherapy (CRT) for intermediate (IR) and high-risk (HR) local-regionally advanced head and neck squamous cell carcinoma (HNSCC): RTOG Foundation 3504. J Clin Oncol 2019;37. https://doi.org/10.1200/JCO.2019.37.15_suppl.6073. suppl; abstr 6073.

[119] Lee NY, Ferris RL, Beck JT, Harrington K, Haddad R, Bourhis J, et al. JAVELIN head and neck 100: a phase 3 trial of avelumab in combination with chemoradiotherapy (CRT) vs CRT for 1st-line treatment of locally advanced squamous cell carcinoma of the head and neck (LA SCCHN). J Clin Oncol 2017;35. https://doi.org/10.1200/JCO.2017.35.15_suppl.TPS6093. suppl; abstr TPS6093.

[120] Machiels J-PH, Licitra L, Tao Y, Yen C-J, Rischin D, Waldron J, et al. Pembrolizumab plus chemoradiation vs chemoradiation alone for locally advanced head and neck squamous cell carcinoma: the phase 3 KEYNOTE-412 study. J Clin Oncol 2018;36. https://doi.org/10.1200/JCO.2018.36.15_suppl.TPS6094. suppl; abstr TPS6094.

[121] Xiao R, Allen CT, Tran L, Patel P, Park SJ, Chen Z, et al. Antagonist of cIAP1/2 and XIAP enhances anti-tumor immunity when combined with radiation and PD-1 blockade in a syngeneic model of head and neck cancer. Oncoimmunology 2018;7(9):e1471440.

[122] Gomez-Roca C, Even C, Le Tourneau C, Rotllan NB, Delord JP, Sarini J, Vergez S, et al. Open-label, non-randomized, exploratory pre-operative window-of-opportunity trial to investigate the pharmacokinetics

and pharmacodynamics of the smac mimetic Debio 1143 in patients with resectable squamous cell carcinoma of the head and neck. In: Proceedings of the annual meeting of the American Association for Cancer Research; 2019 Mar 29–Apr 3; Atlanta, GA. Philadelphia (PA): AACR; 2019. Abstract nr 5001.
[123] Mendez E, Rodriguez CP, Kao MC, Raju S, Diab A, Harbison RA, et al. A phase I clinical trial of AZD1775 in combination with neoadjuvant weekly docetaxel and cisplatin before definitive therapy in head and neck squamous cell carcinoma. Clin Cancer Res 2018;24(12):2740–8.
[124] Friedman J, Morisada M, Sun L, Moore EC, Padget M, Hodge JW, et al. Inhibition of WEE1 kinase and cell cycle checkpoint activation sensitizes head and neck cancers to natural killer cell therapies. J Immunother Cancer 2018;6(1):59.
[125] Patel P, Sun L, Robbins Y, Clavijo PE, Friedman J, Silvin C, et al. Enhancing direct cytotoxicity and response to immune checkpoint blockade following ionizing radiation with Wee1 kinase inhibition. Oncoimmunology 2019;8(11):e1638207.
[126] Kang SH, Keam B, Ahn YO, Park HR, Kim M, Kim TM, et al. Inhibition of MEK with trametinib enhances the efficacy of anti-PD-L1 inhibitor by regulating anti-tumor immunity in head and neck squamous cell carcinoma. Oncoimmunology 2019;8(1):e1515057.
[127] Alexandrov LB, Nik-Zainal S, Wedge DC, Aparicio SA, Behjati S, Biankin AV, et al. Signatures of mutational processes in human cancer. Nature 2013;500(7463):415–21.
[128] Hanna GJ, Lizotte P, Cavanaugh M, Kuo FC, Shivdasani P, Frieden A, et al. Frameshift events predict anti-PD-1/L1 response in head and neck cancer. JCI Insight 2018;3(4).
[129] Weng YM, Peng M, Hu MX, Yao Y, Song QB. Clinical and molecular characteristics associated with the efficacy of PD-1/PD-L1 inhibitors for solid tumors: a meta-analysis. Onco Targets Ther 2018;11:7529–42.
[130] Concha-Benavente F, Gillison ML, Blumenschein GJ, Harrington K, Fayette J, Colevas AD, Licitra L, et al. Characterization of potential predictive biomarkers of response to nivolumab in CheckMate 141 in patients with squamous cell carcinoma of the head and neck (SCCHN). J Clin Oncol 2017;35. https://doi.org/10.1200/JCO.2017.35.15_suppl.6050. suppl; abstr 6050.
[131] Prat A, Navarro A, Pare L, Reguart N, Galvan P, Pascual T, et al. Immune-related gene expression profiling after PD-1 blockade in non-small cell lung carcinoma, head and neck squamous cell carcinoma, and melanoma. Cancer Res 2017;77(13):3540–50.
[132] Ayers M, Lunceford J, Nebozhyn M, Murphy E, Loboda A, Kaufman DR, et al. IFN-gamma-related mRNA profile predicts clinical response to PD-1 blockade. J Clin Invest 2017;127(8):2930–40.
[133] Ladak K, Bass AR. Checkpoint inhibitor-associated autoimmunity. Best Pract Res Clin Rheumatol 2018; 32(6):781–802.
[134] Puzanov I, Diab A, Abdallah K, Bingham 3rd CO, Brogdon C, Dadu R, et al. Managing toxicities associated with immune checkpoint inhibitors: consensus recommendations from the Society for Immunotherapy of Cancer (SITC) Toxicity Management Working Group. J Immunother Cancer 2017;5(1):95.
[135] Burtness B. The role of cetuximab in the treatment of squamous cell cancer of the head and neck. Expert Opin Biol Ther 2005;5(8):1085–93.
[136] Teulings HE, Limpens J, Jansen SN, Zwinderman AH, Reitsma JB, Spuls PI, et al. Vitiligo-like depigmentation in patients with stage III-IV melanoma receiving immunotherapy and its association with survival: a systematic review and meta-analysis. J Clin Oncol 2015;33(7):773–81.

CHAPTER

9

Combination therapy and outcomes in head and neck cancer

Donovan Eu[a], *Ali Hosni*[b], *Anna Spreafico*[c], *Jonathan Irish*[a]

[a]Department of Otolaryngology-Head and Neck Surgery/Surgical Oncology, Princess Margaret Cancer Centre/University Health Network, University of Toronto, Toronto, ON, Canada [b]Department of Radiation Oncology, Princess Margaret Cancer Centre, University of Toronto, Toronto, ON, Canada [c]Division of Medical Oncology and Hematology, Princess Margaret Cancer Centre, University of Toronto, Toronto, ON, Canada

Abstract

Multimodality treatment has long been adopted in the management of head and neck cancers and is currently the standard of care in treatment of locally advanced cancers. The principle is that the synergistic properties of the different modalities complement each other, which translates to an improvement in treatment efficacy while minimizing toxicity. This multidisciplinary approach has dramatically altered the landscape of head and neck cancers. In a similar context, future developments in treatment are unlikely to be isolated to one particular modality or subspecialty but rather will continue to employ a wide spectrum of approaches to improve both survival and quality of life. This chapter will focus on the current developments of selected multimodal therapies, namely, the combination of established treatment modalities and immuno-oncology, adjuncts in surgical and radiation treatment, and window-of-opportunity trials and microbiome studies that may hold future prospects in the therapeutics of head and neck cancers.

Abbreviations

CBCT cone beam computed tomography
CTLA-4 cytotoxic T-lymphocyte–associated protein 4
EGFR epidermal growth factor receptor
PD-1 programmed cell death protein 1
PD-L1 programmed death-ligand 1
PDT photodynamic therapy

Conflict of interest

No potential conflicts of interest were disclosed.

https://doi.org/10.1016/B978-0-12-820679-9.00009-8

Introduction

Management of head and neck cancers is often intricate in view of the complex functional demands within the upper aerodigestive tract. Following the landmark study by the Veterans' Administration Laryngeal Cancer Study Group, the concept of combination therapy for organ preservation changed the paradigm of head and neck oncology [1]. Since then, combination therapies have been instrumental for the treatment of locally advanced head and neck cancers and have resulted in improved oncological outcomes [2]. As we look to the future, it should be recognized that current treatment outcomes and overall survival while improved still remain far from ideal. Understanding that head and neck cancers represent a complex entity of several subsites, each with its own prognostic nuances, it is difficult to compare outcomes among this broad group. A general overview however can be appreciated, with reported rates of only 40% local control over 5 years in patients with locally advanced nonhuman papilloma virus (HPV)–driven cancers [3]. Aside from treatment efficacy, there are also concerns over the functional outcomes and subsequent quality of life especially following completion of successful multimodality treatment. Thus it would only be expected that any new future advances should not only improve survival but also improve quality of life for patients with head and neck cancer.

The explosion in developmental technology and research has been rapid over the recent years. With techniques such as next-generation sequencing, the understanding of the molecular and immunological mechanisms responsible for head and neck cancers has expanded. Personalized medicine and precision oncology are now becoming practice. Recent trials such as the MATCH screening trials (NCT02465060)—a trial designed to stratify treatment plans based on genetic signatures of different cancers—are a prime example of the marriage of advanced technologies and precision therapy. As we uncover more of the basis of oncogenesis, our ability to treat cancers continues to evolve. Similar to the expanding banks of knowledge, the rate of development of novel combination therapies is exponential, and a comprehensive review of all these trials is beyond the scope of this chapter. As such the chapter aims to discuss selected combination therapies that have shown significant potential, which in the near future may be integrated into clinical practice. As we review the current host of novel treatments that are reported, it is important to address the challenges of describing the potential of future treatment paradigms, where many emerging therapeutic options either are in the early stages of clinical trials or are yet to be tested in human trials. Critical to any "change in practice" paradigm is that it must satisfy high levels of evidence prior to implementation. Thus the discussion here does not aim to validate these treatment modalities but rather to discuss the potential for future advances.

Immuno-oncology and radiation medicine

In the recent decade, immunotherapy has been at the forefront of oncological research. Landmark studies such as CheckMate 141, KEYNOTE-012, and KEYNOTE-048 have demonstrated a definite role for use of immunotherapy in treatment of recurrent or metastatic head and neck cancers and have brought with it a new dimension in treatment options [4–6].

The mechanisms of immunotherapy and immune escape have been discussed in prior chapters and thus will not be covered in detail here. Suffice to say, profound developments in the understanding of the tumor microenvironment have since been achieved, and the local tumor immune environment has emerged as a critical target for modulation of cancers within the head and neck.

This notion of harnessing the immune system for cancer treatment is however not new. Radiation and its immune-modulating effects have been suggested since 1979 and more recently have been attributed to the underlying effects of antitumor T-cell activation following radiation treatment [7, 8]. The synergistic effects of immunotherapy agents and radiation have also been noted in the clinical setting. The secondary analysis of KEYNOTE-001 has implied improved outcomes in patients with lung cancer who had undergone both radiation and immunotherapy when compared with single modality treatment with immunotherapy only [9]. Concurrent treatment with immunotherapy and radiation in particular has also further shown improved control in lung cancer and melanoma patients with brain metastasis, depicting longer local control rates and overall survival in this group when compared with sequential treatment approaches with the same agents [10]. These studies have all reiterated the strength of concurrent treatment over monotherapy or sequential multimodal treatment to improve overall treatment efficacy. This is thought to arise from direct synergistic effects of immune activation when both modalities are employed together.

Paradoxically, radiation is also a potent immune suppressor. This is most evidently seen in the immunosuppressive effects in patients undergoing whole body radiation for the treatment of leukemia. For the optimization of immune activation, a delicate balance must be achieved to avoid a counterintuitive response to immunotherapy agents. Marciscano et al. demonstrated the attenuation of immune activation following elective nodal irradiation in a preclinical mouse model undergoing stereotactic radiation therapy and immunotherapy, which cautions against the suppressive effects of radiation treatment following combination with immunotherapy [11]. In the clinical setting, lymphopenia is a recognized effect that has been reported following radiotherapy for treatment of nasopharyngeal cancers. In their study, Cho et al. reported an overall poorer 5-year disease-specific survival in patients who developed lymphopenia with a minimum absolute lymphocyte count of less than 245 [12]. It is apparent that the immune effects of radiation are still not fully understood, and more studies will be required to comprehend the complex effects of radiation on the host immune system to optimize the treatment response in combined treatments involving radiation and immunotherapy.

The systemic effects of the immune activation by combination of immunotherapy and radiotherapy are also of interest. The abscopal effect, which represents tumoricidal effects at sites distal to the irradiated volume of radiation, was first coined by Mole in 1953 [13]. These "off-target effects" of radiation therapy were later determined to be secondary to systemic immune activation via alteration of the tumor microenvironment with activation of cytotoxic T-cell function [14, 15].

In the clinical setting the abscopal effects have been reported in several case reports. In a widely cited report, Postow et al. documented a notable response to metastatic deposits in the hilar lymph nodes and spleen following palliative radiation of a metastatic paraspinal mass in a patient with metastatic melanoma who was treated with ipilimumab (CTLA-4 inhibitor) and radiation [16]. Clearly the ability to consistently replicate these effects is desired especially in patients with advanced and metastatic head and neck cancers. Yet, while the knowledge of immune-mediated effects is well established, trials attempting to employ these effects have been less promising [17]. Preclinical studies have shown that multiple variables including the fractionation and dose of radiation play significant importance to achieving a meaningful immune-stimulatory response [18]. In addition to radiation parameters, other methods such as intratumoral injection of oncolytic viruses have also been studied to improve the abscopal effects [19]. To date, however, there is yet a reliably replicable utility of the abscopal effect to allow for adoption to clinical practice. The future does look promising as more trials and research seek to optimize the variables to achieve a reliable off-target immune response with radiation treatment.

Aside from its synergistic effects, the allure of immunotherapy also stems from the relatively low adverse events that are generally encountered. Compared with concurrent chemoradiation, which is fraught with toxicities such as severe mucositis and long-term functional impairment necessitating gastrostomy tube feeding, immunotherapy has been proposed as an alternative treatment modality that may yield similar survival outcomes with possibly an improved quality of life. It is thus not surprising that there is an impetus for the adoption of immunotherapy in combination treatments of selected subsites to minimize adverse events that are more prevalent with conventional chemotherapy agents.

In the context of primary treatment, there is to date no high-level evidence for treatment efficacy of concurrent immunotherapy and radiotherapy for head and neck squamous cell carcinomas. Several studies are currently ongoing to further evaluate the efficacy of these trials. Early reports from the GORTEC 2015-01 (PembroRad trial, NCT02707588) have shown immunotherapy to be tolerable when compared with cetuximab in groups of patients with locally advanced head and neck cancers who are unfit for platinum-based chemoradiation. Final analysis of this trial and others like it, such as the NCT03258554, will shed light on the benefits of addition of immunotherapy to radiation treatment for a subset of patients who are unable to tolerate standard chemotherapy.

The addition of immunotherapy to chemoradiation for the purpose of treatment intensification for high-risk patients is also currently being evaluated. Synergistic effects of trimodality treatment with immunotherapy, chemotherapy, and radiation have yielded favorable results in other oncological practices such as nonsmall cell lung carcinomas. The recent published PACFIC trial demonstrated significantly prolonged progression-free survival with addition of durvalumab (anti-PD-L1 antibody) following chemoradiation. Within the head and neck, studies are ongoing, and early results have demonstrated the safety and feasibility of this approach [20, 21]. Treatment efficacy of the addition of immunotherapy to chemoradiation for curative intent in patients with high-risk locally advanced head and neck cancers is currently being evaluated. Studies such as KEYNOTE-412 (NCT03040999) and Javelin head and neck 100 (NCT02952586) are randomized phase III studies that incorporate immunotherapy to concurrent chemoradiation and may change the treatment approach for these high-risk patients.

Conversely, comparison of immunotherapy and radiation with standard of care chemoradiation is ongoing as well. The KEYCHAIN trial (NCT03383094) and CCTG HN9 (NCT03410615) are studies designed to determine the efficacy of immunotherapy agents and radiation compared with chemoradiation in patients with HPV-related head and neck cancers. Other studies involving deescalation protocols in patients with HPV-related head and neck cancers are also ongoing and will be discussed in a separate chapter. Overall, several studies are ongoing to determine if immunotherapy is a viable alternative to conventional treatment and if this adoption translates to improved survival and functional outcomes.

In the field of metastatic cancers, concurrent immunotherapy and radiation also show promise. While not limited to head and neck cancers, some early phase I studies and case reports have reported favorable results following combined radiation and immunotherapy in the palliative management of patients with distal metastasis [22, 23]. Certainly, these reports are limited to small numbers, and more work will be required to determine the clinical benefits of such a combined treatment in the palliative setting. Similarly, the utility of immunotherapy in combination with radiation for recurrent disease is also currently undergoing review with clinical trials studying the effects of intensity-modulated radiotherapy and nivolumab (anti-PD1 agent) with recurrent/secondary head and neck cancers (NCT03521570).

Pending these results and future trials, it is not unforeseen that combined immunotherapy and radiation may take on a more significant role in treatment of both primary and recurrent/metastatic head and neck cancers. While combination therapy brings about exciting prospects for head and neck oncology, one should be cautious about the possible complications that may be associated with it. Some complications, such as higher incidence of symptomatic radiation necrosis during combined immunotherapy and stereotactic radiation of brain metastasis, have already been reported [24]. Formal clinical trials will be required to allow for judicious recommendations in clinical adoption.

Immuno-oncology and chemotherapy

Despite the advances in management of head and neck cancers, recurrent and metastatic tumors of the upper aerodigestive tracts have consistently poised a challenge, and prognosis remains dismal when treated with systemic monotherapy. Preferably, recurrent tumors should be identified early and surgical ablation offered at an opportune moment. However, multiple factors such as tumor extent, existing functional compromise, or patient's fitness may render these patients prohibitive to surgery. Conventional approaches for management of recurrent/metastatic patients often result in poor outcomes—with a median overall survival of approximately 10–11 months [25, 26]. Following the recent reports of KEYNOTE-048, the use of immunotherapy as a single agent or as an addition to current regimes of doublet chemotherapy (EXTREME regime) showed promising results, extending the overall survival from 11.0 to 14.7 months in selected patients [27]. These findings have propelled pembrolizumab as an FDA-approved first-line standard of care agent in treatment of recurrent/metastatic cancers within the head and neck. The pace at which immunotherapy has been adopted into clinical practice demonstrates its potential in changing the paradigm of head and neck practice, and more trials are underway to expand its application in the presurgical arena.

Immuno-oncology and surgery

Attempts to incorporate immunotherapy in the setting of a neoadjuvant prior to surgery have shown promise in early phase I/II trials. A significant number of trials aimed to evaluate the efficacy of immunotherapy prior to surgery are designed as window-of-opportunity trials and will be presented in the latter part of this chapter. Formal neoadjuvant trials such as NCT03021993 differ from these window trials to allow for an intentional regimented treatment protocol, often delaying surgery to evaluate possible complete pathological response. Clearly, this is not feasible in window trials where time to surgery remains a priority. In comparison, it is probable that formal neoadjuvant trials allow for a more comprehensive understanding of the effects of immunotherapy in the neoadjuvant setting. Concerns about the effects of immunotherapy in the operative and postoperative phases have also been raised; however, it is the general consensus that surgery remains feasible and not overtly associated with increased morbidity and complications in these settings [28].

Immunotherapy as an adjuvant following surgical ablation is also currently being explored. The use of atezolizumab (anti-PD-L1 antibody) is currently being evaluated in a double-blind phase III trial following definitive treatment of locally advanced head and neck cancers via the IMvoke010 trial (NCT03452137). The PATHWay study (NCT02841748) is a similar double-blinded phase II trial that compares progression-free survival following administration of adjuvant pembrolizumab against a placebo. Addition of immunotherapy to current standard of adjuvant platinum-based chemoradiation in high-risk patients (positive margins and extranodal extension) is also currently being evaluated by several ongoing trials such as KEYNOTE-689 (NCT03765918), NCT02775812, and NCT02777385. KEYNOTE-689, in particular, incorporates immunotherapy agents in the neoadjuvant and adjuvant setting and is an ongoing phase III trial that was designed with its primary outcome measure to evaluate pathological response in addition to event-free survivals compared with the current standard of care [29]. Results from these trials may possibly lead to changes in practice to allow for appropriate exploitation of immunotherapy in the perioperative setting.

Conventional treatment of head and neck cancers is anchored on three pillars: surgery, radiation, and systemic therapy. Immunotherapy has certainly established itself as a formidable treatment option in systemic therapy approaches, and its further combination to our established modalities has so far shown great promise. Certainly, more high-level evidence in the form of randomized control trials will be required to further evaluate the treatment efficacy, functional outcomes, and toxicities that may arise prior to broad clinical implementation.

Surgery and adjunct treatment

The field of surgery has also undergone several developments over the past decade. Transoral robotic surgery has dramatically changed the options available for management of oropharyngeal cancers [30, 31]. The gaining practice of superselective neck dissections and image-guided surgery for head and neck cancer has all further elevated surgical practices [32–34]. The fundamental principles behind these advances are centered on the ability to deliver precise surgery with complete extirpation of the tumor while minimizing collateral damage to the surroundings, thereby maximizing both clinical and functional outcomes.

Image-guided surgery in head and neck cancer

Combined applications of diagnostic imaging and surgery have been long and successful particularly in areas such as endoscopic sinus and skull base surgeries. Image-guided surgery for head and neck cancers has however been infrequently adopted, due to issues of inaccuracies following manipulation of soft tissue and lack of intraoperative image acquisition to guide real-time surgical planning. To mitigate these limitations, advances in intraoperative imaging have allowed for improved on-demand imaging within the operating room. The use of the cone beam CT (CBCT), in particular, allows for high-quality imaging of the surgical field to allow for improved accuracy in intraoperative planning, thereby translating to greater precision during surgical ablation [35, 36]. With the addition of surgical navigation devices, procedures such as mandibular osteotomies can further be performed with increased accuracy [37]. While still a developing field, the use of intraoperative navigation allows for an added guide, enabling greater confidence during surgical ablation.

Current innovations in surgical navigation and in optical imaging include the use of fluorescent-guided surgery that allows for direct tumor delineation of the surgical site, improving visualization of tumor margins to allow for complete extirpation. This has been achieved by several methods, including dynamic optical contrast imaging (DOCI) [38], the evaluation of endogenous fluorophores following excitation by ultraviolet waves, and use of intravenous fluorescence agents such as indocyanine green and panitumumab-IRDye800CW [39, 40]. From these proof-of-principle studies, monoclonal antibody fluorescent-guided surgery has the potential to impact intraoperative decisions.

Pushing the boundaries further, the combination of both fluorescent-guided surgery and photodynamic therapy (PDT) has more recently been explored. Photodynamic therapy has been in use for the management of head and neck cancers for a protracted period of time, and results have shown promise in both primary and intraoperative "adjuvant" treatment [41, 42]. The lack of specificity and often protracted phototoxic effects has, however, led to its aversion in clinical practice. In attempts to overcome the disadvantages as described earlier, the specificity of photosensitizers has been enhanced by targeted cell therapy via surface receptors such as epidermal growth factor receptor (EGFR) and incorporation of nanomedical technologies [43, 44].

Nanoparticles, defined as microscopic particle matter usually less than 100nm, are fast becoming available for clinical care. Current treatment in oncology has focused on the use of nanoparticles for purpose of drug delivery and diagnostic imaging [45, 46]. Novel combination of an agent with both fluorescent properties and photodynamic capabilities was recently described, following formulation of a non-PEGlayted porphyrin-derived nanoparticle with imaging, fluorescent, and photodynamic properties [47]. This advancement in nanomedicine forges a single theranostic particle that exhibits multimodal and multifunctional properties for both therapeutic and diagnostic application. In view of its ultrasmall particulate size (~30nm), these nanoparticles exhibit rapid accumulation within the tumors secondary to the enhanced permeability and retention (EPR) effect commonly displayed in cancers due to neovascularization and angiogenesis. Biological tumor delineation for surgery can then be readily achieved using fluorescent imaging systems that are commercially available, and photodynamic ablation may be sequentially initiated after surgical resection by means of laser activation.

Preclinical studies using buccal carcinomas in the rabbit model have shown tremendous potential in the setting of oral cavity cancers [48, 49]. Preliminary results demonstrate porphysome concentration in the tumor volume, allowing for visualization of tumor extent that is effective for fluorescent-guided surgery (Fig. 1). Following resection, employment of a laser over the surgical bed allows for adjuvant PDT treatment in the operating theater. Pathology correlation confirms destruction of tumor residue by means of apoptosis with preservation of surrounding mucosa. Furthermore, no adverse events or photosensitivity was observed despite there being no light precaution protocols. Formal evaluation of the surrounding oral cavity tissue and systemic biodistribution likewise showed no significant uptake or damage aside from the residual tumor bed. In addition to its local effects, notable lymph node regression was also visualized following PDT administration to the primary tumor in some rabbits when compared with control studies. The probability of a porphysome-derived abscopal effect is currently being evaluated, which may allow for systemic effects of PDT in the near future.

While these data are currently in the preclinical phase, it brings about a new combined treatment modality that can be utilized for oral cavity cancers. Hypothetically, its use is particularly advantageous in destruction of possible microscopic residual disease in the surgical bed following resection, which may then obviate the need for adjuvant radiation and its

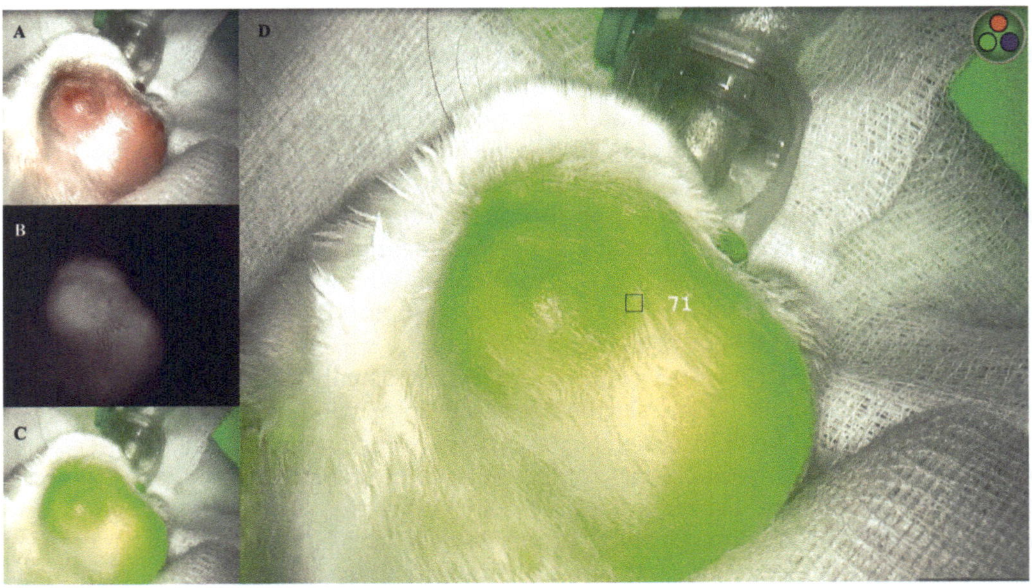

FIG. 1 (A) Lateral view of orthotopic buccal tumor in a rabbit VX-2 model. The fur overlying the right cheek has been shaved for visualization of the tumor. (B) Fluorescent imaging system delineating tumor volume and signal-to-noise ratio. (C and D) Fluorescent imaging overlay on surgical field to allow for real-time surgical navigation.

associated morbidity following surgery. Furthermore, as this can be performed at the same setting of surgery, the overall duration of treatment and recovery time would be shorter. Studies are also ongoing to explore nanoparticle imaging and ablation as minimally invasive treatment for papillary thyroid cancers and to similarly determine if there is a possible role for the ablation of anaplastic thyroid cancers. Certainly, greater refinement will be required prior to its implementation for clinical trials. Nonetheless the ability to incorporate primary surgery and a highly selective "adjuvant" treatment in a single setting is promising especially for patients with locally advanced head and neck cancers.

Image-guided radiation therapy

Similar to surgical ablation the precision of radiation has also been evolving over the years. With the combination of advances in diagnostic imaging, radiation treatment plans can be modified to allow for improved targeting and dosimetry for treatment of head and neck cancers. Among the many advances, the use of molecular image-guided radiation treatment has been suggested to bring about a new paradigm to the manner in which radiation treatment is planned. Some studies have shown that the use of functional imaging such as PET-CT allows for greater accuracy in determining gross tumor volume following comparison with the pathological specimen, while MRI and CT tended to overcall tumor volumes, subjecting them to unnecessary radiation [50]. The use of PET-CT further aids to distinguish tumor from surrounding inflammation, thereby limiting unnecessarily wide radiation volumes. Increasingly the notion of PET-CT as a prognostic tool for treatment failure following radiation has also been proposed [51, 52]. The ability to reliably predict sites of failure using PET-CT data potentiates it as an integral part of radiation planning.

Approximately 20%–30% of patients tend to recur within the radiated volume [53–55]. With the advances in radiation planning (via intensity-modulated radiation therapy (IMRT) or volumetric modulated arc therapy (VMAT)) and radiation delivery through image-guided radiation (e.g., MR-Linac), precise delivery of higher doses of radiation aims to achieve better local control without increasing toxicity to surrounding organs at risk. The delivery of higher doses of radiation however should be implemented in a selective manner to avoid unnecessary side effects on the surrounding tissue.

Functional molecular imaging can be used as a tool to identify areas within the tumor that may be at high risk for radiation resistance in view of tumor proliferation, vascularity, and tumor hypoxia, which are generally associated with aggressive/radio-resistant tumors [56, 57]. Multiple modalities have been utilized for the purpose of defining imaging biomarkers that may accurately predict tumor biology. While ^{18}fluorodeoxyglucose PET-CT is currently most commonly utilized in clinical practice, functional imaging is certainly not limited to this platform. Several other modalities such as PET-MRI and functional MRI incorporating techniques such as apparent diffusion coefficient, diffusion-weighted imaging, and dynamic contrast-enhanced images have all been proposed as tools for prognostication of tumor response. In addition to these techniques, novel tracers such as ^{18}F-fluorothymidine and ^{18}F-fluoromisonidazole as indicators of tumor proliferation and hypoxia have also been studied as functional imaging tools to further stratify patients at risk of radio-resistant tumors.

The continual development and understanding of functional imaging may in the future allow for judicious selection of patients who are likely to benefit from higher radiation doses.

To date, several phase I trials have demonstrated the feasibility of delivering higher radiation doses (up to 80–90 Gy) via dose painting [58, 59]. Early studies by Liu et al. have demonstrated some promising results with 99% complete response rates in locally advanced nasopharyngeal carcinoma patients treated with PET/CT-guided dose painting as opposed to 92.9% response in those treated with conventional CT-based IMRT. Long-term results in this study also demonstrated greater locoregional control, disease-free survival, and overall survival in patients treated with PET/CT-guided dose painting.[60] While the tolerability of increased radiation dosage has been established and some early studies show promise, there is still yet a consensus as to the methodology of dose painting using the data from functional imaging nor is there sufficient high degree of evidence to suggest that dose painting allows for a superior outcome compared with conventional IMRT.

Furthermore, regardless of the advanced molecular imaging capabilities during the planning phase, limitations in treatment delivery are expected due to the lack of "near real-time" target acquisition necessary for precision radiotherapy. Dynamic changes to the position of both tumor and the surrounding structures/organs at risk are not unexpected during the course of radiation treatment. Just by clinical observation, it is not uncommon to see some patients demonstrate remarkable shrinkage in the midst of their planned radiotherapy. Following serial thrice weekly imaging during the course of radiation, Barker et al. demonstrated a progressive decrease of gross tumor volumes at a median rate of 0.2 cm^3 per treatment day, representing a 1.8% decrease in volume [61]. At the final day of treatment, the overall proportion of gross tumor volume was reported to be 69.5% of the initial volume on planning. During this period the contour and shape of the tumor also underwent significant changes. In addition to tumor shrinkage, displacement of the surrounding structures such as the parotid gland was also evident during the course of radiation. These findings would suggest that interval review and possible refinement of the radiation plans may be required to achieve greater precision especially in patients who undergo high-dose radiation.

Adaptive radiotherapy utilizes intratreatment imaging for replanning to correct for the dynamic changes during the course of treatment. This recalibration aims to achieve greater precision with the expectations of improving oncological outcomes and decrease adverse events, such as xerostomia and swallowing dysfunction. In a review by Castelli et al., most studies reported a dose reduction to the parotid glands ranging from 0.6 to 4.1 Gy [62]. In the same review, 2-year locoregional control rates were documented between 88% and 97% compared with historical data of 80% in patients not undergoing adaptive radiotherapy. Improved locoregional control (88% vs 79%) was also achieved by Chen et al. in their study comparing adaptive radiotherapy with conventional IMRT plans despite higher T and N categories in their adaptive radiotherapy group [63]. Using PET/CT-adapted radiation in a phase II clinical trial by Kong et al., the authors were able to remarkably improve locoregional control rates of patients with locally advanced nonsmall cell lung cancers from historical control rates of 34% to 62% [64]. Within the head and neck, there is still no consensus as to the role of adaptive radiotherapy. Favorable studies such as that of Yang et al. show benefits in quality-of-life scores and 2-year locoregional control to patients undergoing adaptive radiation for treatment of nasopharyngeal carcinoma [65]. Other studies however failed to demonstrate significant benefit in dose sparing to the organs at risk when compared with conventional pretreatment planning [66]. While the rationale of intratreatment surveillance and replanning is logical, more evidence is required prior to its implementation into clinical practice. For the purpose of adaptive radiation, molecular/functional imaging is thought to be advantageous in view of its association with tumor viability that can be inferred as a measure of treatment response, thereby prognosticating final outcomes at the end of treatment. However, limitations in its clinical translation arise from lack of uniformity in signal segmentation, high background signal interference during radiation, false-positive avidity due to tissue inflammation during radiation, and significant resources required for repeated imaging and subsequent replanning. All these issues will need to be addressed prior to its initiation into standard practice.

Dose painting and adaptive planning are complementary, and incorporating both techniques is expected to improve outcomes in treatment. Phase I trials utilizing adaptive ^{18}FDG-PET-based dose painting by numbers showed feasibility in their cohort of patients [58, 67]. Using PET-CT images for adaptive treatment, dose titration midway through treatment and further sparing of the organs at risk such as the parotid gland was possible, and no patients required breaks in between their treatment. While feasibility is established, there is still a considerable amount of research to be done prior to incorporation into standard practice. A review of the current literature shows significant heterogeneity in the selection of patients, differences in interval of imaging, standardized uptake value (SUV) cutoffs, interpretation of PET-CT findings, and consistency of PET-CT. Certainly, more robust data are required to guide the implementation of adaptive dose painting radiation protocols into clinical practice. This is hopefully achieved by randomized clinical trials such as the ongoing NCT01341535, a two-arm randomized study comparing adaptive biological imaging-based radiotherapy against standard radiotherapy.

Radiomics is another exciting field of radiation oncology that utilizes artificial intelligence to further extract features such as texture, shape, wavelet, tumor intensity, shape, and Laplace-Gaussian features from a large collection of images in the aim to identify potential radiomic biomarkers that may allow for prognostication and risk stratification of patients [68, 69]. These radiomic biomarkers may allow for greater understanding of failure patterns and outcome variations between groups of patients with similar tumors. In their analysis of MRI-based radiomics for 118 patients with nasopharyngeal carcinoma, Zhang et al. reported a significant relationship with radiomic features and progression-free survival [70]. Several publications have similarly demonstrated the utility of radiomics analysis for prognostication capabilities in patients with other subsites of the head and neck [71, 72]. The ability to harness this body of data may lead to significant changes in practice that may allow for a targeted prescription of individual treatment plans, thereby improving the precision of oncological treatment. Radiomics has certainly been touted as a potential field of exponential advancement in imaging science, and the potential for translation to clinical decision-making is enormous. However, the uniformity of image acquisition and reproducibility across different centers and populations will require further validation before its implementation to clinical practice.

Window-of-opportunity trials

Window-of-opportunity trials have quickly caught on in head and neck oncology as a platform for the study of novel agents in between planned surgery. Unlike formal neoadjuvant studies, window-of-opportunity trials are often shorter in duration and do not delay scheduled treatment. Benefits of window-of-opportunity trials allow for evaluation of the agents' pharmacokinetics and responses in target modulation, which is not feasible in preclinical studies. These trials are usually shorter with its initial aims to identify biomarkers that may predict for tumor response to allow better selection of patients. More recently, numerous window trials have been performed with almost a neoadjuvant purpose, with the introduction of novel agents in the aim of attaining clinical or pathological effects prior to surgery. Some early results from these window trials have reported excellent responses that have warranted further formal investigations.

Window-of-opportunity trials—Immunotherapy

Window trials have been critical to determine biomarkers and predictors of treatment response in patients with head and neck cancers following immunotherapy. In their study, Colevas et al. employed a novel molecular imaging by means of PET tracer [18F]F-AraG that showed potential for early predictor of immunological response following administration of anti-PD-1 immunotherapy in patients planned for surgical resection [73]. These findings are clinically relevant in dichotomizing patient disease response to immunotherapy, in view of the relatively low (~20%) response rates in head and neck cancers [74]. Early identification of patients with good response may allow for further deescalation of treatment and vice versa for patients who exhibit a poor response.

Employed as a neoadjuvant-like setting in window-of-opportunity trials, immunotherapy agents have shown dramatic responses in some reports. CheckMate 358 (NCT02488759) is an ongoing phase II study employing neoadjuvant nivolumab in the setting of resectable HPV-positive and HPV-negative head and neck carcinomas [75]. Preliminary results presented thus far show promise with approximately 48% of patients with documented response on imaging following two doses of nivolumab prior to surgery. A further 14% of this cohort reported a greater than 40% tumor reduction on imaging. In another ongoing trial (NCT02296684), combinations of neoadjuvant and adjuvant immunotherapy with surgery for locally advanced head and neck cancers have also been explored, with preliminary reports demonstrating up to 43% of patients attaining pathological response [76]. Furthermore, early reports from this trial seem to suggest excellent controls pertaining to locoregional recurrence and distal metastasis after 1 year. Preliminary results from these small cohorts demonstrated significant tumor responses to neoadjuvant immunotherapy. Another window-of-opportunity trial using a combination of biologic and immunotherapy neoadjuvant agents is studying the effects of sitravatinib and nivolumab in the Oral Cavity Cancer Window Opportunity Study (SNOW) (NCT03575598). The SNOW trial will evaluate the addition of multiple agents for immune and tumor effects, as well as reliable markers of signal activity that may predict for treatment response. With such remarkable effects on the tumor volumes prior to surgery as described earlier, it is conceivable that this may in the future translate to limited surgical ablation in an effort to maximize postoperative function. This however will need to be supported by long-term results in well-defined clinical trials.

Window-of-opportunity trials—Biologics

Epidermal growth factor receptor inhibitors (EGFR) have commonly been employed in the setting of head and neck cancers. The popularity of EGFR inhibitors is not surprising, in view of the high incidence of EGFR expression in head and neck cancers. In his landmark study, Bonner et al. demonstrated the benefits in survival and locoregional control following combination therapy with cetuximab (EGFR inhibitor) and radiation when compared with radiation alone [77]. Although this trial showed much promise, it has been suggested that response rates to cetuximab in head and neck cancers are limited and subsequent development of resistance further diminishes sustained response; thus meticulous selection is required to optimize treatment response for these patients [78, 79].

Several window trials have attempted to use EGFR inhibitors in the preoperative setting. Some of these trials have reported appreciable objective responses that can be measured with the use of imaging, such as dynamic contrast enhancement (DCE), positron emission tomography (PET), or plain computed tomography (CT). Ferris et al. successfully demonstrated a $\geq$10% reduction in tumor diameter on CT imaging in 33% of patients with locally advanced head and neck cancers following neoadjuvant cetuximab [80]. Of note an increase in EGFR-specific T cell following neoadjuvant cetuximab was reported and has been postulated as a key biomarker for cetuximab response. In another study utilizing DCE-CT and PET-CT, Schmitz et al. reported partial metabolic response on PET-CT in 90% of patients. A greater than 50% decreased cellularity in 71% of their patients treated with neoadjuvant cetuximab and increased peritumoral space on DCE-CT was also noted [81].

Aside from cetuximab, other targeted agents such as trametinib (MEK inhibitor) have also been subjected to window trials. Uppaluri et al. following administration of trametinib in a window trial managed to attain clinical tumor response (defined as measurements in two dimensions) in 65% of patients, partial metabolic response in 46%, and clinical to pathologic tumor downstage in 53% [82]. Afatinib (an oral ErbB inhibitor) has also been evaluated in the EORTC-90111 study, a multinational window trial, and has shown partial metabolic response on FDG PET in 69.5% of the patients [83]. Similar to these trials, several ongoing studies are currently evaluating targeted biologics in the settings of window trials.

Overall, window trials allow for an excellent platform for evaluation of novel agents on a treatment-naive patient awaiting surgery or other treatment. By virtue of the manner in which these trials are preformed, it naturally allows for assessment of these agents in the neoadjuvant setting. While biomarkers are essential for prediction of tumor response, objective measurements of tumor size and volumes on imaging allow for easy appreciation of the effects of these agents on the primary tumor. There is no consensus on the appropriate surgical plan following neoadjuvant therapy, namely, if the resection margins should be based upon the tumor volumes before or after administration of neoadjuvant therapy. What can be appreciated is that the decrease of a bulky tumor makes surgical resection technically easier and may potentially allow for sparing of adjacent healthy tissue translating to better functional outcomes. Metabolic responses and pathological studies all further reinforce the potential effects of these agents on the tumor and may further predict the biological responses in the adjuvant setting. Its direct corelation to patient outcomes is difficult to ascertain from window trials. These findings however allow for continual monitoring of treatment efficacy and further aids prognostication and selection of appropriate agents, thereby improving precision in treatment.

While the advantages in window trials are clear, it is prudent to discuss the limitations of these trials. Certainly the risk of adverse events that may prolong planned surgery is a serious problem that may be encountered. To mitigate this the safety profiles of these agents should ideally be well established prior to the design of a window trial. In the context of window trials involving immunotherapy, the effects of pseudoprogression that is commonly seen following immune activation may obscure its true biological behavior. This is in part due to the short treatment duration when compared with formal neoadjuvant trials. Furthermore, there are still inconsistencies in evaluation of pathological responses that may make interpretation of immunotherapy effects challenging. With regard to surgical ablation, the effects of some neoadjuvant treatment may lead to dramatic reduction in tumor volumes that may make planned surgery difficult due to the lack of tumor delineation or in some cases, the absence of obvious remnant tumor. Detailed documentation and planning are thus required prior to initiation of neoadjuvant treatment if planned surgery encompassing the original tumor extent is to be achieved. As window trials in general are performed in small numbers; they are generally not randomized, and thus results should be interpreted cautiously. Finally, as these trials are not designed for clear evaluation of benefits in terms of survival and disease control, reported benefits even in the context of objective responses do not necessarily equate to improved clinical outcomes. Despite this, window trials allow for a rapid evaluation of novel combination treatments in the surgical setting and act as almost a "pilot" study prior to formal clinical trials. As such, they are extremely useful for effective allocation of study resources to achieve maximal gains.

Regardless of the reported objective responses, window-of-opportunity trials are designed with the primary focus of evaluation for biomarkers of response. Several biomarkers such as PD-L1 expression, tumor immune-cell infiltration, circulating immune cells and patient microbiota are currently being evaluated within the parameters of these trials. Comprehension of these biomarkers will allow for further stratification of patients that are most likely to benefit from specific treatment agents, thereby improving oncological precision.

Microbiome in head and neck cancer

The upper aerodigestive tract is unique due to its persistent exposure to the external environment. The oral flora is quickly gaining attention in research due to enhanced tools that have been available such as 16S rRNA gene sequencing. These techniques allow for greater sensitivity in detection and establishment of the normal oral cavity flora [84–86]. Its use also allows for a deeper understanding of the interplay between the oral cavity microorganisms and pathological processes. As mentioned earlier, current literature has proposed the use of microbiomes as a potential biomarker for response to immunotherapy [87]. However, currently, there are no reliable clinical data to suggest a correlation between microbiome and immunotherapy response in head and neck cancers. The understanding of the oral microbiota and its potential implications is valuable for further design of novel treatment especially for oral cavity cancers. It is foreseeable that the manipulation of oral flora may allow for clinical impact in the fields of prevention, treatment modulation, and prognostication of treatment effect in patients with cancers of the oral cavity.

From earlier studies, the bacterial composition of the oral cavity in patients with oral cancer has been established to be significantly different from healthy individuals [88, 89]. More recently, certain bacteria such as *Porphyromonas gingivalis* and *Fusobacterium nucleatum* have been thought to be potential drivers of carcinogenesis [90–92]. Identification of probable oncogenic bacteria may potentially allow for a noninvasive approach to screening for oral cavity cancers. Furthermore, early detection and possible eradication of these bacteria may allow for a preventive strategy in the care of oral cavity cancers.

In the setting of combined therapies, microbiomes have been shown to influence the outcomes of treatment efficacy following immunotherapy. Preclinical studies have established the direct correlation between gut microbiomes and immunotherapy response. Furthermore, it has been shown that manipulation of the gut microbiome via fecal microbiome transplantation in the mice model allowed for successful modulation of the immune response [93]. In the clinical setting, ROMA LA-OPSCC has demonstrated the variations in microbiota of the oropharynx between patients of different stages and following chemoradiation, which allows for its plausible evaluation as a biomarker for immunotherapy [94]. The understanding of the patient microbiota also allows for attempts to manipulate the native flora as described in the aforementioned preclinical study. As such, studies such as ROMA LA-OPSCC2 (NCT03838601) attempt to similarly evaluate the feasibility of altering the microbiome via administration of live cultures of intestinal bacteria from healthy donors in patients with locally advanced oropharyngeal squamous cell carcinoma undergoing chemoradiation. Early studies such as these show potential for a separate pathway of intervention that may lead to improved response when combined with other modalities. This may improve the overall rates of disease control and survival, and may further decrease morbidity via deescalation of existing treatments.

Conclusion

Advances in treatment approaches to the head and neck are likely to expand from the existing combination therapies that are currently employed. From preliminary reports in the current literature, there is a great potential for further expansion of current findings to improve patient's outcomes in both survivorship and function. The versatility of immunotherapy as a combination agent with surgery, chemotherapy, and radiation and in both neoadjuvant and adjuvant settings has been associated with generally favorable responses in several cancer types and is under investigation in head and neck cancers. Advances in surgical and radiation precision are also likely to establish new treatment paradigms for oncologically sound ablation with minimal injury to critical structures such as the swallowing apparatus. Window trials have rapidly unveiled a plethora of novel treatment approaches that have in some cases shown remarkable results and should in time be followed up with formal trials to validate findings. Finally the comprehension of predictive biomarker of responsiveness, including oral and intestinal microbiomes, and its impact on immunotherapy allow for potential manipulation to improve immune response and thereby treatment outcomes. On the horizon, these exciting approaches all show significant promise but will need to undergo rigorous corroboration by clinical trials to truly establish its benefits in patients with head and neck cancers before clinical implementation.

References

[1] Department of Veterans Affairs Laryngeal Cancer Study Group, Wolf GT, Fisher SG, Hong WK, Hillman R, Spaulding M, Laramore GE, Endicott JW, McClatchey K, Henderson WG. Induction chemotherapy plus radiation compared with surgery plus radiation in patients with advanced laryngeal cancer. N Engl J Med 1991;324:1685–90.

[2] Pignon J-P, le Maître A, Maillard E, Bourhis J, MACH-NC Collaborative Group. Meta-analysis of chemotherapy in head and neck cancer (MACH-NC): an update on 93 randomised trials and 17,346 patients. Radiother Oncol 2009;92:4–14.

[3] Lo Nigro C, Denaro N, Merlotti A, Merlano M. Head and neck cancer: improving outcomes with a multidisciplinary approach. Cancer Manag Res 2017;9:363–71.

[4] Ferris RL, Blumenschein G, Fayette J, Guigay J, Colevas AD, Licitra L, Harrington K, Kasper S, Vokes EE, Even C, Worden F, Saba NF, Iglesias Docampo LC, Haddad R, Rordorf T, Kiyota N, Tahara M, Monga M, Lynch M, Geese WJ, Kopit J, Shaw JW, Gillison ML. Nivolumab for recurrent squamous-cell carcinoma of the head and neck. N Engl J Med 2016;375:1856–67.

[5] Seiwert TY, Burtness B, Mehra R, Weiss J, Berger R, Eder JP, Heath K, McClanahan T, Lunceford J, Gause C, Cheng JD, Chow LQ. Safety and clinical activity of pembrolizumab for treatment of recurrent or metastatic squamous cell carcinoma of the head and neck (KEYNOTE-012): an open-label, multicentre, phase 1b trial. Lancet Oncol 2016;17:956–65.

[6] Burtness B, Harrington KJ, Greil R, Soulières D, Tahara M, De Castro G, Psyrri A, Baste Rotllan N, Neupane PC, Bratland Å, Fuereder T, Hughes BGM, Mesia R, Ngamphaiboon N, Rordorf T, Wan Ishak WZ, Roy A, Cheng J, Jin F, Rischin D. LBA8_PRKEYNOTE-048: phase III study of first-line pembrolizumab (P) for recurrent/metastatic head and neck squamous cell carcinoma (R/M HNSCC). Ann Oncol 2018;29:viii729.

[7] Stone HB, Peters LJ, Milas L. Effect of host immune capability on radiocurability and subsequent transplantability of a murine fibrosarcoma. J Natl Cancer Inst 1979;63:1229–35.

[8] Formenti SC, Demaria S. Combining radiotherapy and cancer immunotherapy: a paradigm shift. J Natl Cancer Inst 2013;105:256–65.

[9] Shaverdian N, Lisberg AE, Bornazyan K, Veruttipong D, Goldman JW, Formenti SC, Garon EB, Lee P. Previous radiotherapy and the clinical activity and toxicity of pembrolizumab in the treatment of non-small-cell lung cancer: a secondary analysis of the KEYNOTE-001 phase 1 trial. Lancet Oncol 2017;18:895–903.

[10] Schapira E, Hubbeling H, Yeap BY, Mehan WA, Shaw AT, Oh K, Gainor JF, Shih HA. Improved overall survival and locoregional disease control with concurrent PD-1 pathway inhibitors and stereotactic radiosurgery for lung cancer patients with brain metastases. Int J Radiat Oncol Biol Phys 2018;101:624–9.

[11] Marciscano AE, Ghasemzadeh A, Nirschl TR, Theodros D, Kochel CM, Francica BJ, Muroyama Y, Anders RA, Sharabi AB, Velarde E, Mao W, Chaudhary KR, Chaimowitz MG, Wong J, Selby MJ, Thudium KB, Korman AJ, Ulmert D, Thorek DLJ, DeWeese TL, Drake CG. Elective nodal irradiation attenuates the combinatorial efficacy of stereotactic radiation therapy and immunotherapy. Clin Cancer Res 2018;24:5058–71.

[12] Cho O, Oh Y-T, Chun M, Noh O-K, Hoe J-S, Kim H. Minimum absolute lymphocyte count during radiotherapy as a new prognostic factor for nasopharyngeal cancer. Head Neck 2016;38:E1061–7.

[13] Mole RH. Whole body irradiation—radiobiology or medicine? Br J Radiol 1953;26:234–41.

[14] Demaria S, Ng B, Devitt ML, Babb JS, Kawashima N, Liebes L, Formenti SC. Ionizing radiation inhibition of distant untreated tumors (abscopal effect) is immune mediated. Int J Radiat Oncol Biol Phys 2004;58:862–70.

[15] Demaria S, Formenti SC. Role of T lymphocytes in tumor response to radiotherapy. Front Oncol 2012;2:95.

[16] Postow MA, Callahan MK, Barker CA, Yamada Y, Yuan J, Kitano S, Mu Z, Rasalan T, Adamow M, Ritter E, Sedrak C, Jungbluth AA, Chua R, Yang AS, Roman R-A, Rosner S, Benson B, Allison JP, Lesokhin AM, Gnjatic S, Wolchok JD. Immunologic correlates of the abscopal effect in a patient with melanoma. N Engl J Med 2012;366:925–31.

[17] McBride SM, Sherman EJ, Tsai CJ, Baxi SS, Aghalar J, Eng J, Zhi WI, McFarland DC, Michel LS, Spielsinger D, Zhang Z, Flynn J, Dunn L, Ho AL, Riaz N, Pfister DG, Lee NY. A phase II randomized trial of nivolumab with stereotactic body radiotherapy (SBRT) versus nivolumab alone in metastatic (M1) head and neck squamous cell carcinoma (HNSCC). J Clin Oncol 2018;36:6009.

[18] Dewan MZ, Galloway AE, Kawashima N, Dewyngaert JK, Babb JS, Formenti SC, Demaria S. Fractionated but not single-dose radiotherapy induces an immune-mediated abscopal effect when combined with anti-CTLA-4 antibody. Clin Cancer Res 2009;15:5379–88.

[19] Fend L, Yamazaki T, Remy C, Fahrner C, Gantzer M, Nourtier V, Préville X, Quéméneur E, Kepp O, Adam J, Marabelle A, Pitt JM, Kroemer G, Zitvogel L. Immune checkpoint blockade, immunogenic chemotherapy or IFN-α blockade boost the local and abscopal effects of oncolytic virotherapy. Cancer Res 2017;77:4146–57.

[20] Gillison ML, Ferris RL, Harris J, Colevas AD, Mell LK, Kong C, Jordan RC, Moore K, Truong MT, Kirsch C, Clump DA, Ohr J, He K, Blakaj D, Deeken JF, Machtay M, Curran WJ, Le Q-T. Safety and disease control achieved with the addition of nivolumab (Nivo) to chemoradiotherapy (CRT) for intermediate (IR) and high-risk (HR) local-regionally advanced head and neck squamous cell carcinoma (HNSCC): RTOG Foundation 3504. J Clin Oncol 2019;37:6073.

[21] Powell SF, Gitau MM, Sumey CJ, Reynolds JT, Lohr M, McGraw S, Nowak RK, Terrell AM, Jensen AW, Blanchard MJ, Ellison C, Black LJ, Thompson PA, Gold KA, Cohen EEW, Lee JH, Spanos WC. Safety of pembrolizumab with chemoradiation (CRT) in locally advanced squamous cell carcinoma of the head and neck (LA-SCCHN). J Clin Oncol 2017;35:6011.

[22] Levy A, Massard C, Soria J-C, Deutsch E. Concurrent irradiation with the anti-programmed cell death ligand-1 immune checkpoint blocker durvalumab: single centre subset analysis from a phase 1/2 trial. Eur J Cancer 2016;68:156–62.

[23] Golden EB, Demaria S, Schiff PB, Chachoua A, Formenti SC. An abscopal response to radiation and ipilimumab in a patient with metastatic non-small cell lung cancer. Cancer Immunol Res 2013;1:365–72.

[24] Martin AM, Cagney DN, Catalano PJ, Alexander BM, Redig AJ, Schoenfeld JD, Aizer AA. Immunotherapy and symptomatic radiation necrosis in patients with brain metastases treated with stereotactic radiation. JAMA Oncol 2018;4:1123–4.

[25] Vermorken JB, Mesia R, Rivera F, Remenar E, Kawecki A, Rottey S, Erfan J, Zabolotnyy D, Kienzer H-R, Cupissol D, Peyrade F, Benasso M, Vynnychenko I, De Raucourt D, Bokemeyer C, Schueler A, Amellal N, Hitt R. Platinum-based chemotherapy plus cetuximab in head and neck cancer. N Engl J Med 2008;359:1116–27.

[26] Janinis J, Papadakou M, Xidakis E, Boukis H, Poulis A, Panagos G, Lefantzis D. Combination chemotherapy with docetaxel, cisplatin, and 5-fluorouracil in previously treated patients with advanced/recurrent head and neck cancer: a phase II feasibility study. Am J Clin Oncol 2000;23:128–31.

[27] Rischin D, Harrington KJ, Greil R, Soulieres D, Tahara M, de Castro G, Psyrri A, Baste N, Neupane PC, Bratland A, Fuereder T, Hughes BGM, Mesia R, Ngamphaiboon N, Rordorf T, Wan Ishak WZ, Zhang Y, Jin F, Gumuscu B, Burtness B. Protocol-specified final analysis of the phase 3 KEYNOTE-048 trial of pembrolizumab (pembro) as first-line therapy for recurrent/metastatic head and neck squamous cell carcinoma (R/M HNSCC). J Clin Oncol 2019;37:6000.

[28] Elias AW, Kasi PM, Stauffer JA, Thiel DD, Colibaseanu DT, Mody K, Joseph RW, Bagaria SP. The feasibility and safety of surgery in patients receiving immune checkpoint inhibitors: a retrospective study. Front Oncol 2017;7:121–3.

[29] Uppaluri R, Lee NY, Westra W, Cohen EEW, Haddad RI, Temam S, Le Tourneau C, Chernock R, Safina S, Klochikhin A, Meirovitz A, Braña I, Ge JY, Swaby RF, Bidadi B, Adkins D. KEYNOTE-689: phase 3 study of adjuvant and neoadjuvant pembrolizumab combined with standard of care (SOC) in patients with resectable, locally advanced head and neck squamous cell carcinoma. J Clin Oncol 2019;37:TPS6090.

[30] O'Malley Jr BW, Weinstein GS, Snyder W, Hockstein NG. Transoral robotic surgery (TORS) for base of tongue neoplasms. Laryngoscope 2006;116:1465–72.

[31] Weinstein GS, O'Malley BW, Snyder W, Sherman E, Quon H. Transoral robotic surgery: radical tonsillectomy. Arch Otolaryngol Head Neck Surg 2007;133:1220.

[32] Robbins KT, Doweck I, Samant S, Vieira F. Effectiveness of superselective and selective neck dissection for advanced nodal metastases after chemoradiation. Arch Otolaryngol Head Neck Surg 2005;131:965.

[33] Suárez C, Rodrigo JP, Robbins KT, Paleri V, Silver CE, Rinaldo A, Medina JE, Hamoir M, Sanabria A, Mondin V, Takes RP, Ferlito A. Superselective neck dissection: rationale, indications, and results. Eur Arch Otorhinolaryngol 2013;270:2815–21.

[34] Keereweer S, Sterenborg HJCM, Kerrebijn JDF, Van Driel PBAA, Baatenburg de Jong RJ, Löwik CWGM. Image-guided surgery in head and neck cancer: current practice and future directions of optical imaging. Head Neck 2012;34:120–6.

[35] Daly MJ, Siewerdsen JH, Moseley DJ, Jaffray DA, Irish JC. Intraoperative cone-beam CT for guidance of head and neck surgery: assessment of dose and image quality using a C-arm prototype. Med Phys 2006;33:3767–80.

[36] Muhanna N, Douglas CM, Daly MJ, Chan HHL, Weersink R, Qiu J, Townson J, de Almeida JR, Goldstein D, Gilbert R, Yu E, Kucharczyk W, Jaffray DA, Irish JC. The image-guided operating room-Utility and impact on surgeon's performance in the head and neck surgery. Head Neck 2019;41:3372–82.

[37] Hasan W, Daly MJ, Chan HHL, Qiu J, Irish JC. Intraoperative cone-beam CT-guided osteotomy navigation in mandible and maxilla surgery. Laryngoscope 2019;37:81–90.

[38] Tajudeen BA, Taylor ZD, Garritano J, Cheng H, Pearigen A, Sherman AJ, Palma-Diaz F, Mishra P, Bhargava S, Pesce J, Kim I, Sebastian C, Razfar A, Papour A, Stafsudd O, Grundfest W, St John M. Dynamic optical contrast imaging as a novel modality for rapidly distinguishing head and neck squamous cell carcinoma from surrounding normal tissue. Cancer 2017;123:879–86.

[39] Baik FM, Hansen S, Knoblaugh SE, Sahetya D, Mitchell RM, Xu C, Olson JM, Parrish-Novak J, Méndez E. Fluorescence identification of head and neck squamous cell carcinoma and high-risk oral dysplasia with BLZ-100, a chlorotoxin-indocyanine green conjugate. JAMA Otolaryngol Head Neck Surg 2016;142:330–8.

[40] van Keulen S, Nishio N, Fakurnejad S, Birkeland A, Martin BA, Lu G, Zhou Q, Chirita SU, Forouzanfar T, Colevas AD, van den Berg NS, Rosenthal EL. The clinical application of fluorescence-guided surgery in head and neck cancer. J Nucl Med 2019;60:758–63.

[41] Biel MA. Photodynamic therapy of head and neck cancers. Methods Mol Biol 2010;635:281–93.

[42] Rigual NR, Shafirstein G, Frustino J, Seshadri M, Cooper M, Wilding G, Sullivan MA, Henderson B. Adjuvant intraoperative photodynamic therapy in head and neck cancer. JAMA Otolaryngol Head Neck Surg 2013; 139:706–11.

[43] El-Sayed IH, Huang X, El-Sayed MA. Selective laser photo-thermal therapy of epithelial carcinoma using anti-EGFR antibody conjugated gold nanoparticles. Cancer Lett 2006;239:129–35.

[44] Master A, Malamas A, Solanki R, Clausen DM, Eiseman JL, Sen Gupta A. A cell-targeted photodynamic nanomedicine strategy for head and neck cancers. Mol Pharm 2013;10:1988–97.

[45] Nune SK, Gunda P, Thallapally PK, Lin Y-Y, Forrest ML, Berkland CJ. Nanoparticles for biomedical imaging. Expert Opin Drug Deliv 2009;6:1175–94.

[46] Singh R, Lillard JW. Nanoparticle-based targeted drug delivery. Exp Mol Pathol 2009;86:215–23.

[47] Cui L, Lin Q, Jin CS, Jiang W, Huang H, Ding L, Muhanna N, Irish JC, Wang F, Chen J, Zheng G. A PEGylation-free biomimetic porphyrin nanoplatform for personalized cancer theranostics. ACS Nano 2015;9:4484–95.

[48] Muhanna N, Jin CS, Huynh E, Chan H, Qiu Y, Jiang W, Cui L, Burgess L, Akens MK, Chen J, Irish JC, Zheng G. Phototheranostic porphyrin nanoparticles enable visualization and targeted treatment of head and neck cancer in clinically relevant models. Theranostics 2015;5:1428–43.

[49] Muhanna N, Cui L, Chan H, Burgess L, Jin CS, MacDonald TD, Huynh E, Wang F, Chen J, Irish JC, Zheng G. Multimodal image-guided surgical and photodynamic interventions in head and neck cancer: from primary tumor to metastatic drainage. Clin Cancer Res 2016;22:961–70.

[50] Daisne J-F, Duprez T, Weynand B, Lonneux M, Hamoir M, Reychler H, Grégoire V. Tumor volume in pharyngolaryngeal squamous cell carcinoma: comparison at CT, MR imaging, and FDG PET and validation with surgical specimen. Radiology 2004;233:93–100.

[51] Allal AS, Slosman DO, Kebdani T, Allaoua M, Lehmann W, Dulguerov P. Prediction of outcome in head-and-neck cancer patients using the standardized uptake value of 2-[18F]fluoro-2-deoxy-D-glucose. Int J Radiat Oncol Biol Phys 2004;59:1295–300.

[52] Lee S, Nam SY, Im KC, Kim JS, Choi EK, Ahn SD, Park SH, Kim SY, Lee B-J, Kim JH. Prediction of prognosis using standardized uptake value of 2-[(18)F] fluoro-2-deoxy-d-glucose positron emission tomography for nasopharyngeal carcinomas. Radiother Oncol 2008;87:211–6.

[53] Oksuz DC, Prestwich RJ, Carey B, Wilson S, Senocak MS, Choudhury A, Dyker K, Coyle C, Sen M. Recurrence patterns of locally advanced head and neck squamous cell carcinoma after 3D conformal (chemo)-radiotherapy. Radiat Oncol 2011;6:54.

[54] Johansen S, Norman MH, Dale E, Amdal CD, Furre T, Malinen E, Evensen JF. Patterns of local-regional recurrence after conformal and intensity-modulated radiotherapy for head and neck cancer. Radiat Oncol 2017;12:87.

[55] Raktoe SAS, Dehnad H, Raaijmakers CPJ, Braunius W, Terhaard CHJ. Origin of tumor recurrence after intensity modulated radiation therapy for oropharyngeal squamous cell carcinoma. Int J Radiat Oncol Biol Phys 2013;85:136–41.

[56] Cliffe H, Patel C, Prestwich R, Scarsbrook A. Radiotherapy response evaluation using FDG PET-CT-established and emerging applications. Br J Radiol 2017;90(1071):20160764.

[57] Rasey JS, Koh WJ, Evans ML, Peterson LM, Lewellen TK, Graham MM, Krohn KA. Quantifying regional hypoxia in human tumors with positron emission tomography of [18F]fluoromisonidazole: a pretherapy study of 37 patients. Int J Radiat Oncol Biol Phys 1996;36:417–28.
[58] Berwouts D, Olteanu LAM, Duprez F, Vercauteren T, De Gersem W, De Neve W, Van de Wiele C, Madani I. Three-phase adaptive dose-painting-by-numbers for head-and-neck cancer: initial results of the phase I clinical trial. Radiother Oncol 2013;107:310–6.
[59] Madani I, Duprez F, Boterberg T, Van de Wiele C, Bonte K, Deron P, De Gersem W, Coghe M, De Neve W. Maximum tolerated dose in a phase I trial on adaptive dose painting by numbers for head and neck cancer. Radiother Oncol 2011;101:351–5.
[60] Liu F, Xi X-P, Wang H, Han Y-Q, Xiao F, Hu Y, He Q, Zhang L, Xiao Q, Liu L, Luo L, Li Y, Mo Y, Ma H-Z. PET/CT-guided dose-painting versus CT-based intensity modulated radiation therapy in locoregional advanced nasopharyngeal carcinoma. Radiat Oncol 2017;12:15.
[61] Barker JL, Garden AS, Ang KK, O'Daniel JC, Wang H, Court LE, Morrison WH, Rosenthal DI, Chao KSC, Tucker SL, Mohan R, Dong L. Quantification of volumetric and geometric changes occurring during fractionated radiotherapy for head-and-neck cancer using an integrated CT/linear accelerator system. Int J Radiat Oncol Biol Phys 2004;59:960–70.
[62] Castelli J, Simon A, Lafond C, Perichon N, Rigaud B, Chajon E, De Bari B, Ozsahin M, Bourhis J, de Crevoisier R. Adaptive radiotherapy for head and neck cancer. Acta Oncol 2018;57:1284–92.
[63] Chen AM, Daly ME, Cui J, Mathai M, Benedict S, Purdy JA. Clinical outcomes among patients with head and neck cancer treated by intensity-modulated radiotherapy with and without adaptive replanning. Head Neck 2014;36:1541–6.
[64] Kong F-M, Ten Haken RK, Schipper M, Frey KA, Hayman J, Gross M, Ramnath N, Hassan KA, Matuszak M, Ritter T, Bi N, Wang W, Orringer M, Cease KB, Lawrence TS, Kalemkerian GP. Effect of midtreatment PET/CT-adapted radiation therapy with concurrent chemotherapy in patients with locally advanced non-small-cell lung cancer: a phase 2 clinical trial. JAMA Oncol 2017;3:1358–65.
[65] Yang H, Hu W, Wang W, Chen P, Ding W, Luo W. Replanning during intensity modulated radiation therapy improved quality of life in patients with nasopharyngeal carcinoma. Int J Radiat Oncol Biol Phys 2013;85:e47–54.
[66] Geets X, Tomsej M, Lee JA, Duprez T, Coche E, Cosnard G, Lonneux M, Grégoire V. Adaptive biological image-guided IMRT with anatomic and functional imaging in pharyngo-laryngeal tumors: impact on target volume delineation and dose distribution using helical tomotherapy. Radiother Oncol 2007;85:105–15.
[67] Duprez F, De Neve W, De Gersem W, Coghe M, Madani I. Adaptive dose painting by numbers for head-and-neck cancer. Int J Radiat Oncol Biol Phys 2011;80:1045–55.
[68] Arimura H, Soufi M, Kamezawa H, Ninomiya K, Yamada M. Radiomics with artificial intelligence for precision medicine in radiation therapy. J Radiat Res 2019;60:150–7.
[69] Li YQ, Tan JSH, Wee JTS, Chua MLK. Adaptive radiotherapy for head and neck cancers: fact or fallacy to improve therapeutic ratio? Cancer Radiother 2018;22:287–95.
[70] Zhang B, Tian J, Dong D, Gu D, Dong Y, Zhang L, Lian Z, Liu J, Luo X, Pei S, Mo X, Huang W, Ouyang F, Guo B, Liang L, Chen W, Liang C, Zhang S. Radiomics features of multiparametric MRI as novel prognostic factors in advanced nasopharyngeal carcinoma. Clin Cancer Res 2017;23:4259–69.
[71] Aerts HJWL, Velazquez ER, Leijenaar RTH, Parmar C, Grossmann P, Carvalho S, Cavalho S, Bussink J, Monshouwer R, Haibe-Kains B, Rietveld D, Hoebers F, Rietbergen MM, Leemans CR, Dekker A, Quackenbush J, Gillies RJ, Lambin P. Decoding tumour phenotype by noninvasive imaging using a quantitative radiomics approach. Nat Commun 2014;5:4006.
[72] Leijenaar RTH, Carvalho S, Hoebers FJP, Aerts HJWL, van Elmpt WJC, Huang SH, Chan B, Waldron JN, O'sullivan B, Lambin P. External validation of a prognostic CT-based radiomic signature in oropharyngeal squamous cell carcinoma. Acta Oncol 2015;54:1423–9.
[73] Colevas AD, Bedi N, Chang S, Nieves UYM, Chatterjee S, Davidzon GA, Srinivas S, Le Q-T, Gambhir A, Sunwoo JB. A study to evaluate immunological response to PD-1 inhibition in squamous cell carcinoma of the head and neck (SCCHN) using novel PET imaging with [18F]F-AraG. J Clin Oncol 2018;36:6050.
[74] Chow LQ, Burtness B, Weiss J, Berger R, Eder JP, Gonzalez EJ, Pulini J, Johnson J, Dolled-Filhart M, Emancipator K, Lunceford JK, Pathiraja K, Gause C, Cheng JD, Seiwert T. LBA31A phase IB study of pembrolizumab (PEMBRO; MK-3475) in patients (PTS) with human papiilloma virus (HPV)-positive and negative head and neck cancer (HNC). Ann Oncol 2014;25v1.

[75] Ferris RL, Gonçalves A, Baxi SS, Martens UM, Gauthier H, Langenberg M, Spanos WC, Leidner RS, Kang H, Russell J, Chiosea S, Soumaoro I, Rao S, Cao ZA, Topalian S. LBA46An open-label, multicohort, phase 1/2 study in patients with virus-associated cancers (CheckMate 358): safety and efficacy of neoadjuvant nivolumab in squamous cell carcinoma of the head and neck (SCCHN). Ann Oncol 2017;28:v604–59.

[76] Uppaluri R, Zolkind P, Lin T, Nussenbaum B, Jackson RS, Rich J, Pipkorn P, Paniello R, Thorstad W, Michel L, Mudianto T, Oppelt PJ, Wildes TM, Dunn GP, Piccirillo JF, Kallogjeri D, Rodig S, Hagemann IS, Chernock R, Adkins D. Neoadjuvant pembrolizumab in surgically resectable, locally advanced HPV negative head and neck squamous cell carcinoma (HNSCC). J Clin Oncol 2017;35:6012.

[77] Bonner JA, Harari PM, Giralt J, Azarnia N, Shin DM, Cohen RB, Jones CU, Sur R, Raben D, Jassem J, Ove R, Kies MS, Baselga J, Youssoufian H, Amellal N, Rowinsky EK, Ang KK. Radiotherapy plus cetuximab for squamous-cell carcinoma of the head and neck. N Engl J Med 2006;354:567–78.

[78] Shayan G, Kansy BA, Gibson SP, Srivastava RM, Bryan JK, Bauman JE, Ohr J, Kim S, Duvvuri U, Clump DA, Heron DE, Johnson JT, Hershberg RM, Ferris RL. Phase Ib study of immune biomarker modulation with neoadjuvant cetuximab and TLR8 stimulation in head and neck cancer to overcome suppressive myeloid signals. Clin Cancer Res 2018;24:62–72.

[79] Wheeler DL, Huang S, Kruser TJ, Nechrebecki MM, Armstrong EA, Benavente S, Gondi V, Hsu K-T, Harari PM. Mechanisms of acquired resistance to cetuximab: role of HER (ErbB) family members. Oncogene 2008;27:3944–56.

[80] Ferris RL, Kim S, Trivedi S, Srivastava RM, Concha-Benavente F, Heron DE, Clump DA, Duvvuri U, Gooding WE, Johnson JT, Ohr J, Bauman JE. Correlation of anti-tumor adaptive immunity with clinical response in a phase II "window" trial of neoadjuvant cetuximab in patients with resectable stage III-IV head and neck squamous carcinoma (HNSCC). J Clin Oncol 2016;34:6060.

[81] Schmitz S, Rommel D, Michoux N, Lhommel R, Hanin F-X, Duprez T, Machiels J-P. Dynamic contrast-enhanced computed tomography to assess early activity of cetuximab in squamous cell carcinoma of the head and neck. Radiol Oncol 2015;49:17–25.

[82] Uppaluri R, Winkler AE, Lin T, Law JH, Haughey BH, Nussenbaum B, Paniello RC, Rich JT, Diaz JA, Michel LP, Wildes T, Dunn GP, Zolkind P, Kallogjeri D, Piccirillo JF, Dehdashti F, Siegel BA, Chernock RD, Lewis JS, Adkins DR. Biomarker and tumor responses of oral cavity squamous cell carcinoma to trametinib: a phase II neoadjuvant window-of-opportunity clinical trial. Clin Cancer Res 2017;23:2186–94.

[83] Machiels J-PH, Menis J, Lia M, Fortpied C, Lhommel R, Lemort M, Guzzo M, Bianchi R, Crippa F, Quattrone P, Bossi P, Schmitz S, Duprez T, Lalami Y, Quiriny M, de Saint Aubain N, Clement PM, Coropciuc R, Hauben E, Licitra LF. Activity of afatinib administered in a window pre-operative study in squamous cell carcinoma of the head and neck (SCCHN): EORTC-90111. J Clin Oncol 2016;34:6049.

[84] Aas JA, Paster BJ, Stokes LN, Olsen I, Dewhirst FE. Defining the normal bacterial flora of the oral cavity. J Clin Microbiol 2005;43:5721–32.

[85] Preza D, Olsen I, Willumsen T, Grinde B, Paster BJ. Diversity and site-specificity of the oral microflora in the elderly. Eur J Clin Microbiol Infect Dis 2009;28:1033–40.

[86] Chen T, Yu W-H, Izard J, Baranova OV, Lakshmanan A, Dewhirst FE. The human oral microbiome database: a web accessible resource for investigating oral microbe taxonomic and genomic information. Database (Oxford) 2010;2010:baq013.

[87] Routy B, Le Chatelier E, Derosa L, Duong CPM, Alou MT, Daillère R, Fluckiger A, Messaoudene M, Rauber C, Roberti MP, Fidelle M, Flament C, Poirier-Colame V, Opolon P, Klein C, Iribarren K, Mondragón L, Jacquelot N, Qu B, Ferrere G, Clémenson C, Mezquita L, Masip JR, Naltet C, Brosseau S, Kaderbhai C, Richard C, Rizvi H, Levenez F, Galleron N, Quinquis B, Pons N, Ryffel B, Minard-Colin V, Gonin P, Soria J-C, Deutsch E, Loriot Y, Ghiringhelli F, Zalcman G, Goldwasser F, Escudier B, Hellmann MD, Eggermont A, Raoult D, Albiges L, Kroemer G, Zitvogel L. Gut microbiome influences efficacy of PD-1–based immunotherapy against epithelial tumors. Science 2018;359:91–7.

[88] Mager DL, Haffajee AD, Devlin PM, Norris CM, Posner MR, Goodson JM. The salivary microbiota as a diagnostic indicator of oral cancer: a descriptive, non-randomized study of cancer-free and oral squamous cell carcinoma subjects. J Transl Med 2005;3:27.

[89] Schmidt BL, Kuczynski J, Bhattacharya A, Huey B, Corby PM, Queiroz ELS, Nightingale K, Kerr AR, DeLacure MD, Veeramachaneni R, Olshen AB, Albertson DG, Teh M-T. Changes in abundance of oral microbiota associated with oral cancer. PLoS One 2014;9.

[90] Olsen I, Yilmaz Ö. Possible role of Porphyromonas gingivalis in orodigestive cancers. J Oral Microbiol 2019;11:1563410.

[91] Lafuente Ibáñez de Mendoza I, Maritxalar Mendia X, García de la Fuente AM, Quindós Andrés G, Aguirre Urizar JM. Role of *Porphyromonas gingivalis* in oral squamous cell carcinoma development: a systematic review. J Periodontal Res 2020;55(1):13–22.

[92] Gholizadeh P, Eslami H, Kafil HS. Carcinogenesis mechanisms of Fusobacterium nucleatum. Biomed Pharmacother 2017;89:918–25.

[93] Gopalakrishnan V, Spencer CN, Nezi L, Reuben A, Andrews MC, Karpinets TV, Prieto PA, Vicente D, Hoffman K, Wei SC, Cogdill AP, Zhao L, Hudgens CW, Hutchinson DS, Manzo T, Petaccia de Macedo M, Cotechini T, Kumar T, Chen WS, Reddy SM, Szczepaniak Sloane R, Galloway-Pena J, Jiang H, Chen PL, Shpall EJ, Rezvani K, Alousi AM, Chemaly RF, Shelburne S, Vence LM, Okhuysen PC, Jensen VB, Swennes AG, McAllister F, Marcelo Riquelme Sanchez E, Zhang Y, Le Chatelier E, Zitvogel L, Pons N, Austin-Breneman JL, Haydu LE, Burton EM, Gardner JM, Sirmans E, Hu J, Lazar AJ, Tsujikawa T, Diab A, Tawbi H, Glitza IC, et al. Gut microbiome modulates response to anti-PD-1 immunotherapy in melanoma patients. Science 2018;359:97–103.

[94] Oliva Bernal M, Schneeberger PHH, Taylor R, Rey V, Hansen AR, Taylor K, Bayley A, Hope A, Hosni A, Bratman SV, Ringash J, Gilbert RW, Weinreb I, Perez-Ordonez B, Waldron J, Xu W, Guttman D, Siu LL, Coburn B, Spreafico A. Role of the oral and gut microbiota as a biomarker in locoregionally advanced oropharyngeal squamous cell carcinoma (ROMA LA-OPSCC). J Clin Oncol 2019;37:6045.

CHAPTER

10

Targeted radiation: Specificity and deescalation

Rebecca Levin-Epstein, Neil Parikh, Ritchell van Dams, John V. Hegde, Robert K. Chin, Amar U. Kishan

Department of Radiation Oncology, University of California Los Angeles, Los Angeles, CA, United States

Abstract

The identification of human papillomavirus (HPV) as a unique risk factor for development of oropharyngeal squamous cell carcinoma (OPSCC) marked a paradigm shift in the understanding and treatment of the disease. HPV-associated OSPCC is an entirely distinct disease entity with markedly improved prognosis compared with HPV-negative OSPCC. In the context of the morbidity associated with treatment of OPSCC and the improved prognosis among patients with HPV-associated OPSCC, attempts have been made to identify which patients could potentially benefit from further deescalated treatment. This has led to numerous trials of deescalated therapy including radiation dose reduction in the definitive and adjuvant settings, systemic therapy modification (substitution of chemotherapy for less toxic regimens or biologic agents, reduction in chemotherapy dose, and elimination of concurrent systemic therapy altogether), response-driven treatment deintensification after induction therapy, RT target volume reduction, and replacing RT with minimally invasive surgery. Despite these efforts, there remains no consensus treatment paradigm specific to patients with HPV-associated OPSCC. Enrollment of all eligible patients in any of the several ongoing clinical trials described earlier is highly encouraged, and through these efforts, such a rational guideline may be established.

Abbreviations

^{18}F-FMISO	^{18}F-fluoromisonidazole
AE	adverse event
AJCC	American Joint Committee on Cancer
BID	twice daily
cCR	clinical complete response
chemoRT	chemoradiation
CSS	cancer-specific survival
CT	computed tomography
DC	distant control

https://doi.org/10.1016/B978-0-12-820679-9.00010-4

DFMS	distant metastasis-free survival
ENE	extranodal extension
HPV	human papillomavirus
IMPT	intensity-modulated proton therapy
IMRT	intensity-modulated radiation therapy
LRC	locoregional control
LVSI	lymphovascular space invasion
MDADI	MD Anderson Dysphagia Inventory
NCT	National Clinical Trial
ND	neck dissection
OPSCC	oropharyngeal squamous cell carcinoma
OS	overall survival
pCR	pathologic complete response
PET-CT	positron emission tomography-computed tomography
PFS	progression-free survival
PRO	patient-reported outcome
QOL	quality of life
RECIST	response evaluation criteria in solid tumors
RT	radiotherapy
RTOG	Radiation Therapy Oncology Group
SIB	simultaneous integrated boost
TORS	transoral robotic surgery
VMAT	volumetric arc therapy

Conflict of interest

Dr. Kishan has received honoraria and consulting fees from Varian Medical Systems, Inc.; ViewRay, Inc.; and Intelligent Automation, Inc. Dr. Kishan has also served on an advisory board for Janssen, Inc.

Introduction

The identification of human papillomavirus (HPV) as a unique risk factor for the development of oropharyngeal squamous cell carcinoma (OPSCC) marked a paradigm shift in the understanding and treatment of the disease. HPV-associated OSPCC is an entirely distinct disease entity from HPV-negative OSPCC, with separate underlying oncogenetic pathways and a substantially improved prognosis [1–4]. Recognition of these differences led to the codification of p16 status as a key prognostic indicator and to the development of a new staging system specific to HPV-associated OPSCC [5]. In tandem with these findings, the Radiation Therapy Oncology Group (RTOG) trial 0022 found that select patients could achieve high tumor control rates with reduced long-term toxicity through the use of intensity-modulated radiotherapy (IMRT), a technique that allows for complex shaping of the radiation dose to spare adjacent organs at risk [6]. In the context of the morbidity associated with treatment of OPSCC and the improved prognosis among patients with HPV-associated OPSCC, attempts have been made to identify which patients could potentially benefit from further deescalated treatment. O'Sullivan et al. [7] used recursive partitioning analysis and found that HPV-associated OPSCC patients with T1-3N0-2c disease had the lowest risk of distant metastasis (DM), with the caveat that patients with N2c disease had reduced distant control (DC) with radiotherapy (RT) alone. Another risk schema was formed from a secondary analysis of RTOG 0129, which investigated altered fractionation RT versus standard fractionation for OPSCC [8]. Based on overall survival (OS) rates, Ang and colleagues classified patients as low risk if they had HPV-associated OPSCC, with the exception of those with N2b-3 disease and a > 10 pack-year smoking history, who were instead classified as intermediate risk. These risk strata formed the foundation for subsequent trials investigating the efficacy and safety of various modes of deescalation among patients with HPV-associated OPSCC.

Completed deescalation trials

Numerous approaches to treatment deintensification and improved targeting have been investigated for HPV-related OPSCC. The major classes of these approaches include RT dose reduction in the definitive and adjuvant settings, systemic therapy modification (substitution of chemotherapy for less toxic regimens or biologic agents, reduction in chemotherapy dose, and elimination of concurrent systemic therapy altogether), response-driven treatment deintensification after induction therapy, RT target volume reduction, and replacing RT with minimally invasive surgery. This section describes prominent completed trials in these arenas, with an accompanying summary available in Table 1.

TABLE 1 Completed clinical trials involving treatment deescalation for HPV-associated OPSCC.

Trial	Trial purpose	Patient details	Trial details	Results
UNC [9,10] Chera et al.	Assess pCR rates after deescalated definitive chemoRT (reduction in both chemotherapy and RT)	$N = 43$ HPV-related OPSCC (95%) or unknown primary (5%) Stage: T0-3, N0-N2c (AJCC 7th) Mostly ≤10 pack-year smoking history	Phase II single-arm Accrual: 2012–2014 IMRT 60 Gy/6 weeks (5 days per week, 2gy/fx), 54 Gy to subclinical sites Concurrent cisplatin, 30 mg/m^2 weekly x6 → Planned surgical evaluation (biopsy or resection, if N+ at presentation, also SLND)	Median follow-up 36 months Clinical and pathologic response: cCR rate 64% (98% primary, 60% neck) pCR rate[a] 86% (98% primary, 84% neck) Three-year oncologic outcomes: LC: 100% RC: 100% CSS: 100% DMFS: 100% OS: 95% Treatment completion: 95% received ≥four doses of cisplatin, 70% full six weekly doses Toxicity: - Feeding tube rate 39%, none long term - Acute grade ≥3: xerostomia 75%, dysphagia 55%, mucositis 45% - No late grade ≥3 - PROs: Global QOL similar, xerostomia worse versus baseline

Continued

TABLE 1 Completed clinical trials involving treatment deescalation for HPV-associated OPSCC—cont'd

Trial	Trial purpose	Patient details	Trial details	Results
UNC Chera et al. [11]	Reassess same deescalated definitive chemoRT without posttreatment planned surgery	$N=114$ HPV-related OPSCC Stage: T0-T3, N0-N2c (AJCC 7th) or T0-3 N0-2 (AJCC 8th) Minimal remote smoking history	Phase II single-arm Accrual: 2014–2017 IMRT 60Gy/6weeks (5days per week, 2Gy/fx)/54Gy subclinical sites Concurrent weekly cisplatin (30mg/m^2) for 6weeks; no chemotherapy for T0-2 N0-1 (AJCC 7th)	Median follow-up 31.8months Two-year oncologic outcomes: Two-yr PFS[a]: 86% Two-yr LRC: 95% Two-yr DMFS: 91% Two-yr OS: 95%s Toxicity: - Feeding tube rate 34%, none permanent - No late grade ≥3 toxicities - 2-year PROs: global OQL better, xerostomia worse
De-ESCALaTE HPV [12] Mehanna et al.	Compare concurrent cisplatin versus cetuximab with standard dose RT	$N=334$ Low-risk p16+ OPSCC Stage: T3-4N0, T1N1-T4N3 (AJCC seventh) Nonsmokers or <10 pack-years	Phase III RCT Accrual: 2012–2016 IMRT 70Gy/35fx plus either: Bolus cisplatin q3 weeks × three cycles versus Weekly cetuximab (loading dose then weekly ×7weeks)	Median follow-up 25.9months All outcomes are concurrent cisplatin versus cetuximab: Two-year oncologic outcomes: Recurrence: 6.0% versus 16.1% ($P=.0007$) DM: 3% versus 9% ($P=.0092$) OS: 97.5% versus 89.4% ($P=.0012$) Subset analysis: cisplatin superior for all stages (I–III, AJCC 8th) Treatment completion: - No significant differences - 38% received all three cisplatin cycles, median total dose 200mg/m^2 - 79% received all eight cetuximab cycles - Full RT dose: 95%–97% Toxicity[a]: - No differences in any grade or grade ≥3 acute or late toxicity; though significantly more hospitalizations with cisplatin - PROs: Overall function favors cisplatin

TABLE 1 Completed clinical trials involving treatment deescalation for HPV-associated OPSCC—cont'd

Trial	Trial purpose	Patient details	Trial details	Results
RTOG 1016 [13]	Compare concurrent cisplatin versus cetuximab with accelerated RT	$N=849$ p16+ OPSCC Stage: T1-2, N2a-3 OR T3-4 with any N (AJCC 7th) Any pack-year smoking history (38% >10 pack-years)	Phase III RCT, noninferiority design Accrual: 2011–2014 Accelerated IMRT 70 Gy/6 weeks (35 fx) plus either: Bolus cisplatin × two cycles versus Weekly cetuximab (loading dose then weekly x 7 weeks)	Median follow-up 4.5 years All outcomes are concurrent cisplatin versus cetuximab: Five-year oncologic outcomes: OS[a]: 84.6% versus 77.9% ($P=.016$) PFS: 78.4% versus 67.3% ($P=.0002$) LRF: 9.9% versus 17.3% ($P=.0005$) DM: 8.5% versus 11.7% ($P=.09$) -Subset: Zubrod 1 still did better with cisplatin Treatment completion: - 93% received two cycles of cisplatin - 85% received at least seven cycles of cetuximab - RT per protocol 83%–86% Toxicity: Acute grade ≥3: 81.7% versus 77.4% ($P=.16$); using T score cisplatin significantly higher acute toxicity Late grade ≥3: 20.4% versus 16.5% ($P=.12$) PROs similar
NRG HN002 [16] Yom et al.	Compare two different forms of deintensification (no standard intensity arm): ChemoRT dose deescalation versus accelerated RT without chemotherapy	$N=306$ p16+ OPSCC Stage: T1-2, N1-2b, T3N0-N2b[b] ≤10 pack-year smoking history	Phase III RCT Accrual: 2014–2017 IMRT 60 Gy/6 weeks with weekly cisplatin at 40 mg/m^2 (IMRT+C) versus accelerated RT alone: 60 Gy/5 weeks (6 fx/week)	Median follow-up 2.6 years All outcomes are IMRT +C versus IMRT: Acute toxicity (≤180 days): - Grade 4: 15.1% versus 2.0% - Grade 3: 64.5% versus 50.3% Late toxicity (>180 days): - Grade 4: 1.3% versus 1.4% - Grade 3: 20.0% versus 16.7%

Continued

TABLE 1 Completed clinical trials involving treatment deescalation for HPV-associated OPSCC—cont'd

Trial	Trial purpose	Patient details	Trial details	Results
				- MDADI at 1 year[a]: 85.3 versus 81.76 (both met threshold for acceptability) Oncologic outcomes: Two-year PFS: 90.5% versus 87.6%, measured against historical rate PFS 85% only IMRT+C met superiority threshold (P =.035) Two-year OS: 96.7% versus 97.3%
ECOG 1308 [17] Marur et al.	Evaluate RT dose deescalation based on induction chemotherapy response	N =80 HPV-associated OPSCC Stage: III–IV, surgically resectable	Phase II Accrual: 2010–2011 Induction chemotherapy: cisplatin, paclitaxel, and cetuximab × three cycles →If cCR, IMRT 54 Gy/27 fx with weekly cetuximab →If no cCR, 69.3 Gy/33 fx with weekly cetuximab	Median follow-up 35.4 months Primary tumor cCR: 70% Outcomes reported for de-escalated vs. standard RT dose arm: Oncologic outcomes: Two-year PFS[a]: 81% versus 67% Two-year OS: 93% versus 87% Subset among cCR: If ≤T3, ≤N2b, and ≤ 10 pack-year smoking and received deescalated RT: Two-year PFS: 96% Two-year OS: 96% Toxicity: - Difficulty swallowing solids: 40% versus 89% (P =.0111) - Impaired nutrition: 10% versus 44% (P =.025)
CCRO-022 (UCLA/UC Davis) [18] Chen et al.	Evaluate RT dose deescalation based on induction chemotherapy response	N =45 p16+ OPSCC Stage: III–IV (AJCC 7th) Any smoking history	Phase II Accrual: 2012–2015 Induction: carboplatin/paclitaxel × two cycles →PR or CR (by RECIST): IMRT 54 Gy/27 fx →Nonresponder: 60 Gy/30 fx with weekly paclitaxel - All RT volumes based on	Median follow-up 30 months Oncologic outcomes: Responder (PR or CR): 55% (11% CR) Two-year PFS[a]: 92% Two-year OS: 98% Toxicity: - Any grade ≥3 toxicity: 39% (grade ≥3 dysphagia and mucositis rates 9.1%)

TABLE 1 Completed clinical trials involving treatment deescalation for HPV-associated OPSCC—cont'd

Trial	Trial purpose	Patient details	Trial details	Results
			prechemotherapy disease extent	- Feeding tube rate: 15%, none long term - Late grade ≥3 toxicity: 5%
OPTIMA [23] Siewert et al.	Two aims: Target volume deescalation (neck) and deescalation based on response to induction chemotherapy and risk classification	$N=62$ HPV-positive OPSCC Staging: Defined low or high risk: - Low risk: ≤T3, ≤N2b, ≤10 pack-year smoking - High risk: T4 or ≥N2c, or >10 pack-year (AJCC 7th)	Phase II Accrual: 2014–2016 Induction: carboplatin/nab-paclitaxel × 3 cycles →Response-based stratification (>50%, 30%–50%, or <50% response) incorporating low- or high-risk classification: IMRT 50 Gy/5 weeks alone, versus ChemoRT 45 Gy in 1.5 Gy/fraction BID with concurrent TFHX, versus ChemoRT 75 Gy in 1.5 Gy/fraction BID with concurrent TFHX All RT to primary disease plus only first echelon of uninvolved nodes	Median follow-up 29 months Of low-risk patients, 71% received RT alone Of high-risk patients, 71% received 45 Gy chemoRT Oncologic outcomes: Low risk versus high risk: 2-year PFS[a]: 95% versus 94% 2-year OS: 100% versus 97% Toxicity: RT alone versus 45 Gy ChemoRT versus 75 Gy ChemoRT: - Acute grade ≥3 mucositis: 30% versus 63% versus 91% ($P=.004$) - Feeding tube use: 0% versus 31% versus 82% ($P<.0001$)
ORATOR [38] Nichols et al.	Compare QOL after definitive RT versus TORS	$N=68$ OPSCC (88% p16+) with likely achievable negative surgical margins Stage: early stage, T1-2, N0-2 (nodes ≤4 cm)	Phase II RCT Accrual: 2012–2017 IMRT 70 Gy + concurrent bolus cisplatin (83%) or cetuximab or carboplatin, +/− salvage neck dissection if gross residual nodes versus TORS +/− adjuvant RT (60–64 Gy) +/− adjuvant chemotherapy depending on pathology risk	Median follow-up 27 months Three-year oncologic outcomes: No significant differencesin PFS or OS between RT and TORS - PFS for p16+ patients (either treatment): 93.0% - OS for p16+ patients (either treatment): 93.1% Toxicity: RT versus TORS:

Continued

TABLE 1 Completed clinical trials involving treatment deescalation for HPV-associated OPSCC—cont'd

Trial	Trial purpose	Patient details	Trial details	Results
				- MDADI score at 1 year[a]: 86.9 versus 80.1 (P=.042) - Clinically meaningful decline (≥10 points) significantly worse for TORS in global and emotional domains - Grade ≥2 toxicity: 91% versus 97% (not significant - More pain and bleeding in TORS group, including one bleeding-related death
MC1273 [39] Ma et al.	Evaluate adjuvant RT dose deescalation	N = 80 p16+ OPSCC after curative-intent surgery with negative margins and ≥1 intermediate risk factor[b] or ENE Stage: III–IV (AJCC 7th ed.) ≤10 pack-year smoking history	Phase II single arm Accrual: 2013–2016 Intermediate risk: Adjuvant 30 Gy in 1.5 Gy/fx BID + concurrent weekly docetaxel + if ENE, SIB to nodal levels with ENE, 36 Gy in 1.8 Gy/fx BID	Median follow-up 25 months Oncologic outcomes: Two-year LRC[a]: 96.2% Two-year PFS 91.1% Two-year OS: 98.7% Two-year DMFS: 94.9% Toxicity: Grade ≥3 toxicity: 0% Swallowing function and diet normalcy improved from pre-RT to post-RT (P = .01) Xerostomia at baseline by 1 year

[a] *Indicates primary outcome.*
[b] *N2c not included due to higher distant metastasis risk with de-intensification per O'Sullivan et al. [7].*
Abbreviations: *pCR*, pathologic complete response; *ChemoRT*, chemoradiation; *HPV*, human papillomavirus; *OPSCC*, oropharyngeal squamous cell carcinoma; *AJCC*, American Joint Committee on Cancer; *IMRT*, intensity-modulated radiation therapy; *RT*, radiation therapy; *cCR*, clinical complete response; *PR*, partial clinical response; *TORS*, Transoral robotic surgery; *Fx*, fraction; *N+*, node positive; *SLND*, selective neck dissection; *LC*, local control; *RC*, regional control; *CSS*, cancer-specific survival; *DMFS* Distant metastasis–free survival; *OS*, overall survival; *PROs*, patient-reported outcomes; *QOL*, quality of life; *LRC*, locoregional control; *PFS*, progression-free survival; *RCT*, randomized controlled trial; *MDADI*, MD Anderson Dysphagia Inventory; higher score signifies better swallowing function. *RECIST*, Response Evaluation Criteria in Solid Tumors; *BID*, twice daily; *TFHX*, paclitaxel, 5-fluorouracil, hydroxyurea; *ENE*, extranodal extension.

Deescalation of concurrent chemoradiation

Chera et al. [9] **(UNC Chapel Hill)** published one of the early investigations of deintensified chemoradiation (chemoRT) with deescalation of both RT dose and systemic therapy intensity. In this single-arm phase II trial, 43 patients with p16+ T0-3, N0-2c (AJCC seventh ed.) OPSCC, or unknown primary (the latter comprising only 5%) received definitive intent RT to 60 Gray (Gy) (54 Gy for subclinical disease) over 6 weeks with concurrent weekly reduced-dose cisplatin (30 mg/m^2). Given minimal preceding proof of efficacy of deintensification, all patients underwent planned surgical evaluation at 7- to 14-week postcompletion of chemoRT. The degree of surgery was based on degree of response via clinical evaluation with computed tomography (CT) and laryngoscopy at 4- to 8-week posttreatment. Patients underwent transoral robotic surgery (TORS) resection if they only had a partial clinical response, but those with clinical complete responses (cCRs) had only a biopsy of the primary site with optional resection if the biopsy showed residual disease. All patients who were clinically node positive on initial staging also underwent planned selective neck dissection. Grade $\geq$3 acute toxicities were common (grade $\geq$3 xerostomia 75%, dysphagia 55%, and mucositis 45%), though there were no late grade $\geq$3 adverse events. The primary endpoint, pathologic complete response (pCR) rate at a mean 9 weeks, was 86% overall (96% at primary and 84% in the neck). With longer-term follow-up of 36 months [10], 3-year locoregional control (LRC), distant metastasis-free survival (DFMS), and cancer-specific survival (CSS) were all 100%, and 3-year OS was 95%. High pathologic response rates and promising 3-year oncologic outcomes with a deintensified chemoRT regimen were important early steps in the path toward future deescalation trials for HPV-associated OPSCC. An important caveat is that these high disease control and survival outcomes occurred with trimodality therapy involving surgery, and further investigation of definitive deintensified chemoRT without surgery was the logical next step.

Thus a follow-up phase II study from the **UNC group (Chera et al.)** [11] investigated the same deescalated regimen without the addition of surgery in the same patient population, with the exception that very early-stage patients (T0-2, N0-1 by American Joint Committee on Cancer [AJCC] 7th ed.) received RT alone without chemotherapy. At median follow-up of 31.8 months, 2-year LRC was 95%, 2-year DMFS was 91%, 2-year progression-free survival (PFS) was 86%, and 2-year OS was 95%. Importantly, global quality of life returned to baseline over time, with patients reporting minimal to no worsening of their swallowing at later time points.

Concurrent systemic therapy deescalation

Two major trials, **De-ESCALaTE HPV** [12] (the United Kingdom/Ireland/the Netherlands) and **RTOG 1016** [13] (the United States/Canada), investigated the use of concurrent cetuximab with RT as a potentially less toxic radiosensitizing agent in place of cisplatin. This strategy was largely based on the favorable safety profile of this treatment combination in the Bonner trial [14], in which the toxicity experienced by patients treated with concurrent RT + cetuximab versus RT alone was similar. Though the Bonner trial was not specifically designed to test these two treatment strategies for HPV-related OPSCC, a retrospective analysis of this trial published by Rosenthal et al. [15] demonstrated superior disease control and OS with the addition of cetuximab.

De-ESCALaTE HPV [12] was a phase III trial of 334 patients with low-risk p16+ OPSCC with 1:1 randomization to IMRT (70 Gy/35 fractions over 7 weeks) with either concurrent bolus cisplatin (100 mg/m^2 q3 weeks x3 cycles) or concurrent cetuximab (loading dose 400 mg/m^2 followed by seven weekly infusions of 250 mg/m^2). The primary endpoint was overall (acute and late) severe grade $\geq$3 toxicity. Secondary endpoints included 2-year recurrence rates, 2-year OS, other toxicity metrics, and patient-reported outcomes (PROs) for quality of life (QOL). With median follow-up of 25.9 months, the RT + cisplatin arm showed significantly better 2-year recurrence rates (6.0% with cisplatin vs. 16.1% with cetuximab, $P = .0007$) and OS (97.5% with cisplatin vs. 89.4% with cetuximab, $P = .0012$). The vast majority of patients (95%–97%) were able to receive the full 70 Gy dose of RT on either arm, though systemic therapy completion differed, as only 38% of patients received the full three cycles of cisplatin compared with 79% receiving the full eight cetuximab infusions. There were no major differences in all-grade or grade ≥3 acute and late toxicity, although there were significantly more events qualifying as serious adverse events (AEs) leading to hospitalizations with cisplatin. PROs were generally equivalent to slightly favoring cisplatin (e.g., overall functioning). The study authors conclude that RT + cisplatin should continue to be the standard of care in HPV-associated OPSCC.

RTOG 1016 [13] was a randomized multicenter noninferiority trial of 849 patients with either low-risk (79%) or intermediate-risk (29%) HPV-associated OPSCC. Patients were randomized to either concurrent cisplatin or concurrent cetuximab, both with accelerated IMRT (70 Gy/35 fractions over 6 weeks, with six fractions per week involving delivery of two fractions on 1 day). In addition to the difference in RT schedule, the cisplatin arm entailed only two (rather than three) scheduled bolus infusions of 100 mg/m^2 each on days 1 and 22 of RT. The cetuximab dosing was identical to De-ESCALaTE. The primary endpoint was OS, and with median follow-up 4.5 years, noninferiority of cetuximab was rejected. For cetuximab + RT versus cisplatin + RT, 5-year OS estimates were 77.9% versus 84.6% ($P = .0163$), 5-year PFS was 67.3% versus 78.4% ($P = .0002$), and 5-year locoregional failure rates were 17.3% versus 9.9% ($P = .0005$). Rates of DM did not differ significantly (11.7% with cetuximab vs. 8.6% with cisplatin, $P = .09$). Like De-ESCALaTE, there were no significant differences in acute or late toxicity as measured by CTCAE grades; however, using a T-score approach, cisplatin was associated with a 40% higher acute toxicity burden. Notably, acute grade ≥3 toxicity was considerable in both arms at 77.4% (cetuximab) and 81.7% (cisplatin). PROs were also similar between groups. Receipt of the full radiation therapy course as prescribed was notably lower with the accelerated regimen, with 83%–86% of patients receiving RT per protocol. A subset analysis of patients with Zubrod 1 (as opposed to 0) performance status demonstrated that these patients actually did worse with cetuximab compared with cisplatin, and patients with Zubrod 1 received a higher mean overall cisplatin dose of 192 mg/m^2 compared with the overall cisplatin arm average of 184.7 mg/m^2, eschewing the notion that these patients are less able to tolerate cisplatin or have compromised outcomes with the drug. Notably, about 2/3 of patients had feeding tubes placed. Though direct comparisons are not possible, it is notable that 9% of patients were feeding tube dependent at 1 year on the RTOG trial, versus only 1% on the second-generation phase II UNC trial [11]; the frequency of feeding tube use on the UNC trial was also only 39%. It is thus plausible that a modest dose of radiation (from 70 Gy to 60 Gy) may lead to more marked toxicity improvements than switching from cytotoxic chemotherapy to a biological agent. The authors conclude that RT + cisplatin should remain standard of care for HPV-associated OPSCC.

Attempts to eliminate concurrent systemic therapy

Results from the phase II RCT **NRG HN-002** [16] were presented at the annual meeting for the American Society for Radiation Oncology in September 2019 and as of yet are only available in abstract form. This trial randomized 306 patients with p16+, T1-2N1-2b/T3N0-N2b (AJCC 7th ed.) OPSCC, and ≤10 pack-years of smoking history to either concurrent chemoRT with IMRT to 60 Gy over 6 weeks with concurrent weekly cisplatin at 40 mg/m^2 (IMRT+C) or accelerated IMRT alone (60 Gy over 5 weeks, with six fractions per week involving twice-daily [BID] fractions 1 day per week). Notably, both arms of the study can be considered experimental, as the dose of RT is low in both arms. Almost all patients (97%) received the full 60 Gy, and in the cisplatin arm, 80.9% of patients received at least five cycles (though some with dose reductions). With a median follow-up of 2.6 years, 2-year PFS was 90.5% with IMRT+C and 87.6% with accelerated IMRT. Each arm was measured against the prespecified null hypothesis 2-year PFS of 85%, and by this measure, IMRT+C achieved significantly improved PFS ($P = .0350$), but IMRT alone did not ($P = .2284$). Two-year OS was similar (96.7% and 97.3% for IMRT+C and IMRT, respectively). Swallowing function measured by the MD Anderson Dysphagia Inventory (MDADI) at 1 year met acceptability criteria in both arms. Grade 3 and 4 acute toxicities were 64.5% and 15.1%, respectively, in the IMRT+C arm, compared with 50.3% and 2.0%, respectively, in the IMRT arm; late toxicity rates were similar (16.7%–20.0% grade 3 and 1.3%–1.4% grade 4). Though acute severe toxicity was higher, IMRT+C met prespecified acceptability criteria for both PFS and swallowing function, while dose-reduced accelerated IMRT without concurrent systemic therapy met only swallowing endpoints. Similar to RTOG 1016 and De-ESCALaTE, these results suggest that omission of cisplatin is unwise. The IMRT+C arm was thus selected as the experimental arm in the newly opened NRG HN-005 trial (NCT03952585), which will compare this dose deescalated IMRT+C arm against a control arm of 70 Gy over 6 weeks with cisplatin (which was also the control arm of RTOG 1016).

Induction chemotherapy response-driven deescalation

ECOG 1308 [17] was a phase II single-arm trial that investigated deescalation of the definitive RT dose based on response to induction chemotherapy. In this trial, 80 patients with surgically resectable stage III–IV (AJCC 7th ed.) HPV-associated OPSCC received three cycles of induction cisplatin, paclitaxel, and cetuximab, followed by clinical assessment including endoscopy for primary tumor response. Those who achieved primary-site cCR (70% of patients who completed induction therapy) received 54 Gy in 27 fractions to gross disease and 51.3 Gy/27 fractions to elective nodal volumes with concurrent cetuximab. Those who did not achieve cCR after induction chemotherapy received 69.3 Gy in 33 fractions and the same subclinical volume dosing, with concurrent cetuximab. There was considerable crossover, with 8/18 patients intended to receive 69.3 Gy crossing over to the deescalated arm and 5/56 patients intended for the deescalated arm crossing over to the high-dose arm. Grade 4 toxicities were rare (<5% in both arms). The dose deescalated arm had less grade 3 mucositis (30% vs. 47%) and dysphagia (15% vs. 29%). Swallowing function and nutrition were significantly better at 12 months following RT in the deescalated arm.

Among all patients treated to ≤54 Gy, regardless of actual clinical response ($n = 62$ patients), 2-year PFS and OS were 81% and 93%, respectively. Two-year PFS and OS for patients treated to 69.3 Gy were 67% and 87%, respectively. Importantly, this trial was composed of a relatively higher-risk population compared with other deescalation studies, with 85% of patients staged IVA/B and 39% with >10 pack-year smoking history, including 16% current smokers. On subset analysis of patients with ≤10 pack-year smoking history, ≤T3 tumor, and ≤N2b disease who received deescalated RT, 2-year PFS was substantially higher at 96%. The authors concluded that response to induction chemotherapy can be used as a marker of responsiveness for patients who may achieve equivalent oncologic outcomes with dose-reduced RT while benefitting from reduced toxicity and that more favorable patient selection (lower tumor volume and minimal smoking history) may be the ideal population for definitive deescalation.

CCRO-022 (Chen, UCLA/UC Davis) [18] was a phase II single-arm trial of 45 patients with p16+ stage III–IV (AJCC 7th ed.) OPSCC treated at two institutions investigating radiation dose deescalation based on response to induction chemotherapy. Patients received two cycles of induction carboplatin/paclitaxel followed by CT scan at approximately 2 weeks following completion of chemotherapy. Those who exhibited a complete or partial response by RECIST criteria (55% of patients, with 11% achieving complete radiologic response of all disease) received a deintensified definitive chemoRT regimen of 54 Gy in 27 fractions (43 Gy to subclinical sites) with concurrent weekly paclitaxel × five cycles. Patients who did not meet at least partial response criteria received 60 Gy in 30 fractions (48 Gy to subclinical sites) with weekly paclitaxel. Treatment volumes were based on prechemotherapy extent of disease. With median follow-up of 30 months, 2-year PFS (primary endpoint) was 92%, and 2-year OS was 98%. Grade 3 toxicity was 39% (20% for nonhematologic toxicity), with notably low rates of grade 3 mucositis and dysphagia (9.1%, compared with 15%–30% with similar RT doses in ECOG 1308), as well as temporary feeding tube dependence (14%, over half of which were placed prophylactically). The authors concluded that induction chemotherapy to select patients who might benefit the most from deintensified definitive therapy showed promise.

Functional imaging-driven deescalation

Though HPV-associated OPSCC as a group demonstrates superior radiosensitivity, some HPV-associated tumors clearly do still recur. Specifically, intratumoral hypoxia is known to confer increased radioresistance and reduced local and distant disease control, and this has been shown to hold true for HPV-associated tumors [19,20]. A pilot study from Memorial Sloan Kettering [21] investigated incorporation of metabolic imaging to help identify patients most likely to have successful treatment deescalation. Using a previously validated PET-based tracer for detection of hypoxia, ^{18}F-fluoromisonidazole (^{18}F-FMISO) [22], Lee and colleagues [21] piloted incorporation of ^{18}F-FMISO to identify ideal candidates for treatment deintensification among patients with p16-positive OPSCC, in this case, specifically for deescalation in the treatment of grossly involved lymph nodes. Thirty-three patients underwent ^{18}F-FMISO PET at baseline pretreatment and again at 1 week into chemoRT. All tumors demonstrated a high baseline degree of hypoxia, and at repeat imaging 1 week into concurrent chemoRT, 10 patients (30%) demonstrated sufficient resolution of hypoxia and received a 10-Gy reduction from 70 Gy to 60 Gy. Two-year LRC was 100%, with 2-year DMFS of 97% and 2-year OS of 100%.

Target volume deescalation

OPTIMA [23] was a phase II trial exploring both induction therapy response–based treatment deescalation and target volume deescalation for HPV-associated OPSCC. Patients were classified as low risk ($n = 28$) or high risk ($n = 34$) based on AJCC 7th ed. T stage, N stage, and smoking status, in a manner slightly different from Ang risk stratification. Patients underwent three cycles of induction carboplatin/nab-paclitaxel and were assessed for response using RECIST criteria. Low-risk patients with ≥50% response to induction therapy received 50Gy RT alone over 5weeks. High-risk patients with ≥50% response and low-risk patients with 30%–50% response received concurrent chemoRT using a hyperfractionated RT regimen of 45Gy in 1.5-Gy fraction delivered BID, with concurrent TFHX (paclitaxel, 5-FU, and hydroxyurea). All other patients were treated with concurrent hyperfractionated chemoRT to 75Gy in 1.5-Gy fraction-delivered BID with concurrent TFHX (considered standard of care). Importantly, target volumes included only gross tumor and the first echelon of uninvolved nodes. At 4–8weeks after completion of therapy, all patients underwent neck dissection and/or biopsy or surgical resection of the primary tumor. Pathologic CR was observed in 90% of patients overall and 92% of patients who received deescalated therapy. At median follow-up of 29months, 2-year PFS was 95% for low-risk and 94% for high-risk disease, both meeting the noninferiority threshold for comparison to historic controls (85%). Deescalation from standard of care 75Gy chemoRT was associated with substantial reductions in grade ≥3 mucositis, dermatitis, and feeding tube requirement, with the lowest rates in the RT alone arm.

Deintensification of toxicity with radiation modality

Photon-based radiation planning and delivery techniques have improved dramatically over the past decade, with continued improvements in delivery of intensity modulation techniques. In particular, incorporation of volumetric arc therapy (VMAT) allows for highly conformal intensity-modulated dose distributions using continuous beam shaping with dynamic multileaf collimators and a significantly greater variety of gantry angles compared with static-beam IMRT [24,25]. This technique allows for improved healthy tissue sparing while maintaining coverage of target volumes [26]. Optimization of intensity modulation techniques based on better understanding of dosimetric predictors of xerostomia and dysphagia has improved the ability to prioritize toxicity-related objectives to reduce toxicity without compromising target coverage [27,28]. For example, in a study of 45 patients with OPSCC (39 with HPV-associated OPSCC) treated with IMRT focused on swallowing function preservation planning objectives, at median 6.5year follow-up, patients had stable or improved health-related QOL compared with pretreatment baseline and no late detriments compared with QOL at 2-year posttreatment [29].

Proton therapy has been proposed as an attractive alternative to photon therapy, due to its sharp dose falloff (Bragg peak) [30] compared with the slower decline and exit dose produced by photons. This is particularly relevant in OPSCC, for which critical tissue structures involved in swallowing function and salivary production are often in close proximity to tumor, creating the need for a rapid slope of dose decline between target volumes and healthy neighboring tissues.

Intensity-modulated proton therapy (IMPT) may further reduce dose to critical swallowing function structures and clinically reduce severe swallowing dysfunction compared with IMRT [31]. In a planning study comparison of IMPT versus IMRT with the objective to reduce dose to swallowing function organs at risk, Laan et al. [31] published a predicted 9% reduction in grade ≥2 predicted swallowing dysfunction compared with IMRT with the same tissue-sparing objectives. Much of the actual dose reductions from proton-based planning were achieved in the mean dose to the ipsilateral and contralateral parotid glands and the supraglottic larynx, with small improvements in mean submandibular gland dose. Importantly, target location with respect to critical structures remains the greatest predictor of dose to those tissues [31].

A prospective study of 50 patients with OPSCC (89% p16+) treated with IMPT at MD Anderson investigated toxicity and disease control outcomes for this modality, which was combined with concurrent chemotherapy in 64% of patients [32]. There were no grade ≥4 CTCAE toxicities, and while 58% of patients experienced grade 3 mucositis, acute and late grade 3 xerostomia rates were both only 2%, and acute and late dysphagia rates were 24% and 12%, respectively. Overall, 2-year PFS was 88.6% with a crude failure rate of 8% [32].

Deescalation with surgical techniques

The advent of minimally invasive surgical techniques such as TORS for the management of resectable OPSCC has brought significant reductions in treatment-related morbidity compared with previously used open surgical techniques [33]. However, improvements in conformal dose delivery and the ability to spare critical structures with IMRT and more advanced modulation techniques have also led to dramatic reductions in treatment toxicity and concomitant improvements in cancer-specific survival [34–36]. A metaanalysis examining cancer-specific and functional outcomes after TORS or IMRT found similar oncologic outcomes between the two approaches but possibly superior functional outcomes with TORS; however, there was substantial heterogeneity in the population and an imbalance in patient factors between patients treated surgically versus nonsurgically [37]. Thus with excellent outcomes for HPV-associated early-stage OPSCC, the question of whether one modality is truly superior still remains an important unanswered question.

Published in October 2019, **ORATOR** [38] (radiotherapy vs. transoral robotic surgery and neck dissection for oropharyngeal squamous cell carcinoma) was the first randomized trial comparing definitive IMRT versus TORS and was designed to fill this gap in the knowledge base regarding the optimal treatment of early-stage OPSCC. This phase II trial randomized 68 patients with T1-2 and N0-2 (maximum nodal size ≤4cm) OPSCC to either standard definitive IMRT (if N0) or chemoRT (if N1-2) to 70Gy/35 fractions versus TORS plus selective neck dissection (ND), with the option for adjuvant therapy (RT or chemoRT) based on pathology. Though not specifically directed at HPV-associated OPSCC, 88% of enrolled patients had p16 + disease. In the definitive RT arm, concurrent chemotherapy was given for all node-positive patients (72% of patients), mostly with bolus cisplatin q3 weeks (83% and 17% with carboplatin or cetuximab). Only 29% of patients received TORS without adjuvant therapy. Based on imaging performed at 8–10weeks after completion of RT demonstrating evidence of residual gross disease (general, neck nodes >1cm), 12% of patients in the RT underwent subsequent salvage neck dissection. Adjuvant RT was delivered after TORS for node positivity, lymphovascular space invasion (LVSI), pT3-4, or close margin of <2mm (adjuvant RT alone delivered to 47% of patients), and concurrent chemotherapy (cisplatin or carboplatin) was added for extranodal extension (ENE) or positive margin (24% of patients).

The primary endpoint was swallowing-related QOL as measured by the MDADI at 1year, with a 10-point change in score designated as a clinically meaningful change. Total MDADI scores resulted in favor of RT, although not meeting the 10-point clinically meaningful threshold. However, global and emotional domains did meet the threshold for clinical significance in favor of RT. Though the rates of grade ≥2 toxicities were similar between arms (91% and

97% with RT and TORS, respectively), side-effect profiles differed, with more neutropenia and otologic toxicities in the RT arm versus worse trismus, dysphagia, and bleeding (18% grade ≥2 bleeding events, including one bleeding-related death) in the TORS arm. Though not quite reaching significance, 100% of patients in the RT arm returned to total oral diet at 1 year compared with 84% with TORS/ND ($P = .055$). With 27-month median follow-up, 3-year PFS and OS were similar between the two treatment arms. On subgroup analysis of p16+ versus p16– patients, p16 positivity was associated with markedly superior 3-year PFS (93.0% vs. 30.0%) and OS (93.1% vs. 30.0%), though the number of patients with p16– tumors was small. The investigators have launched a subsequent randomized phase II trial, the ORATOR II study (NCT03210103), to compare deescalated radiotherapy versus TORS with deescalated adjuvant therapy in patients with HPV-associated OPSCC.

Radiation dose deescalation in the adjuvant setting

MC1273 [39] was a phase II single-arm nonrandomized trial of markedly dose deescalated adjuvant chemoRT in p16+ OPSCC after curative-intent surgery (mostly TORS) with negative margins. Patients were stratified by the presence of ENE. All patients received 30 Gy in 1.5-Gy fraction delivered BID with concurrent weekly docetaxel at 15 mg/m^2. Patients with ENE ($n = 43$) also received simultaneous integrated boost (SIB) to 36 Gy (1.8 Gy/fraction, BID) to nodal levels with ENE. The impetus for choosing this dose was derived from the original Nigro protocol for anal cancer [40]. There was a 100% completion rate of chemoRT with no major treatment breaks. Patients were stage III–IV by AJCC 7th ed. (or, by AJCC 8th ed., 72% stage I, 24% stage II, and 3.8% stage III) with a ≤10 pack-year smoking history. With median follow-up of 3 years, 2-year LRC was 96.2%, 2-year DMFS was 94.9%, and 2-year PFS was 91.1%. The majority of local and distant failures occurred in the ENE+ arm, and multiple excisions required to achieve negative margins appeared to be a risk factor for recurrence. Late toxicity rates were low: grade 2 rate 1.4% at 1 year and 6.7% at 2 years. All-grade ≥3 toxicities were acute, with 16.5% grade 3 toxicities at 3 months and none at 6-month follow-up or later. A subset of patients was included in a cost analysis, which showed a 33% reduction in cost with this deintensified regimen. Overall, results were favorable compared with prior reports of outcomes for standard therapy.

Ongoing deescalation trials

Currently, there are numerous ongoing clinical trials studying deescalation in HPV-associated OPSCC. A total of 26 National Clinical Trial (NCT)-registered studies investigating this subject are underway—including 18 in the definitive setting (Table 2A) and 8 in the adjuvant setting (Table 2B)—representing accrual goals of 2291 and 1678 patients, respectively. Overall the phase of study for the 26 trials varies: phase II (16), phase II/III (1), phase III (5), and not classified (4). Staging inclusion criteria for each trial are based on AJCC 7th ed. (9), AJCC 8th ed. (12), or not specified (5). Regarding trial study design, there are 10 randomized studies and 7 single-arm studies, with the remainder composed of studies in which treatment is decided based on stratification criteria.

TABLE 2A Ongoing clinical trials involving treatment deescalation for HPV-associated OPSCC in the definitive setting.

Title (NCT number)	Phase	Enrollment goal	Study completion date	Selected inclusion criteria	Study design/interventions	Primary measure
Chemotherapy and locoregional therapy trial (surgery or radiation) for patients with head and neck cancer (OPTIMA-II) (NCT03107182) [42]	2	56	06/2020	- T3-T4/N2-N3 (AJCC 7th)	After induction nab-paclitaxel/carboplatin/nivolumab × 3: - Low-risk and small-volume tonsillar/BOT disease (T1–2, nonbulky N2A-N2B) with ≥50% reduction following induction: TORS +/− selective nodal dissection (plus adjuvant RT for adverse path); if patient not candidate or refuses TORS, then RT alone (50 Gy) - Low risk with 30%–50% reduction OR high-risk features (T4, bulky N2B, N2C-N3, >10 pack years) with ≥50% reduction: ChemoRT (50 Gy) with cisplatin or ChemoRT (45 Gy) with TFHX - Low-risk features and <30% reduction OR high-risk disease with <50% reduction or any patients with progressive disease during induction chemo: ChemoRT (70 Gy) with cisplatin or ChemoRT (75 Gy) with TFHX (Adjuvant nivolumab offered to all patients for 6-month postcompletion)	Two-year deep response rate (DRR) measured by tumor shrinkage (%)

Major radiation reduction for HPV+ oropharyngeal carcinoma (NCT03323463) [41]	2	76	10/2020	- p16+ SCC of OP or UP (w/PI consent) - cT1-2/N1-2c/M0 (AJCC 7th)	Single-arm study - ChemoRT (30 Gy over 3 weeks) with cisplatin (or carboplatin/5-FU if cisplatin not tolerated)	Two-year effectiveness of treatment by standard CT (or MRI) or tumor site and PET scan
Deescalation protocols in HPV-related oropharyngeal carcinoma in Chinese populations (NCT04012502) [47]	N/A	83	12/2020	- T1-2/N1-3 (except T1N1 and single LN < 3 cm) or T3–4/N0–3/M0 (AJCC 8th)	After induction chemo (TP × 2): - <50% PR after induction: ChemoRT (70 Gy) with cisplatin - >50% PR after induction: RT alone (60 Gy)	Two-year PFS
Adaptive treatment deescalation in favorable risk HPV-positive oropharyngeal carcinoma (NCT03215719) [56]	2	54	04/2021	- T1-T2/N1-N2b or T3/N1-N2b (AJCC 7th)	Stratification based on nodal shrinkage at 4-week interval scan - >40% nodal shrinkage: Dose deescalated treatment[a] - ≤40% nodal shrinkage: Standard treatment[a]	Two-year PFS
Use of a proliferation saturation index to determine personalized radiotherapy for HPV+oropharyngeal cancers (NCT03656133) [58]	N/A	60	08/2021	- T1-3/N0-1 (AJCC 8th)	Stratification based on proliferation saturation index (PSI) - RT[a] (Standard fractionation at 2Gy once daily) - RT[a] (Hyperfractionation at 1.2 Gy twice daily [≥6 h apart])	% of response at week 4
The quarterback trial: reduced dose radiotherapy for HPV+ oropharynx cancer (NCT01706939) [44]	3	24 (Actual)	12/2021	- p16+ SCC of OP/UP/NP - Stage III–IV disease without evidence of distant metastases	Randomized - ChemoRT (70 Gy) with carboplatin - ChemoRT (56 Gy) with carboplatin	Three-year PFS

Continued

TABLE 2A Ongoing clinical trials involving treatment deescalation for HPV-associated OPSCC in the definitive setting—cont'd

Title (NCT number)	Phase	Enrollment goal	Study completion date	Selected inclusion criteria	Study design/interventions	Primary measure
Weekly cetuximab/RT versus weekly cisplatin/RT in HPV-associated oropharyngeal squamous cell carcinoma (HPV oropharynx) (TROG 12.01) (NCT01855451) [53]	3	189 (Actual)	06/2022	- ≤10 pack-year: Stage III (excluding T1-2N1) or stage IV (excluding T4, N3, and distant metastasis) - >10 pack-year: N0-N2a (AJCC 7th)	Randomized - ChemoRT (70 Gy in 35 fractions over 7 weeks; weekly cetuximab) - ChemoRT (70 Gy in 35 fractions over 7 weeks; weekly cisplatin)	Twenty-week symptom severity
Ipilimumab, nivolumab, and radiation therapy in treating patients with HPV-positive advanced oropharyngeal squamous cell carcinoma (NCT03799445) [50]	2	180	08/2022	- T1/N2a-N2c, T2/N1-N2c, or T3N0-N2c (AJCC 7th)	Single-arm study - Nivolumab + Ipilimumab + IMRT (50–66 to low-intermediate volume in 30 fractions)	- Baseline to 1-month dose-limiting toxicity (safety lead in) - 6-month pCR rate - Baseline to 2-year PFS
Cisplatin or immunotherapy in association with definitive radiotherapy in HPV-related oropharyngeal squamous cell carcinoma: a randomized phase II trial (CITHARE) (NCT03623646) [51]	2	66	11/2022	- T1/N1-2 or T2-T3/N0-2 (AJCC 8th)	Randomized - ChemoRT (70 Gy) with cisplatin - RT (70 Gy) with durvalumab	One-year PFS
Individualized adaptive deescalated radiotherapy for HPV-related oropharynx cancer (NCT03416153) [43]	2	75	11/2022	- Stage I–II (AJCC 8th)	Stratification based on response at midtreatment PET-CT - If not meeting pPET-CT/iPET-CT parameters: ChemoRT (70 Gy in 35 fractions) with concurrent carboplatin/paclitaxel	One-year LRR

					- If meets pPET-CT/iPET-CT parameters: ChemoRT (54/43.2Gy to PTV-HR/PTV-LR in 27 fractions) with concurrent carboplatin/paclitaxel	
Phase II treatment stratification trial using neck dissection-driven selection to improve quality of life for low-risk patients with HPV+ oropharyngeal squamous cell cancer (NCT02784288) [57]	2	100	12/2022	- T1-3/N0-2c (AJCC 7th)	Stratification based on pathology from neck dissection - 0-1 LN <6cm, no ENE, no PNI/PVI): TOS alone - 2 nodes without ENE, or if PNI/PVI: RT[a] - ENE or positive margins: ChemoRT[a] w/carboplatin/paclitaxel	One-year QOL posttreatment (vs. historical controls)
PEARL PET-based adaptive radiotherapy clinical trial (PEARL) (NCT03935672) [55]	2	50	02/2023	- cT1-3/N0-1/M0 (UICC 8th)	Single-arm study - ChemoRT[a] (33 fractions) with concurrent cisplatin; rescan after 2 weeks of treatment helps delineate new GTV from which RT is replanned subsequently thereafter	Two-year PFS
p16+ oropharyngeal cancer radiation optimization trial reducing elective treatment volumes (PROTEcT) (NCT04104945) [48]	N/A	32	10/2023	- T1-3/N1-2 (AJCC 8th)	Single-arm study: - ChemoRT (60/54Gy in 30 fractions to gross disease/subclinical regions at risk with level IB excluded) with concurrent cisplatin or cetuximab (at medical oncologist's discretion)	Xerostomia QOL Toxicity (CTCAE)

Continued

TABLE 2A Ongoing clinical trials involving treatment deescalation for HPV-associated OPSCC in the definitive setting—cont'd

Title (NCT number)	Phase	Enrollment goal	Study completion date	Selected inclusion criteria	Study design/interventions	Primary measure
Deintensification of radiation and chemotherapy for low-risk HPV-related oropharyngeal SCC: follow-up study (NCT02281955) [45]	2	115 (Actual)	11/2024	- T0-3/N0-2c/M0 (AJCC 7th)	Single-arm study - ChemoRT (60/54 Gy in 30 fractions to PTV-HR/PTV-SR, respectively) with cisplatin (chemo omitted if T0-2/N0-1 disease × 6, ≤10 pack-year smoking history; PET-CT 10–16 weeks posttreatment completion and if positive patients undergo surgical evaluation)	Two-year PFS
Deescalation radiotherapy in patients with low-risk HPV-related oropharyngeal squamous cell carcinoma (EVADER) (NCT03822897) [54]	2	100	12/2024	- T1-3/N0-1/M0 (AJCC 8th)	Single-arm study - Volume-reduced elective nodal irradiation with two options: chemoRT (70/56 Gy in 35 fractions; 5 fractions/week with cisplatin) or RT alone (70 Gy/56 Gy in 35 fractions, 6 fractions/week)	Five-year event-free survival
Deintensified radiation therapy with chemotherapy (cisplatin) or immunotherapy (nivolumab) in treating patients with early-stage, HPV-positive, nonsmoking-associated oropharyngeal cancer (NRG-HN005) (NCT03952585) [49]	2/3	711	02/2025	- cT1-2/N1 or T3/N0-1 (AJCC 8th)	Randomized - ChemoRT (70 Gy in 6 weeks; 6 fractions/week) with cisplatin - ChemoRT (60 Gy in 6 weeks; 5 fractions/week) with cisplatin ChemoRT (60 Gy in 5 weeks; 6 fractions/week) with nivolumab	PFS QOL (MDADI)

Cisplatin plus radiotherapy versus durvalumab plus radiotherapy followed by durvalumab versus durvalumab plus radiotherapy followed by tremelimumab and durvalumab in intermediate risk HPV-positive oropharyngeal SCC (CCRG HN.9) (NCT03410615) [52]	2	180	07/2026	- If ≥10 pack-year: T1-2/N1 or T3/N0-1 - Any smoking history: T1-3/N2 (AJCC 8th)	Randomized - ChemoRT (70 Gy in 35 fractions; q3w cisplatin) - RT (70 Gy in 35 fractions) with concurrent durvalumab followed by adjuvant durvalumab - RT (70 Gy in 35 fractions) with concurrent durvalumab followed by adjuvant durvalumab/tremelimumab	Three-year event-free survival
Primary radiotherapy versus primary surgery for HPV-associated oropharyngeal cancer (ORATOR2) (NCT03210103) [46]	2	140	08/2028	- T1-2/N0-2 (AJCC 8th)	Randomized - RT (60 Gy) +/− chemo - Transoral surgery + neck dissection (plus 50 Gy RT, if required)	Two-year OS

[a] *Radiation dose not explicitly specified in clinical trial protocol.*

TABLE 2B Ongoing clinical trials involving treatment deescalation for HPV-associated OPSCC in the adjuvant setting.

Title (NCT number)	Phase	Enrollment goal	Study completion date	Selected inclusion criteria	Study design/ interventions	Primary measure
Transoral surgery followed by low-dose or standard-dose radiation therapy with or without chemotherapy in treating patients with HPV-positive stage III–IVA oropharyngeal cancer (ECOG E3311) (NCT01898494) [64]	2	511 (Actual)	02/2020	- Low risk: T1-T2/N0-1 and clear (≥ 3 mm) margins, and no ENE/PNI/LVI - Intermediate-risk: any 1 of following: <3 mm margin, <1 mm ENE, N2a-b (AJCC 7th), PNI or LVI - High risk: any 1 of following: positive margin, >1 mm ENE, or 5+ positive LNs	Parallel assignment/ randomized: - Low risk: observation - Intermediate risk: - IMRT (50 Gy in 25 fractions) - IMRT (60 Gy in 30 fractions) - High risk: ChemoRT (66 Gy in 33 fractions) with cisplatin	Two-year PFS (intermediate risk)
Postoperative adjuvant therapy deintensification trial for human papillomavirus-related, p16+ oropharynx cancer (ADEPT) (NCT01687413) [67]	3	41	08/2021	- S/p TOS/ neck dissection for T1–4a OP primary to negative margin - pN+ with extracapsular extension	Randomized - IMRT alone (60 Gy in 30 fractions) - ChemoRT (60 Gy in 30 fractions) with weekly cisplatin	Five-year DFS Five-year LRC
Adjuvant deescalated radiation + adjuvant nivolumab for intermediate high risk P16+ oropharynx cancer (NCT03715946) [62]	2	135	08/2021	- If <10 pack-year: T0–3 plus either N2b-3 (AJCC 7th), +ENE >1 mm, or +margin - If >10 pack-year: T0-3 plus any N2-3, +ENE >1 mm, or + margin	Single-arm study - RT (45 or 50 Gy depending on tumor volume in 25 fractions, 6 fractions per week) with concurrent and adjuvant nivolumab	Three-year PFS One-year PEG tube dependence
Surgery followed by risk-directed postoperative adjuvant therapy for HPV-related oropharynx squamous cell carcinoma: "The Minimalist Trial (MINT) (NCT03621696) [61]	2	40	07/2024	- p16+ SCC of OP/UP - Stage I-III (AJCC 8th)	Stratification based on pathologic features - No ENE and no + margin: RT (42 Gy in 21 fractions) alone	Percentage weight loss in patients during postoperative adjuvant chemoRT

					- ENE or + margin: ChemoRT (42 Gy in 21 fractions) with cisplatin - pT4 / cN3 disease: ChemoRT (60 Gy in 30 fractions) with cisplatin	
Evaluation of deescalated adjuvant radiation therapy for human papillomavirus (HPV)-associated oropharynx cancer (DART-HPV) (NCT02908477) [60]	3	214	09/2024	- GTR of primary and at least unilateral neck dissection - Absence of distant metastases - At least one of: LN >3 cm; 2+ LNs; PNI; LVSI; T3-T4 primary; ENE	Parallel assignment/ randomized: - Active comparator: RT (60 Gy in 30 fractions over 6 weeks) if intermediate risk; add concurrent cisplatin for high risk - Experimental: ChemoRT (30 Gy if intermediate risk or 36 Gy if high risk; in 20 BID fractions) with weekly docetaxel	Two-year adverse events rate (grade 3–5 toxicities)
Postoperative adjuvant treatment for HPV-positive tumors (PATHOS) (NCT02215265) [65]	3	242	11/2026	- T1-3/N0-2b (UICC 7th), or T1-T3/N0-1 (UICC 8th)	Initial stratification based on pathology; arms subsequently randomized - No adverse histological features: no adjuvant treatment - T3, T1-T2 (with additional risk factors), pN2a/pN2b (AJCC 7th), PNI/PVI, 1–5 mm margin:	One-year MDADI One-year OS

Continued

TABLE 2B Ongoing clinical trials involving treatment deescalation for HPV-associated OPSCC in the adjuvant setting—cont'd

Title (NCT number)	Phase	Enrollment goal	Study completion date	Selected inclusion criteria	Study design/ interventions	Primary measure
					- PORT (60 Gy in 30 fractions) - PORT (50 Gy in 25 fractions) - <1 mm margin (but negative marginal biopsies) and/or ENE: - POCRT (60 Gy in 30 fractions with cisplatin) - PORT (60 Gy in 30 fractions)	
Study of postoperative radiation therapy for HPV-associated oropharyngeal squamous cell carcinoma (NCT03875716) [66]	2	111	12/2027	- Low risk: (pT1-2/N0-1 (AJCC 8th); ≥15 LNs dissected; ≤2+LNs confined to level 2/3; no ENE; clear margins) - Intermediate risk (pT1-2/N0-2 (AJCC 8th) and either >2+LNs, <15 LNs dissected, +LNs in ipsilateral level IB/IV/V or anywhere contralaterally, <1 mm ENE, or close margins) - High-risk (pT1-4/N0-2 (AJCC 8th) and >1-mm ENE or microscopic +margins)	Stratification based on pathologic features - Low risk: No intervention - Intermediate risk: Reduced-dose adjuvant RT[a] - High risk: Adjuvant RT[a] without chemo	Two-year DFS

Deescalation of adjuvant radio (chemo) therapy for HPV-positive head-neck squamous cell carcinomas (DELPHI) (NCT03396718) [63]	N/A	384	01/2029	- Intermediate risk (pT3/R0 and/or LN+ but ENE-) - High risk (R1 and/or pT4 and/or 3+ LNs and/or ENE)	Stratification based on pathologic features - Observational arm, HPV-: RT (60/66 Gy) +/− chemo - Observational arm, HPV+: RT (60/66 Gy) +/− chemo - Experimental arm A, HPV+: RT (54/59.4 Gy) +/− chemo - Experimental arm B, HPV+: RT (48.8/55 Gy) +/− chemo (Chemo given based on # of LNs, ENE, and residual tumor)	Two-year LRR

[a] *Radiation dose not explicitly specified in clinical trial protocol.*

Abbreviations: *AJCC*, American Joint Committee on Cancer; *BOT*, base of tongue; *CT*, computed tomography; *ChemoRT*, chemoradiation; *ENE*, extranodular extension; *GTR*, gross total resection; *GTV*, gross tumor volume; *HPV*, human papillomavirus; *IMRT*, intensity-modulated radiation therapy; *LN*, lymph node; *LRC*, locoregional recurrence; *LRR*, locoregional recurrence; *LVSI*, lymphovascular space invasion; *MDADI*, MD Anderson Dysphagia Inventory; *NP*, nasopharynx; *OP*, oropharynx; *OPSCC*, oropharyngeal squamous cell carcinoma; *OS*, overall survival; *pCR*, pathologic complete response; *PET*, positron emission tomography; *PFS*, progression free survival; *PNI*, perineural invasion; *POCRT*, postoperative chemoradiation; *PORT*, postoperative radiation; *PTV-HR*, high-risk planning treatment volume; *PTV-LR* low-risk planning treatment volume; *PTV-SR*, standard-risk planning treatment volume; *PVI*, perivascular invasion; *Q3W*, every 3 weeks; *QOL*, quality of life; *RT*, radiation therapy; *TFHX*, paclitaxel, 5-FU, hydroxyurea, dexamethasone, famotidine, and diphenhydramine; *TORS*, transoral robotic surgery; *TOS*, transoral surgery; *UICC*, Union for International Cancer Control; *UP*, unknown primary.

Deescalation in the definitive setting

Of the HPV-associated OPSCC deescalation studies in the definitive setting, deescalation has been attempted through various mechanisms. Ten studies involve reduction in radiation dose, with total doses ranging from 30 Gy (Phase II **MSKCC study** [41], NCT03323463), 45–50 Gy (Phase II **OPTIMA-II** [42], NCT03107182), 54 Gy (Phase II **University of Michigan study** [43], NCT03416153), 56 Gy (Phase III **Quarterback Trial** [44], NCT01706939), up to 60 Gy (Phase II **UNC study** [45], NCT02281955; Phase II **ORATOR2** [46], NCT03210103; **Fudan University study** [47], NCT04012502; **PROTEcT** [48], NCT04104945; and Phase II/III **NRG-HN005** [49], NCT03952585), instead of the standard 66–70 Gy usually delivered in the definitive setting. Notably, within this cohort, the randomized **NRG-HN005** [49] (estimated enrollment of 711 patients with cT1-2/N1 or T3/N0-1 p16+ OPSCC) aims to concurrently study deescalation of radiation dose (70 Gy vs. 60 Gy) while also investigating deescalation with concurrent systemic therapy (cisplatin vs. nivolumab).

Beyond **NRG-HN005** [49], other studies in the definitive setting investigating deescalation of concurrent systemic therapies include single-arm Phase II **MDACC study** [50] (NCT03799445) evaluating nivolumab/ipilimumab/IMRT, randomized Phase II **CITHARE** [51] (NCT03623646) evaluating chemoRT with concurrent cisplatin versus durvalumab, randomized Phase II **CCRG HN.9** [52] (NCT03410615) evaluating RT with concurrent cisplatin versus concurrent durvalumab and adjuvant durvalumab versus concurrent durvalumab and adjuvant durvalumab/tremelimumab, and randomized Phase III **TROG 12.01** [53] (NCT01855451) evaluating chemoRT with cisplatin versus cetuximab.

In addition to **PROTEcT** [48], which involves reduction of RT dose and omission of radiation to nodal level IB, two other trials also study the feasibility of reducing total volume irradiated—phase II **EVADER** [54] (NCT03822897) involving volume-reduced elective nodal irradiation and phase II **PEARL** [55] (NCT03935672) involving positron emission tomography-computed tomography (PET-CT) scan 2 weeks after initiation of RT to facilitate delineation of a smaller GTV that is targeted for the remainder of treatment.

While 12 of 18 deescalation trials in the definitive setting are either single-arm or randomized, six involve stratification of patients based on degree of response to initial treatment. Of these, two utilize tumor response to induction chemotherapy to determine whether a patient will receive deescalated RT dose: Phase II **OPTIMA-II** [42] involving induction nab-paclitaxel/carboplatin/nivolumab × 3 cycles and a **Fudan University study** [47] involving induction docetaxel + cisplatin × 2 cycles. Two additional studies seek to determine whether to deescalate radiation dose based on a patient's response to radiation midtreatment: a phase II **NYU study** [56] (NCT03215719) involving scanning at the 4-week treatment interval and a phase II **University of Michigan study** [43] involving midtreatment PET-CT. Another phase II **University of Michigan study** [57] (NCT02784288) aims to stratify patients based on results of an initial neck dissection—with patients subsequently getting either transoral surgery, radiation alone, or chemoradiation dependent on pathology. Finally, a **Moffitt study** [58] (NCT03656133) aims to use the proliferation saturation index (defined as the ratio of tumor volume to the host-influenced tumor-carrying capacity) [59] to determine whether a patient would be better suited for radiation with standard fractionation or hyperfractionation.

Deescalation in the adjuvant setting

Regarding the eight studies investigating deescalation in the adjuvant setting, seven involve a component of reduction in RT dose. The deescalated adjuvant doses range from 30 to 36 Gy depending on intermediate/high risk (Phase III **DART-HPV** [60], NCT02908477, expected to accrue 214 patients), 42 Gy (Phase II **MINT trial** [61], NCT03621696), 45–50 Gy depending on tumor volume (Phase II **University of Pittsburgh study** [62], NCT03715946), 48–54 Gy (**DELPHI** [63], NCT03396718, expected to accrue 384 patients), up to 50 Gy (Phase II **ECOG 3311** [64], NCT01898494, which has enrolled 511 patients; Phase III **PATHOS** [65], NCT02215265, expected to accrue 242 patients).

Of the aforementioned adjuvant studies involving deescalation of RT dose, two also involve simultaneous deescalation of concurrent systemic therapy: **University of Pittsburgh study** [62] (single-arm study in patients with intermediate high risk receive adjuvant RT with concurrent and adjuvant nivolumab) and **DART-HPV** [60] (randomized trial in which the high-risk subset of patients received either docetaxel with low-dose RT vs. cisplatin in standard dose RT arm). An additional three studies involve a component of systemic therapy avoidance in certain adjuvant settings. In the subset of **PATHOS** [65] patients (inclusion criteria T1-3/N0-2b AJCC 7th ed.) with close margin or ENE, patients are randomized to postoperative chemoRT (60 Gy) with cisplatin or postoperative RT alone (60 Gy). In a Phase II **DFCI study** [66] (NCT03875716), patients with pT1-4/N0-2 (AJCC 8th ed.) and either >1-mm ENE or microscopic positive margins are treated with adjuvant radiation alone without systemic therapy. Finally, in the Phase III **ADEPT trial** [67] (NCT01687413), patients with pN+ and ENE are randomized to receive either adjuvant IMRT alone (60 Gy) or adjuvant chemoRT (60 Gy) with cisplatin.

Conclusion

HPV-associated OPSCC is a separate clinical disease entity distinct from HPV-negative OPSCC and is associated with markedly improved treatment outcomes. As a result, numerous efforts have been made to identify patients eligible for treatment deescalation and to ascertain methods of deescalation that preserve excellent prognosis while limiting treatment toxicity. Despite these efforts, there remains no nationally recognized treatment paradigm specific to patients with HPV-associated OPSCC. As a result, it remains imperative at this time to treat all patients off-protocol according to the standard of care developed for all patients with OPSCC regardless of HPV status. Enrollment of all eligible patients in any of the several ongoing clinical trials described earlier is highly encouraged.

References

[1] Mirghani H, Blanchard P. Treatment de-escalation for HPV-driven oropharyngeal cancer: where do we stand? Clin Transl Radiat Oncol 2018;8:4–11. https://doi.org/10.1016/j.ctro.2017.10.005.

[2] Boscolo-Rizzo P, Del Mistro A, Bussu F, Lupato V, Baboci L, Almadori G, DA Mosto MC, Paludetti G. New insights into human papillomavirus-associated head and neck squamous cell carcinoma. Acta Otorhinolaryngol Ital 2013;33(2):77–87.

[3] Arenz A, Ziemann F, Mayer C, Wittig A, Dreffke K, Preising S, Wagner S, Klussmann JP, Engenhart-Cabillic R, Wittekindt C. Increased radiosensitivity of HPV-positive head and neck cancer cell lines due to cell cycle dysregulation and induction of apoptosis. Strahlenther Onkol 2014. https://doi.org/10.1007/s00066-014-0605-5.

[4] Bonilla-Velez J, Mroz EA, Hammon RJ, Rocco JW. Impact of human papillomavirus on oropharyngeal cancer biology and response to therapy. Implications for treatment. Otolaryngol Clin North Am 2013;46(4):521–43. https://doi.org/10.1016/j.otc.2013.04.009.

[5] O'Sullivan B, Huang SH, Su J, Garden AS, Sturgis EM, Dahlstrom K, Lee N, Riaz N, Pei X, Koyfman SA, Adelstein D, Burkey BB, Friborg J, Kristensen CA, Gothelf AB, Hoebers F, Kremer B, Speel EJ, Bowles DW, Raben D, Karam SD, Yu E, Xu W. Development and validation of a staging system for HPV-related oropharyngeal cancer by the International Collaboration on Oropharyngeal cancer Network for Staging (ICON-S): a multicentre cohort study. Lancet Oncol 2016;17(4):440–51. https://doi.org/10.1016/S1470-2045(15)00560-4.

[6] Eisbruch A, Harris J, Garden AS, Chao CKS, Straube W, Harari PM, Sanguineti G, Jones CU, Bosch WR, Ang KK. Multi-institutional trial of accelerated hypofractionated intensity-modulated radiation therapy for early-stage oropharyngeal cancer (RTOG 00-22). Int J Radiat Oncol Biol Phys 2010;76(5):1333–8. https://doi.org/10.1016/j.ijrobp.2009.04.011.

[7] O'Sullivan B, Huang SH, Siu LL, Waldron J, Zhao H, Perez-Ordonez B, Weinreb I, Kim J, Ringash J, Bayley A, Dawson LA, Hope A, Cho J, Irish J, Gilbert R, Gullane P, Hui A, Liu FF, Chen E, Xu W. Deintensification candidate subgroups in human papillomavirus-related oropharyngeal cancer according to minimal risk of distant metastasis. J Clin Oncol 2013;31(5):543–50. https://doi.org/10.1200/JCO.2012.44.0164.

[8] Ang KK, Harris J, Wheeler R, Weber R, Rosenthal DI, Nguyen-Tân PF, Westra WH, Chung CH, Jordan RC, Lu C, Kim H, Axelrod R, Silverman CC, Redmond KP, Gillison ML. Human papillomavirus and survival of patients with oropharyngeal cancer. N Engl J Med 2010. https://doi.org/10.1056/NEJMoa0912217.

[9] Chera BS, Amdur RJ, Tepper J, Qaqish B, Green R, Aumer SL, Hayes N, Weiss J, Grilley-Olson J, Zanation A, Hackman T, Funkhouser W, Sheets N, Weissler M, Mendenhall W. Phase 2 trial of de-intensified chemoradiation therapy for favorable-risk human papillomavirus–associated oropharyngeal squamous cell carcinoma. Int J Radiat Oncol Biol Phys 2015;93(5):976–85. https://doi.org/10.1016/j.ijrobp.2015.08.033.

[10] Chera BS, Amdur RJ, Tepper JE, Tan X, Weiss J, Grilley-Olson JE, Hayes DN, Zanation A, Hackman TG, Patel S, Sheets N, Weissler MC, Mendenhall WM. Mature results of a prospective study of deintensified chemoradiotherapy for low-risk human papillomavirus-associated oropharyngeal squamous cell carcinoma. Cancer 2018;124(11):2347–54. https://doi.org/10.1002/cncr.31338.

[11] Chera BS, Amdur RJ, Green R, Shen C, Gupta G, Tan X, Knowles M, Fried D, Hayes N, Weiss J, Grilley-Olson J, Patel S, Zanation A, Hackman T, Zevallos J, Blumberg J, Patel S, Kasibhatla M, Sheets N, Weissler M, Yarbrough W, Mendenhall W. Phase II trial of de-intensified chemoradiotherapy for human papillomavirus–associated oropharyngeal squamous cell carcinoma. J Clin Oncol 2019;37(29):2661–9. https://doi.org/10.1200/JCO.19.01007.

[12] Mehanna H, Robinson M, Hartley A, Kong A, Foran B, Fulton-Lieuw T, Dalby M, Mistry P, Sen M, O'Toole L, Al Booz H, Dyker K, Moleron R, Whitaker S, Brennan S, Cook A, Griffin M, Aynsley E, Rolles M, De Winton E, Chan A, Srinivasan D, Nixon I, Grumett J, Leemans CR, Buter J, Henderson J, Harrington K, Mcconkey C, Gray A, Dunn J, Group D-EscalHPVT. Radiotherapy plus cisplatin or cetuximab in low-risk human papillomavirus-positive oropharyngeal cancer (De-ESCALaTE HPV): an open-label randomised controlled phase 3 trial. Lancet 2019;393(10166):51–60. https://doi.org/10.1016/S0140-6736(18)32752-1.

[13] Gillison ML, Trotti AM, Harris J, Eisbruch A, Harari PM, Adelstein DJ, Sturgis EM, Burtness B, Ridge JA, Ringash J, Galvin J, Yao M, Koyfman SA, Blakaj DM, Razaq MA, Colevas AD, Beitler JJ, Jones CU, Dunlap NE, Seaward SA, Spencer S, Galloway TJ, Phan J, Dignam JJ, Le QT. Radiotherapy plus cetuximab or cisplatin in human papillomavirus-positive oropharyngeal cancer (NRG Oncology RTOG 1016): a randomised, multicentre, non-inferiority trial. Lancet 2019;393(10166):40–50. https://doi.org/10.1016/S0140-6736(18)32779-X.

[14] Bonner JA, Harari PM, Giralt J, Azarnia N, Shin DM, Cohen RB, Jones CU, Sur R, Raben D, Jassem J, Ove R, Kies MS, Baselga J, Youssoufian H, Amellal N, Rowinsky EK, Ang KK. Radiotherapy plus cetuximab for squamous-cell carcinoma of the head and neck. N Engl J Med 2006;354(6):567–78. https://doi.org/10.1056/NEJMoa053422.

[15] Rosenthal DI, Harari PM, Giralt J, Bell D, Raben D, Liu J, Schulten J, Ang KK, Bonner JA. Association of Human Papillomavirus and p16 status with outcomes in the IMCL-9815 Phase III registration trial for patients with Locoregionally advanced oropharyngeal squamous cell carcinoma of the head and neck treated with radiotherapy with or without C. J Clin Oncol 2016;34(12):1300–8. https://doi.org/10.1200/JCO.2015.62.5970.

[16] Yom SS, Torres-Saavedra P, Caudell JJ, Waldron JN, Gillison ML, Truong MT, Jordan R, Subramaniam R, Yao M, Chung C, Geiger JL, Chan J, O'Sullivan B, Blakaj DM, Mell LK, Thorstad WL, Jones CU, Banerjee RN, Lominska CE, Le QT. NRG-HN002: a randomized phase II trial for patients with p16-positive, non-smoking-associated, locoregionally advanced oropharyngeal cancer. Int J Radiat Oncol Biol Phys 2019;105(3):684–5. https://doi.org/10.1016/j.ijrobp.2019.08.038.

[17] Marur S, Li S, Cmelak AJ, Gillison ML, Zhao WJ, Ferris RL, Westra WH, Gilbert J, Bauman JE, Wagner LI, Trevarthen DR, Balkrishna J, Murphy BA, Agrawal N, Colevas AD, Chung CH, Burtness B. E1308: phase II trial of induction chemotherapy followed by reduced-dose radiation and weekly cetuximab in patients with HPV-associated resectable squamous cell carcinoma of the oropharynx—ECOG-ACRIN cancer research group. J Clin Oncol 2017;35(5):490–7. https://doi.org/10.1200/JCO.2016.68.3300.

[18] Chen AM, Felix C, Wang P-C, Hsu S, Basehart V, Garst J, Beron P, Wong D, Rosove MH, Rao S, Melanson H, Kim E, Palmer D, Qi L, Kelly K, Steinberg ML, Kupelian PA, Daly ME. Reduced-dose radiotherapy for human papillomavirus-associated squamous-cell carcinoma of the oropharynx: a single-arm, phase 2 study. Lancet Oncol 2017;18(6):803–11. https://doi.org/10.1016/S1470-2045(17)30246-2.

[19] Brizel DM, Dodge RK, Clough RW, Dewhirst MW. Oxygenation of head and neck cancer: changes during radiotherapy and impact on treatment outcome. Radiother Oncol 1999;53(2):113–7. https://doi.org/10.1016/S0167-8140(99)00102-4.

[20] Sørensen BS, Busk M, Olthof N, Speel E-J, Horsman MR, Alsner J, Overgaard J. Radiosensitivity and effect of hypoxia in HPV positive head and neck cancer cells. Radiother Oncol 2013;108(3):500–5. https://doi.org/10.1016/j.radonc.2013.06.011.

[21] Lee N, Schoder H, Beattie B, Lanning R, Riaz N, Mcbride S, Katabi N, Li D, Yarusi B, Chan S, Mitrani L, Zhang Z, Pfister DG, Sherman E, Baxi S, Boyle J, LGT M, Ganly I, Wong R, Humm J. Strategy of using intratreatment hypoxia imaging to selectively and safely guide radiation dose de-escalation concurrent with chemotherapy for locoregionally advanced human papillomavirus–related oropharyngeal carcinoma. Int J Radiat Oncol Biol Phys 2016;96(1):9–17. https://doi.org/10.1016/j.ijrobp.2016.04.027.

[22] Rasey JS, Koh WJ, Evans ML, Peterson LM, Lewellen TK, Graham MM, Krohn KA. Quantifying regional hypoxia in human tumors with positron emission tomography of [18F]fluoromisonidazole: a pretherapy study of 37 patients. Int J Radiat Oncol Biol Phys 1996;36(2):417–28. https://doi.org/10.1016/s0360-3016(96)00325-2.

[23] Seiwert TY, Foster CC, Blair EA, Karrison TG, Agrawal N, Melotek JM, Portugal L, Brisson RJ, Dekker A, Kochanny S, Gooi Z, Lingen MW, Villaflor VM, Ginat DT, Haraf DJ, Vokes EE. OPTIMA: a phase II dose and volume de-escalation trial for human papillomavirus-positive oropharyngeal cancer. Ann Oncol 2019; 30(2):297–302. https://doi.org/10.1093/annonc/mdy522.

[24] Chui C-S, Chan MF, Yorke E, Spirou S, Ling CC. Delivery of intensity-modulated radiation therapy with a conventional multileaf collimator: comparison of dynamic and segmental methods. Med Phys 2001;28(12):2441–9. https://doi.org/10.1118/1.1418018.

[25] Verbakel WFAR, Cuijpers JP, Hoffmans D, Bieker M, Slotman BJ, Senan S. Volumetric intensity-modulated arc therapy vs. conventional IMRT in head-and-neck cancer: a comparative planning and dosimetric study. Int J Radiat Oncol Biol Phys 2009;74(1):252–9. https://doi.org/10.1016/j.ijrobp.2008.12.033.

[26] Vanetti E, Clivio A, Nicolini G, Fogliata A, Ghosh-Laskar S, Agarwal JP, Upreti RR, Budrukkar A, Murthy V, Deshpande DD, Shrivastava SK, Dinshaw KA, Cozzi L. Volumetric modulated arc radiotherapy for carcinomas of the oro-pharynx, hypo-pharynx and larynx: a treatment planning comparison with fixed field IMRT. Radiother Oncol 2009;92(1):111–7. https://doi.org/10.1016/j.radonc.2008.12.008.

[27] Tam M, Riaz N, Kannarunimit D, Peña AP, Schupak KD, Gelblum DY, Wolden SL, Rao S, Lee NY. Sparing bilateral neck level IB in oropharyngeal carcinoma and xerostomia outcomes. Am J Clin Oncol 2015;38(4):343. https://doi.org/10.1097/COC.0000000000000064.

[28] van der Laan HP, Gawryszuk A, Christianen MEMC, Steenbakkers RJHM, Korevaar EW, Chouvalova O, Wopken K, Bijl HP, Langendijk JA. Swallowing-sparing intensity-modulated radiotherapy for head and neck cancer patients: treatment planning optimization and clinical introduction. Radiother Oncol 2013; 107(3):282–7. https://doi.org/10.1016/j.radonc.2013.05.004.

[29] Vainshtein JM, Moon DH, Feng FY, Chepeha DB, Eisbruch A, Stenmark MH. Long-term quality of life after swallowing and salivary-sparing chemo–intensity modulated radiation therapy in survivors of human papillomavirus–related oropharyngeal cancer. Int J Radiat Oncol Biol Phys 2015;91(5):925–33. https://doi.org/10.1016/j.ijrobp.2014.12.045.

[30] Moreno AC, Frank SJ, Garden AS, Rosenthal DI, Fuller CD, Gunn GB, Reddy JP, Morrison WH, Williamson TD, Holliday EB, Phan J, Blanchard P. Intensity modulated proton therapy (IMPT)—the future of IMRT for head and neck cancer. Oral Oncol 2019;88:66–74. https://doi.org/10.1016/j.oraloncology.2018.11.015.
[31] van der Laan HP, van de Water TA, van Herpt HE, Christianen MEMC, Bijl HP, Korevaar EW, Rasch CR, van 't Veld AA, van der Schaaf A, Schilstra C, Langendijk JA, Group on behalf of the R cooperative. The potential of intensity-modulated proton radiotherapy to reduce swallowing dysfunction in the treatment of head and neck cancer: a planning comparative study. Acta Oncol 2013;52(3):561–9. https://doi.org/10.3109/0284186X.2012.692885.
[32] Gunn GB, Blanchard P, Garden AS, Zhu XR, Fuller CD, Mohamed AS, Morrison WH, Phan J, Beadle BM, Skinner HD, Sturgis EM, Kies MS, Hutcheson KA, Rosenthal DI, Mohan R, Gillin MT, Frank SJ. Clinical outcomes and patterns of disease recurrence after intensity modulated proton therapy for oropharyngeal squamous carcinoma. Int J Radiat Oncol Biol Phys 2016;95(1):360–7. https://doi.org/10.1016/j.ijrobp.2016.02.021.
[33] Lee SY, Park YM, Byeon HK, Choi EC, Kim S-H. Comparison of oncologic and functional outcomes after transoral robotic lateral oropharyngectomy versus conventional surgery for T1 to T3 tonsillar cancer. Head Neck 2014;36(8):1138–45. https://doi.org/10.1002/hed.23424.
[34] Beadle BM, Liao K-P, Elting LS, Buchholz TA, Ang KK, Garden AS, Guadagnolo BA. Improved survival using intensity-modulated radiation therapy in head and neck cancers: a SEER-medicare analysis. Cancer 2014;120(5):702–10. https://doi.org/10.1002/cncr.28372.
[35] Beadle BM, Liao K-P, Giordano SH, Garden AS, Hutcheson KA, Lai SY, Guadagnolo BA. Reduced feeding tube duration with IMRT for head and neck cancer: a SEER-medicare analysis. Cancer 2017;123(2):283–93. https://doi.org/10.1002/cncr.30350.
[36] Pauloski BR, Rademaker AW, Logemann JA, Discekici-Harris M, Mittal BB. Comparison of swallowing function after IMRT and conventional radiotherapy for head and neck cancer. Head Neck 2015;37(11):1575–82. https://doi.org/10.1002/hed.23796.
[37] Yeh DH, Tam S, Fung K, MacNeil SD, Yoo J, Winquist E, Palma DA, Nichols AC. Transoral robotic surgery vs. radiotherapy for management of oropharyngeal squamous cell carcinoma—a systematic review of the literature. Eur J Surg Oncol 2015;41(12):1603–14. https://doi.org/10.1016/j.ejso.2015.09.007.
[38] Nichols AC, Theurer J, Prisman E, Read N, Berthelet E, Tran E, Fung K, de Almeida JR, Bayley A, Goldstein DP, Hier M, Sultanem K, Richardson K, Mlynarek A, Krishnan S, Le H, Yoo J, SD MN, Winquist E, Hammond JA, Venkatesan V, Kuruvilla S, Warner A, Mitchell S, Chen J, Corsten M, Johnson-Obaseki S, Eapen L, Odell M, Parker C, Wehrli B, Kwan K, Palma DA. Radiotherapy versus transoral robotic surgery and neck dissection for oropharyngeal squamous cell carcinoma (ORATOR): an open-label, phase 2, randomised trial. Lancet Oncol 2019;20:1349–59. https://doi.org/10.1016/S1470-2045(19)30410-3.
[39] Ma DJ, Price KA, Moore EJ, Patel SH, Hinni ML, Garcia JJ, Graner DE, Foster NR, Ginos B, Neben-Wittich M, Garces YI, Chintakuntlawar AV, Price DL, Olsen KD, Van Abel KM, Kasperbauer JL, Janus JR, Waddle M, Miller R, Shiraishi S, Foote RL. Phase II evaluation of aggressive dose de-escalation for adjuvant chemoradiotherapy in human papillomavirus–associated oropharynx squamous cell carcinoma. J Clin Oncol 2019;37(22):1909–18. https://doi.org/10.1200/JCO.19.00463.
[40] Leichman L, Nigro N, Vaitkevicius VK, Considine B, Buroker T, Bradley G, Seydel HG, Olchowski S, Cummings G, Leichman C, Baker L. Cancer of the anal canal. model for preoperative adjuvant combined modality therapy. Am J Med 1985;78(2):211–5. https://doi.org/10.1016/0002-9343(85)90428-0.
[41] Major radiation reduction for HPV+ oropharyngeal carcinoma, https://clinicaltrials.gov/show/NCT03323463.
[42] Chemotherapy and locoregional therapy trial (surgery or radiation) for patients with head and neck cancer, https://clinicaltrials.gov/show/NCT03107182.
[43] Individualized adaptive de-escalated radiotherapy for HPV-related oropharynx cancer, https://clinicaltrials.gov/show/NCT03416153.
[44] The quarterback trial: reduced dose radiotherapy for HPV+ oropharynx cancer, https://clinicaltrials.gov/show/NCT01706939.
[45] De-intensification of radiation and chemotherapy for low-risk HPV-related oropharyngeal SCC: follow-up study, https://clinicaltrials.gov/show/NCT02281955.
[46] Primary radiotherapy versus primary surgery for HPV-associated oropharyngeal cancer, https://clinicaltrials.gov/show/NCT03210103.
[47] De-escalation protocols in HPV-related oropharyngeal carcinoma in Chinese populations, https://clinicaltrials.gov/show/NCT04012502.

[48] p16+ oropharyngeal cancer radiation optimization trial reducing elective treatment volumes (PROTEcT), https://clinicaltrials.gov/show/NCT04104945.
[49] De-intensified radiation therapy with chemotherapy (cisplatin) or immunotherapy (nivolumab) in treating patients with early-stage, HPV-positive, non-smoking associated oropharyngeal cancer, https://clinicaltrials.gov/show/NCT03952585.
[50] Ipilimumab, nivolumab, and radiation therapy in treating patients with HPV positive advanced oropharyngeal squamous cell carcinoma, https://clinicaltrials.gov/show/NCT03799445.
[51] Cisplatin or immunotherapy in association with definitive radiotherapy in HPV-related oropharyngeal squamous cell carcinoma: a randomized phase II trial, https://clinicaltrials.gov/show/NCT03623646.
[52] Cisplatin plus radiotherapy vs durvalumab plus radiotherapy followed by durvalumab vs durvalumab plus radiotherapy followed by tremelimumab and durvalumab in intermediate risk HPV-positive oropharyngeal SCC, https://clinicaltrials.gov/show/NCT03410615.
[53] Weekly cetuximab/RT versus weekly cisplatin/RT in HPV-associated oropharyngeal squamous cell carcinoma, https://clinicaltrials.gov/show/NCT01855451.
[54] De-escalation radiotherapy in patients with low-risk HPV-related oropharyngeal squamous cell carcinoma, https://clinicaltrials.gov/show/NCT03822897.
[55] PEARL PET-based adaptive radiotherapy clinical trial, https://clinicaltrials.gov/show/NCT03935672.
[56] Adaptive treatment de-escalation in favorable risk HPV-positive oropharyngeal carcinoma, https://clinicaltrials.gov/show/NCT03215719.
[57] Phase II treatment stratification trial using neck dissection-driven selection to improve quality of life for low risk patients with HPV+ oropharyngeal squamous cell cancer, https://clinicaltrials.gov/show/NCT02784288.
[58] Use of a proliferation saturation index to determine personalized radiotherapy for HPV + oropharyngeal cancers, https://clinicaltrials.gov/show/NCT03656133.
[59] Prokopiou S, Moros EG, Poleszczuk J, Caudell J, Torres-Roca JF, Latifi K, Lee JK, Myerson R, Harrison LB, Enderling H. A proliferation saturation index to predict radiation response and personalize radiotherapy fractionation. Radiat Oncol 2015;10:159. https://doi.org/10.1186/s13014-015-0465-x.
[60] Evaluation of ee-escalated adjuvant radiation therapy for human papillomavirus (HPV)-associated oropharynx cancer, https://clinicaltrials.gov/show/NCT02908477.
[61] Surgery followed by risk-directed post-operative adjuvanttTherapy for HPV-related oropharynx squamous cell carcinoma: "the minimalisttrial (MINT).", https://clinicaltrials.gov/show/NCT03621696.
[62] Adjuvant de-escalated radiation + adjuvant nivolumab for intermediate-high risk P16+ oropharynx cancer, https://clinicaltrials.gov/show/NCT03715946.
[63] De-escalation of adjuvant radio (chemo) therapy for HPV-positive head-neck squamous cell carcinomas, https://clinicaltrials.gov/show/NCT03396718.
[64] Transoral surgery followed by low-dose or standard-dose radiation therapy with or without chemotherapy in treating patients with HPV positive stage III-IVA oropharyngeal cancer, https://clinicaltrials.gov/show/NCT01898494.
[65] Post-operative adjuvant treatment for HPV-positive tumours (PATHOS), https://clinicaltrials.gov/show/NCT02215265.
[66] Study of postoperative radiation therapy for HPV associated oropharyngeal squamous cell carcinoma, https://clinicaltrials.gov/show/NCT03875716.
[67] Post operative adjuvant therapy de-intensification trial for human papillomavirus-related, p16+ oropharynx cancer, https://clinicaltrials.gov/show/NCT01687413.

CHAPTER

11

Trials and tribulations: Clinical trials and the future

Suchin Khanna, Barbara Burtness

Department of Internal Medicine, Yale Cancer Center and Yale School of Medicine, New Haven, CT, United States

Abstract

There have been many barriers to progress in systemic treatment for head and neck cancer. This is in part due to the pathobiology and heterogeneity of the disease, lack of funding for and poor accrual of studies, dearth of adequate risk stratification tools to properly select candidates for trials, and lack of enough dedicated head and neck specialists. Despite the barriers, there have been recent exciting advances in the field. We discuss historical and more recent landmark trials that have established the current standard of care. We also explore deescalation trials in HPV-associated HNSCC and posit why these studies have largely been unable to show a benefit to deescalation in this setting. We discuss clinical trials that have looked at the role of immunotherapy and targeted therapy in locally advanced and recurrent/metastatic head and neck cancer. We finish with ongoing promising clinical trials.

Abbreviations

ECOG Eastern Cooperative Oncology Group
EGFR epidermal growth factor receptor
EORTC European Organization for Research and Treatment of Cancer
FDA Federal Drug Administration
HNSCC head and neck squamous cell carcinoma
HPV human papilloma virus
NIH National Institutes of Health
RTOG Radiation Therapy Oncology Group

Conflict of interest

No potential conflicts of interest were disclosed.

https://doi.org/10.1016/B978-0-12-820679-9.00011-6

Barriers to head and neck cancer research

Squamous cell carcinomas arising in the head and neck (HNSCC) are often devastating cancers and account for 550,000 cases and 380,000 deaths annually, making them in aggregate the sixth most common cancer type worldwide [1]. Early stage cancers are readily cured with surgery or radiation, but locally advanced cancers require morbid multimodality therapy and nonetheless carry high recurrence rates, while metastatic disease has been only rarely curable to date. One of the major barriers to progress in systemic treatment for advanced head and neck cancers is the underlying pathobiology of each of the predominant forms of HNSCC, which is HPV associated (HPV+) and which is not (HPV −). HPV-negative HNSCC has a high rate of genetic heterogeneity, resulting in loss-of-function mutations in tumor suppressor genes such as *TP53* and *CDKN2A*, while their products, the key tumor suppressors p53, and $p16^{INK4a}$ are targeted by viral oncoproteins in HPV+ cancers. Though our understanding of the biology of this cancer has deepened considerably in recent years, the predominance of undruggable tumor suppressor gene mutations and other noncatalytic abnormalities that drive head and cancer continues to present major barriers to developing novel targeted and other advanced molecular therapies. Despite the predominance of loss-of-function tumor suppressor gene mutations, activation of certain oncogenes is also well known to contribute to HNSCC including EGFR amplification and overexpression and PIK3CA mutation and amplification. EGFR overexpression was the basis for the clinical testing of the anti-EGFR monoclonal antibody cetuximab in head and neck cancer. To date, this drug has been the only targeted drug validated to demonstrate efficacy in both HPV-negative and HPV-associated HNSCC. It has been FDA approved for treatment of head and neck cancer, but efficacy is far from satisfactory with a disappointing single agent response rate of 13% and an absence of validated predictive biomarkers to guide therapy [2].

A further barrier to research and development of novel therapies is the somewhat more limited commercial opportunity presented by head and neck cancer therapies, which must be developed against a background of high comorbidity burden and competing risk of mortality in HPV-negative disease [3] and which will find a smaller market than more common solid tumors. National cooperative groups have also devoted lesser funding to head and neck cancer and supported fewer annual accruals to trials in head and neck cancer than in more common cancers. For instance, in the fiscal year 2017, NIH funding to head and neck cancer was 63.6 million dollars total, whereas it was 545 million dollars for breast cancer, 320 million dollars for lung cancer, 208.4 million dollars for colon cancer, and 233 million dollars for prostate cancer. In fact, head and neck cancer was one of the lowest funded tumor sites of all cancers [4]. While lesser funding for less common cancers would be anticipated, analysis has shown that even when accounting for number of cases, head and neck cancer is disproportionately underfunded. For instance, calculations based on total number of new cases in the United States [5] show that the NIH funding dollars granted per new patient diagnosed with head and neck cancer in fiscal year 2017 was approximately $1000 per new patient, whereas for lung cancer, it was approximately $1450 per new patient. As novel insights into head and cancer biology and the success of immunotherapy open new clinical research questions in the field, advocacy from the patient and research communities will be needed to improve the balance of funding for this difficult disease.

Trials in head and neck cancer have also frequently been underpowered for several reasons. As mentioned, clinical research in HNSCC has historically faced funding limitations in both the industry and cooperative group spheres. A number of trials have proven at analysis to have flawed statistical assumptions regarding outcomes in the control arm: this difficulty is related to the small number of trials in this disease, creating a reliance for hypotheses regarding the control arm on older data or on results from single trials with low sample size. The large number of studies currently accruing in locally advanced and recurrent/metastatic disease to evaluate novel immunotherapy strategies is likely to provide multiple adequately sized cohorts for estimating outcomes in the control arms of future studies. It has also been the case that numerous trials have proven unable to enroll the required numbers of patients and been halted early. A study found that of 694 Phase I-III head and neck cancer trials analyzed, 91 (13.1%) of these were terminated, suspended, or withdrawn early. Most of the closed trials were in patients with locoregionally advanced disease ($n = 47$, 51.6%) or recurrent/metastatic disease ($n = 42$, 46.2%). Of the 91 closed studies, 27.5% did not provide a reason for stopping recruitment early. Insufficient accrual was the most common reason provided for trial closure when this was stated, with other common reasons being loss of funding/support (12.1%), safety concerns (8.8%), and lack of efficacy (6.6%) [6]. Even trials that were not terminated, suspended, or withdrawn experienced recruitment challenges. Early closure leads to reduced patient numbers, resulting in a significant reduction in statistical power for outcome analyses. For example, in the DeCIDE trial (docetaxel-based chemotherapy with or without induction chemotherapy in locoregionally advanced HNSCC patients: NCT 00117572), only 280 (70%) of the planned 400 patients were enrolled between 2004 and 2009; this study completed but did not meet its primary endpoint of overall survival [7,8].

Trials historically and even currently are further limited by the paucity of adequate risk stratification biomarkers and other tools to help properly select trial patients. For instance, patients with HPV-associated HNSCC are now recognized as having a distinct disease with superior response to treatment and better prognosis compared with HPV-negative patients [9]. In the recent past and continuing to some current trials, tumors of a range of subsites of the head and neck and both HPV+ and HPV− cancers have been enrolled, often without stratification for HPV or other molecular markers. The advent of further risk stratification tools in the future will enable research into more personalized treatments. The more stringent eligibility criteria that have been defined for current studies and will be defined for future studies based on ever more sophisticated risk stratification tools will improve our ability to answer specific questions in HNSCC. However, highly specific eligibility criteria will likely only increase the challenge of patient recruitment. International collaboration such as between the Canadian and US cooperative groups, or that facilitated through the Head and Neck International Group, and transnational industry studies will increasingly be required to achieve evidence-based progress in clinical management of HNSCC.

Other potential barriers to effective research in HNSCC include the fact that there are a small number of experienced head and neck physicians available to perform head and neck cancer trials [10]. The very nature of the disease may also present a potential barrier to HNSCC clinical trial recruitment. As noted earlier, comorbidity is common occurrence in patients with HNSCC and has a negative impact on survival and eligibility for clinical trials [11]. Another facet of HNSCC that is often an exclusion criterion in clinical trials is the presence of second primary tumors. Second primary cancer is a leading long-term cause of mortality in smoking or alcohol-related HPV-negative HNSCC [12].

Clinical trials

Despite the barriers, there have been recent exciting advances in the field. We will begin with historical and more recent landmark trials that have established the current standard of care. We will also explore recent deescalation trials for HPV-associated HNSCC and clinical trials in locally advanced and recurrent/metastatic HNSCC looking at immunotherapies and targeted therapies in these settings. We will finish with ongoing promising clinical trials that have not yet completed.

Historical landmark trials

RTOG 90-03 [13]

This study was designed to test whether any of three altered radiation fractionation schemes improved locoregional tumor control rates for patients with HNSCC when compared with standard fractionation of 70 Gy using daily 2-Gy fractions delivered 5 days a week. The study enrolled 1113 patients between 1991 and 1997. Inclusion criteria included nonmetastatic Stage III and Stage IV SCC (AJCC 3rd edition staging manual) of oral cavity, oropharynx, larynx, and hypopharynx. The three experimental arms included (1) 1.2Gy per fraction twice per day to a total dose of 81.6 Gy over 7 weeks (this was termed the hyperfractionation arm), (2) 1.6 Gy twice per day over 6 weeks but with a 2-week break after 38.4 Gy (this was termed the accelerated fractionation arm), and (3) 72 Gy over 6 weeks with 1.8 Gy daily fractions and an additional afternoon 1.5 Gy boost dose for the last 12 treatment days (this was termed the accelerated fractionation with a concomitant boost arm). Both the hyperfractionation and accelerated fractionation with concomitant boost arms improved 5-year local-regional control (LRC) by about 19% compared with standard fractionation (HR = 0.81, $P = .08$). The absolute benefit in 5-year LRC was 6.5% with hyperfractionation and 6.6% with accelerated fractionation with concomitant boost. The hyperfractionation arm appeared to improve overall survival (OS) when patients were initially evaluated at 5 years; however, the difference in OS compared with standard fractionation lost statistical significance with longer follow-up. Toxicities between the standard and experimental arms did not differ significantly at 5 years. There was no significant difference in distant metastasis rates or second malignancy rates between the groups. It is important to note that this trial used radiation therapy as a single modality for the management of locoregionally advanced head and neck cancer as this was the standard of care at the time this trial was designed and conducted. The current standard of care for locoregionally advanced head and neck cancer is in general concurrent chemoradiotherapy or surgery with risk-guided postoperative therapy. Therefore the results of RTOG-90-03 are not directly comparable with more modern studies. However, this landmark trial is worth noting because it was the largest to compare head-to-head several radiation therapy fractionations.

VA larynx study [14]

The historical treatment for patients with locally advanced laryngeal SCC was total laryngectomy with or without postoperative radiation therapy. The absence of a larynx after surgery results in significant morbidity with inability to perform essential functions such as phonation, deglutition, and respiration. The need for a larynx-sparing approach led the VA in 1991 to undertake a prospective, multiinstitutional randomized trial of nonoperative therapy consisting of induction chemotherapy followed by definitive radiation therapy in Stage III and IV SCC of the larynx (AJCC 3rd edition staging manual) as compared with total laryngectomy followed by postoperative radiation therapy. The induction chemotherapy was cisplatin 100 mg/m^2 of body surface area followed by continuous infusion fluorouracil 1000 mg per square meter per day for 5 days starting on days 1, 22, and 43. Clinical response was assessed 18–21 days after the start of cycle 2 of chemotherapy. Responders went on to receive a third cycle of chemotherapy and definitive radiation therapy. Partial response was defined as 50% reduction of tumor, and patients without partial response after two cycles of chemotherapy underwent immediate surgery and postoperative radiation therapy. A partial or complete response was seen in 85% of patients after two cycles of induction chemotherapy. There was no significant difference in OS between arms of the trial, with 2-year estimated survival of 68% in both arms ($P = .98$). Moreover, survival rates were similar among patients who responded to chemotherapy and those who did not respond to chemotherapy ($P = .98$) and when these two groups were compared with patients assigned to upfront surgery ($P = .74$). The overall rate of relapse was similar between the two arms of the trial, but the pattern of relapse was different. There were more patients with locally persistent tumor in the chemotherapy arm of the trial ($P = .042$) and more patients with distant metastasis in the surgery arm of the trial. Of patients assigned to induction chemotherapy, 64% were able to preserve their larynx. There were significantly more salvage laryngectomies required in stage IV versus stage III patients ($P = .0048$) and T4 primary stage of tumor compared with lesser primary stages of tumor ($P = .0001$). This trial helped change the standard of care for patients with locally advanced laryngeal SCC from total laryngectomy to an upfront larynx preservation approach.

RTOG 91-11 [15]

The VA Larynx study established an upfront nonsurgical larynx preserving approach as a reasonable alternative to surgery for patients with laryngeal SCC. The purpose of RTOG 91–11 was to determine the optimal larynx-preserving regimen, specifically the optimum sequencing of chemotherapy in relation to radiotherapy. This prospective, multicenter, randomized trial enrolled 547 patients with locally advanced laryngeal SCC between 1992 and 2000. The patients were randomized to one of three arms: (1) induction chemotherapy followed by radiotherapy, (2) concurrent chemoradiation, and (3) radiation therapy alone. The induction chemotherapy arm was three cycles of cisplatin (100 mg/m^2 on day 1) and fluorouracil (1000 mg/m^2 per day for 5 days) delivered every 3 weeks. The concurrent chemotherapy was cisplatin 100 mg/m^2 on days 1, 22, and 43 of radiation therapy. The radiation therapy arm was 2 Gy per fraction to a dose of 70 Gy. The concurrent chemoradiation arm had the highest larynx preservation rate. The concurrent cisplatin and radiation therapy group demonstrated a relative reduction of 54% compared with the radiation therapy alone group (HR = 0.46, $P < .001$) and 42% relative reduction compared with the induction chemotherapy arm (HR = 0.58, $P = .005$). There was no significant difference in OS among the three arms, but laryngectomy-free survival was significantly worse in the radiation alone arm, while it was not significantly different between the concurrent and induction arms. The best locoregional control was achieved with concurrent cisplatin and radiation therapy. The concurrent arm was associated with a relative risk reduction of locoregional failure of 41% compared with radiation therapy alone and 34% compared with the induction chemotherapy arm. There was no significant difference in cumulative incidence of grades 3–5 late toxicity between arms. This study has changed the standard of care for appropriately selected stage III and IV laryngeal SCC patients to concurrent chemoradiation. Interestingly, although the study was not powered to assess noncancer mortality specifically, results of 10-year follow-up show statistically nonsignificant but clinically meaningful increase in noncancer or cancer treatment-related mortality rates in the concurrent chemoradiation arm (30.8%) as compared with the other arms (20.8% in the induction chemotherapy arm and 16.9% in the RT alone arm). This highlights the need for long-term follow-up in curative intent trials.

Cetuximab + RT for locally advanced HNSCC [16]

Cetuximab is a monoclonal antibody that targets the epidermal growth factor receptor (EGFR) and blocks its downstream signaling functions. High levels of EGFR expression have been detected in over 90% of HNSCC. This prospective randomized trial's purpose was to determine whether the addition of cetuximab to definitive radiotherapy would improve outcomes for patients with locally advanced HNSCC. The study randomized 424 patients between 1999 and 2002 to either definitive radiation therapy alone or definitive radiation therapy and concomitant cetuximab. Median OS was improved from 29.3 months with radiation alone to 49 months with the concurrent cetuximab and radiation regimen. Interestingly, patients who developed a prominent cetuximab-induced rash experienced improved survival compared with those who developed a mild rash, with median OS of 68.8 months versus 25.6 months, respectively. Like RTOG 91-11, this study demonstrated that adding systemic therapy to radiotherapy results in better outcomes than radiotherapy alone.

RTOG 0522 [17]

As it had been established that radiotherapy combined with cisplatin and radiotherapy combined with cetuximab improved outcomes when compared with radiotherapy alone for locally advanced HNSCC, this study was designed to assess whether the addition of cetuximab to concomitant cisplatin and radiotherapy would improve outcomes for these patients. The study enrolled 940 patients with locally advanced HNSCC between 2005 and 2009 and randomized them to receive either radiotherapy plus cisplatin plus cetuximab or radiotherapy plus cisplatin without cetuximab. There were no statistically significant differences between arms for PFS, OS, LRC, or distant metastasis. In addition, there was more toxicity in the cetuximab arm including high rates of grades 3 and 4 skin reactions, mucositis, anorexia, fatigue, and hypokalemia. There was no difference in relapse or overall survival based on level of tumor EGFR expression.

PARADIGM [18]

The PARADIGM trial evaluated whether the addition of induction chemotherapy to concurrent chemoradiotherapy improved OS in patients with locally advanced HNSCC. Patients randomized to the induction chemotherapy arm received three cycles repeated every 3 weeks of docetaxel, cisplatin, and fluorouracil. Following induction chemotherapy, patients were stratified based on response to chemoradiotherapy. Poor responders received concurrent chemoradiation with weekly docetaxel at 20 mg/m^2 for 4 weeks. Favorable responders received concurrent chemoradiotherapy with weekly carboplatin (AUC 1.5) for 7 weeks. Patients randomized to concurrent chemoradiotherapy without induction chemotherapy received two doses of cisplatin 100 mg/m^2 on days 1 and 22 of radiation therapy. There was no difference in PFS or OS between the induction arms and the noninduction arm. OS at 3 years was 73% for the induction arms and 78% for the noninduction arm ($P = .77$). There was a higher rate of neutropenic fever in the group randomized to induction chemotherapy, and a total of 52 serious adverse events were reported in the induction arm as compared with 22 in the noninduction arm. One of the major limitations of this study was that it was only able to enroll 145 patients between 2004 and 2008 and was halted due to slow accrual, so many concluded that the question of whether there was a survival advantage to adding induction chemotherapy to concurrent chemoradiotherapy remained unanswered by this study. However, it should be noted that in addition to be underpowered, another flaw of this study was that the patient population studied was heterogeneous in the sense that the study was not stratified for patient HPV status. Despite these flaws, it is unlikely that full accrual or risk stratification would have changed the conclusions of the study, as 3-year overall survival was 73% (95% CI 60–82) in the induction therapy followed by chemoradiotherapy group and 78% (66–86) in the chemoradiotherapy alone group with a hazard ratio of 1.09, 95% CI 0.59–2.03; $P = .77$.

DECIDE [19]

This trial randomized 285 patients between 2004 and 2009 to concurrent chemoradiation plus or minus induction chemotherapy. To be eligible, patients had to have N2 or N3 disease (AJCC 6th edition staging manual). The concurrent chemoradiation consisted of docetaxel, fluorouracil, and hydroxyurea plus radiation therapy at a dose of 150 cGy delivered twice per day and every other week. The induction chemotherapy consisted of two cycles of docetaxel, cisplatin, and fluorouracil delivered in 21-day cycles. There was no statistically significant difference in OS or relapse-free survival between the two arms of the trial. There were significantly more toxicities in the induction arm as compared with the noninduction arm. Serious adverse events were 47% in the induction arm as compared with 28% in the noninduction arm ($P = .002$). There were significantly higher rates of neutropenia, leukopenia, anemia, hypomagnesemia, and alopecia in the induction arm. While the study was underpowered, the lack of significant difference in survival and the increased toxicities with the induction arm led to the conclusion that induction chemotherapy added to concurrent chemoradiotherapy in locally advanced head and neck cancer with N2 or N3 disease could not be routinely recommended. In addition to the two design flaws this study shares with PARADIGM, it had the further concerning design feature that the induction arm of the study included the use of cisplatin, whereas the concurrent chemoradiation only arm did not include cisplatin, and therefore even if the study had been positive, it would not have been possible to conclude that induction should be added to the current standard of care chemoradiation with a platinum.

MACH-NC [20]

One approach employed to assess clinical research findings in HNSCC, given multiple underpowered studies, has been metaanalysis. The MACH-NC collaborative group, which stands for **M**eta-**A**nalysis of **C**hemotherapy in **H**ead and **N**eck **C**ancer, performs metaanalyses in head and neck cancer with this goal in mind. One such metaanalysis included 93 randomized trials and 17,346 patients treated with curative intent for nonmetastatic HNSCC. Patients were treated with curative intent surgery or radiation therapy or with the same locoregional therapy combined with chemotherapy. Concomitant chemotherapy with locoregional management improved overall survival and event-free survival. The hazard ratio for death was 0.81 ($P < .0001$) with an absolute survival improvement of 6.5% at 5 years in favor of adding concomitant chemotherapy. There was no effect of concomitant chemotherapy on noncancer deaths, and the benefit seen for chemotherapy was due to improvement in head and neck cancer-specific survival. When comparing monochemotherapy regimens, the benefit with cisplatin was significantly higher than other chemotherapy agents. Concurrent chemotherapy significantly improved the rate of local and regional control. There was no significant improvement in LRC with induction chemotherapy.

EORTC 22931 and RTOG 9501 [21]

The landmark EORTC 22931 and RTOG 9501 trials were published in 2004 and demonstrated that the addition of concurrent chemotherapy with cisplatin to radiation therapy in the postoperative setting improved outcomes for selected HNSCC patients, though it should be noted that the specific features that portend better outcomes with the addition of chemotherapy were identified in post-hoc analysis. Both trials enrolled high-risk, locoregionally advanced patients for postoperative treatment either with standard radiation therapy alone or combined cisplatin (100 mg/m^2 on days 1, 22, 43) with radiation therapy. The EORTC trial defined high risk as positive surgical margins (at 5 mm or less), extracapsular extension of nodal metastasis, level 4 or 5 nodal involvement of oral cavity and oropharynx cancer, perineural invasion, or lymphovascular invasion. The RTOG trial defined high risk as positive surgical margin not further specified, extracapsular extension of nodal metastasis and/or two or more positive lymph nodes. In post-hoc analysis combining data from both trials, patients who had positive surgical margins or extracapsular extension experienced worse overall survival than patients who did not have these factors. The addition of cisplatin chemotherapy improved overall survival for patients with positive margins and/or extracapsular extension in both trials. There was no significant survival benefit for the addition of cisplatin to postoperative radiation therapy in either trial in patients that had neither positive margin nor extracapsular extension. The pooled analysis of both trial populations yielded a significant benefit of adding cisplatin with 28% reduction in risk of death. Both the EORTC and the RTOG trials demonstrated a statistically significant improvement in locoregional control and disease-free survival for concurrent adjuvant chemoradiation. The pooled data demonstrated a relative reduction of 42% in locoregional failure and 30% in risk of treatment failure. However, it should be noted that these post-hoc analyses are retrospective in nature and were unplanned at the time of design of these two trials so need to be interpreted within this light. However, despite this, there is consensus that postoperative chemoradiation for patients with surgical positive margins and/or extracapsular nodal extension is better than radiation alone and this has become the standard of care.

De-escalation trials in HPV-associated HNSCC

RTOG 1016 [22]

This noninferiority trial compared cisplatin plus radiotherapy versus cetuximab plus radiotherapy to see whether giving cetuximab could improve toxicities from treatment while maintaining cure rates. The study enrolled 987 patients with T1-T2, N2a-N3, M0, or T3-T4 N0-N3 M0 (AJCC 7th edition staging manual) HPV+ oropharyngeal carcinoma between 2011 and 2014. The radiotherapy plus cetuximab did not meet the noninferiority criteria for OS compared with the radiotherapy plus cisplatin group. The 5-year OS for the cetuximab group was 77.9% and for the cisplatin group was 84.6%, while the 5-year progression-free survival was 67.3% for the cetuximab group and 78.8% for the cisplatin group. One curious finding of this trial was that 5-year survival did not differ among patients with preserved performance status, who represented the majority of enrolled patients. Toxicities were not significantly different between the cetuximab and cisplatin groups, though the overall burden of acute toxicity was greater for the cisplatin group compared with the cetuximab group as measured by the mean number of grades 3 or 4 adverse events per patient. As with other studies of cetuximab in HNSCC, no stratification or patient selection on the basis of putative biomarkers of resistance was employed; however, specimen collection was robust, and the anticipated correlative analyses should have sufficient power to identify possible subgroups for which cisplatin or cetuximab may have greater benefit.

De-ESCALaTE HPV [23]

This trial also compared concurrent cisplatin plus radiotherapy versus concurrent cetuximab plus radiotherapy in patients with low-risk HPV-positive oropharyngeal carcinoma. Low-risk HPV-positive oropharyngeal cancer was defined as p16-positive AJCC 7th edition stages T3-T4 N0 and T1 N1-T4N3 squamous cell carcinoma arising in patients with less than a 10-pack-year tobacco use history. This trial constitutes a further example of the compromises in trial methodology that limit conclusions from HNSCC trials. The target sample size of the De-ESCALaTE HPV trial was 304 patients, allowing a 90% power to detect a 25% reduction in acute toxicity for the use of cetuximab. It is important to note that the study was not powered to detect a survival difference. Treatment groups were balanced with a minimization algorithm according to trial center, stage groupings (AJCC 7th edition T1-T2 vs T3-T4 and N0-1 vs N2-3), unilateral versus bilateral radiation, and planned gastrostomy. No patient selection or stratification for putative markers of cetuximab resistance was employed. Three hundred thirty-four patients were enrolled between 2012 and 2016. The arms were balanced across the stage groupings; however, with 76% of patients in each arm classified as N2-N3 and only 1% in each arm N3, it is not possible to discern whether there is an imbalance in those with bilateral adenopathy or high node number. This trial demonstrated significantly worse tumor control in the cetuximab group compared with the cisplatin group. Two-year OS was 89.4% in the cetuximab group compared with 97.5% in the cisplatin group ($P = .001$), though there were a significant amount of noncancer deaths, without which the arms would not have significantly differed. The 2-year recurrence rate was significantly higher in the cetuximab group at 16.1% compared with the cisplatin group at 6% ($P = .0007$). There was no significant difference in acute or late grade 3–5 toxicity between the cetuximab and cisplatin groups at 24 months.

ECOG 1308 [24]

In a multicenter cooperative group study, ECOG 1308, 80 patients with Stage III or IVA HPV-associated HNSCC (AJCC 7th edition) received induction chemotherapy with three cycles of cisplatin, paclitaxel, and cetuximab. Patients were then reevaluated clinically. Lower dose radiation therapy (RT), 54 Gy in 27 fractions, was given to patients who achieved a clinical complete response to induction chemotherapy at the primary site. Conventional dose RT, 69.3 Gy in 33 fractions, was given to patients with a clinical partial response or stable disease. In both groups, RT was given in conjunction with weekly cetuximab. A total of 56 patients had a primary site clinical complete response, and 51 of these patients proceeded to 54 Gy of radiation. For these 51 patients who had a complete response to initial induction chemotherapy and received a reduced dose of radiation, 2-year PFS was 80%, and 2-year OS was 94%. Among patients with favorable risk, defined as less than T4 primary stage, less than AJCC 7th edition N2c nodal stage, and less than 10 pack-years of tobacco use, 2-year PFS and OS were 96%, with the only death noncancer related. This exploratory trial established that HPV+ oropharynx cancer may be treated for cure with substantially lower radiation dose than had become standard and gave rise to a number of additional single arm studies that aim to refine a reduce radiation dose approach suitable for phase III testing.

UCLA phase II deintensification trial [25]

Similar to the previous study, this study investigated whether chemoradiotherapy with reduced-dose radiation would maintain survival outcomes while improving tolerability for patients with HPV-associated oropharyngeal carcinoma. It was a single-arm, phase 2 trial enrolling patients with newly diagnosed locally advanced SCC of the oropharynx, positive for HPV by p16 testing. Patients received two cycles of induction chemotherapy with 175 mg/m^2 paclitaxel and carboplatin (AUC=6) given 21 days apart, followed by intensity-modulated radiotherapy with daily image guidance plus 30 mg/m^2 paclitaxel per week concomitantly, utilizing a regimen pioneered in E2399 [26]. Complete or partial responders to induction chemotherapy received 54 Gy in 27 fractions and those with less than partial or no responses received 60 Gy in 30 fractions. The primary endpoint was PFS at 2 years. Forty-five patients were enrolled between 2012 and 2015: 55% achieved complete or partial responses to induction chemotherapy and consequently received 54Gy radiation, while the 45% with less than partial responses received 60Gy. Two-year PFS was 92%. No grade 4 events were reported. The conclusion was that chemoradiotherapy with radiation doses reduced by 15%–20% was associated high PFS and improved toxicity profile compared with historical regimens using standard doses.

OPTIMA [27]

Patients with HPV+ oropharyngeal cancer were assigned to dose and volume deescalated radiotherapy or chemoradiotherapy based on response to induction chemotherapy in this study in an effort to limit treatment-related toxicity while preserving efficacy. Patients were classified as low risk if they had equal to or less than T3N2B disease by AJCC 7th Edition Staging Manual and had less or equal to a 10-pack-year smoking history. They were considered high risk otherwise. After three cycles of carboplatin/nab-paclitaxel, response was assessed. Low-risk patients with greater or equal to 50% response received 50 Gy radiotherapy (RT50), while low-risk patients with 30%–50% response or high-risk patients with greater or equal to 50% response received 45 Gy chemoradiotherapy (CRT45). Patients with lesser response received 75 Gy chemoradiotherapy (CRT75). Sixty-two patients were enrolled. Of low-risk patients, 71% received RT50, while 21% received CRT45. Of high-risk patients, 71% received CRT45. Two-year PFS and OS were 95% and 100% for low-risk patients and 94% and 97% for high-risk patients, respectively. The overall 2-year PFS was 94.5% and within the 11% noninferiority margin for the historic control. Grades 3 and 4 mucositis occurred in 30%, 63%, and 91% of the RT50, CRT45, and CRT75 groups, respectively ($P = .004$). This third independent trial again demonstrates that cure rates for the paradigm of induction chemotherapy followed by response-adjusted radiation dosing are higher than those reported for favorable risk patients treated with high dose cisplatin and radiation. Taken together, these exploratory trials create an imperative for phase III comparison of this paradigm to the conventional definitive regimen of high dose cisplatin concurrent with 70 Gy radiation.

Ongoing deintensification trials

A randomized trial comparing weekly cetuximab with radiotherapy versus weekly low-dose cisplatin with radiotherapy in HPV-associated oropharyngeal squamous cell carcinoma is ongoing (NCT01855451). In the phase II ECOG 3311 trial (NCT01898494), patients with HPV-associated HNSCC with stage III or IV (AJCC 7th edition staging manual) underwent transoral resection of their oropharyngeal carcinoma and were assigned to one of four treatment arms based on risk, including Arm (A) transoral surgical resection only, Arm (B) transoral surgical resection followed by low-dose intensity-modulated radiotherapy (IMRT) daily for 5 days a week for 5 weeks, Arm (C) transoral surgical resection followed by IMRT daily 5 days a week for 6 weeks, and Arm (D) transoral surgical resection followed by standard-dose IMRT daily 5 days a week for 6–7 weeks and with concomitant weekly cisplatin or carboplatin. Five hundred eleven patients have been enrolled, but results have not yet been reported. In the phase II trial (NRG HN-002, NCT02254278), patients with stage III or locoregionally advanced stage IV HPV associated oropharyngeal SCC are being randomly assigned to treatment with mildly accelerated but dose-reduced RT alone (60 Gy in 5 weeks) or nonaccelerated, dose-reduced RT (60 Gy in 6 weeks) with weekly concurrent cisplatin. In the PATHOS phase II/III trial, patients with histologically confirmed HPV-associated T1-T3N0-N2b OPSCC and who are fit for surgery undergo transoral surgery and neck dissection. Then, based on adverse risk features of surgical pathology, patients are allocated into three groups. Patients with no adverse risk pathologic features will receive no adjuvant treatment as is the current standard of care. Patients with intermediate-risk pathologic features (defined as T3 and/or N2 tumors, tumors with evidence of perineural invasion or lymphovascular invasion or close margins [1–5 mm] around the primary tumor specimen but with negative marginal biopsies and no evidence of cervical lymph node extranodal extension) will be randomized to receive either standard dose postoperative radiotherapy or reduced dose radiotherapy. Patients with high-risk pathologic features (defined as having positive [<1 mm] margins around the primary tumor and/or evidence of cervical lymph node extranodal extension) will be randomized to receive either standard postoperative chemoradiotherapy or radiotherapy alone. The primary outcome of the study is patient reported swallowing function measured using the MD Anderson Dysphagia Inventory score prior to surgery and at 12 months post treatment [28]. Additional data and longer follow-up will be required from these and other trials before lower-dose RT, substitution of radiation dose by induction chemotherapy, use of potentially less toxic drugs, use of minimally invasive surgery, or radiotherapy alone can be considered as new standard approaches for HPV-positive patients. The safety of such changes will need to be carefully validated so as not to compromise the otherwise excellent prognosis in these patients.

NRG HN005 is a phase II/III open-label randomized trial being done to see if a reduced dose of radiation therapy and/or nivolumab works as well as standard dose radiation therapy and cisplatin in treating patients with locally advanced HPV-associated oropharyngeal cancer. The study began in July 2019 and is estimated to be completed in February 2025. Seven hundred eleven participants are estimated to be enrolled in this study. The control arm of standard dose radiation therapy and standard high-dose cisplatin is being compared with two experimental arms: (1) reduced dose RT and standard high-dose cisplatin and (2) reduced dose RT and nivolumab. Primary outcome measures include PFS and quality of life, whereas secondary outcome measures include locoregional failure, distant failure, OS, and adverse events.

Another notable study that does not specifically look at deintensification in HPV-associated HNSCC but is worth mentioning briefly is the European BEST OF study, a phase III 1:1 randomized study comparing IMRT to transoral surgery for T1 or T2 OPSCC and its effect on swallowing outcomes. MD Anderson Dysphagia Inventory (MDADI) score (the primary outcome) will be reported by the patients prior to treatment and then at months 4.5, 6, 9, and 12 after randomization. Secondary outcomes of this study include treatment response at 6 months after randomization, oncologic outcomes at year 1 and 5, toxicities based on CTCAE, quality of life based on QLQ-C30 and HN43, and out-of-pocket costs [29].

Recurrent/metastatic (R/M) HNSCC, second-line therapy

For purposes of chronology, since the trials that led to the standard of care for second-line therapy occurred before KN-048 (a new landmark clinical trial that has changed standard practice for first-line therapy for R/M HNSCC, discussed in more detail in the succeeding text), we review second-line therapies before the trials in first-line therapy.

KN-040 [30]

In this open-label phase 3 clinical trial, 495 patients who had failed prior platinum therapy were randomly assigned to pembrolizumab or investigator's choice of several different standards of care monotherapy options including methotrexate, docetaxel, or cetuximab. Checkpoint inhibitors were allowed at the time of progression in any treatment group and were more commonly used in those assigned to the standard of care group, 13% versus 5%. Patients in the pembrolizumab arm had fewer adverse events, 63% versus 84%. Although not statistically significant, more patients assigned to pembrolizumab had tumor responses, 15% versus 10%. These responses were more durable with a median duration of response 18 versus 5 months. For the group as a whole, pembrolizumab modestly improved OS with 1-year survival 37% versus 27% and median survival 8.4 versus 6.9 months. In exploratory analysis the survival benefit was demonstrated in 26% of patients who had PD-L1 expression greater or equal to 50%, whereas it was not statistically significant in those with PD-L1 expression less than 50%.

KN-012 [31]

This large nonrandomized study enrolled 192 patients and treated them with pembrolizumab on one of two schedules: 10 mg/kg every 2 weeks or a fixed dose of 200 mg every 3 weeks. This was a heavily pretreated population with 45% having received at least three prior therapies. Objective response rate was 18%. The response rate was similar in HPV-positive and HPV-negative patients. Responses lasted greater or equal to 6 months in 85% of patients, and overall survival at 12 months was 38%.

KN-055 [32]

In this large nonrandomized phase II trial, 171 patients were treated with pembrolizumab 200 mg every 3 weeks. All patients had platinum and cetuximab refractory head and neck cancer. The objective response rate was 16%, and the median duration of response was 8 months. Response rates were similar in patients who were HPV positive and HPV negative and in the various PD-L1 subgroups.

Checkmate-141 [33]

In this phase III trial, 361 patients with platinum-refractory recurrent or metastatic disease were randomly assigned to either nivolumab 3 mg/kg every 2 weeks or a single-agent investigator's choice of therapy: methotrexate, docetaxel, or cetuximab. Overall survival for the entire study population was significantly longer in patients treated with nivolumab with median 7.7 months versus 5.1 months and 1-year survival of 34% versus 19.7%. The objective response rate was also increased with nivolumab, 13.3% versus 5.8%. In exploratory analysis, overall survival was significantly increased with immunotherapy in patients with PD-L1 expression greater or equal to 1%, 8.7 versus 4.6 months. Overall survival was not significantly increased with PD-L1 expression <1%. In a post hoc exploratory analysis, overall survival was significantly increased in patients with HPV-positive tumors treated with immunotherapy with median survival 9.1 versus 4.4 months. The difference in OS was not statistically significant in those with HPV-negative tumors with median 7.5 versus 5.8 months. Patient-reported outcomes on a quality-of-life questionnaire found that there was no clinically meaningful deterioration associated with treatment with nivolumab immunotherapy, in contrast to chemotherapy where quality of life significantly decreased in 8 of 15 domains.

HAWK [34]

This international, multiinstitutional, single-arm phase II study evaluated durvalumab monotherapy in PD-L1 high patients with platinum-refractory recurrent or metastatic HNSCC. Immunotherapy-naïve patients with confirmed PD-L1 tumor expression of greater than or equal to 25% received durvalumab 10 mg/kg intravenously every 2 weeks for up to 12 months. Among 111 patients, objective response rate was 16.2%, 29.4% for HPV-positive patients, and 10.9% for HPV-negative patients. Median progression-free survival and overall survival for treated patients was 2.1 months and 7.1 months, respectively. Progression-free survival and overall survival at 12 months were 14.6% and 33.6%, respectively. Treatment-related adverse events were 57%, and 8% of patients had grade three or four adverse events.

Phase II trial of cetuximab in cisplatin-refractory R/M HNSCC [35]

A total of 103 patients with recurrent or metastatic HNSCC who had progressed following two to six cycles of platinum-based therapy were given cetuximab at an initial dose of 400 mg/m^2 followed by 250 mg/m^2 weekly. The objective response rate was 13%, including 10% who had had progression with cisplatin and 18% who had had stable disease. The median survival was 7.5 months.

Cytotoxic chemotherapy

After failure of initial chemotherapy, objective responses to subsequent lines of cytotoxic chemotherapy are uncommon, particularly when contemporary response criteria are applied, and there is no evidence that second-line treatment prolongs survival. Notably, there are no second-line trials yet reporting response rates or survival rates to chemotherapy after initial immunotherapy or chemoimmunotherapy.

Lenvatinib + pembrolizumab phase IB/II trial [36]

In addition to the trials discussed earlier, there has been more recent interest in researching novel combinations in the second-line setting, such as immunotherapy and antiangiogenic therapy. For instance, in this multicenter open-label phase Ib/II trial, the combination of pembrolizumab and lenvatinib was studied in patients with R/M HNSCC. Initial results of the study were presented at ASCO in 2018. Twenty-two patients had been enrolled by that time with 86% of the patients having had greater or equal to one prior anticancer therapy. ORR at 24 weeks was 36.4% (95% CI 17.2%–59.3%), median duration of response and median PFS at that point were both 8.2 months. PFS rate at 12 months was 37.2% (95% CI 12.8%–62.2%).

Recurrent/metastatic HNSCC, first-line therapy

EXTREME [37]

Drawing on the evidence from E5397 that the addition of cetuximab to cisplatin chemotherapy increased response rates, PFS, and numerical but underpowered improvement in OS [38], the EXTREME phase III trial randomly assigned 442 patients with recurrent or metastatic HNSCC to a platinum (cisplatin 100 mg/m^2 on day 1 or carboplatin AUC = 5 on day 1) plus fluorouracil (1000 mg/m^2 per day for 4 days) every 3 weeks, with or without cetuximab (400 mg/m^2 followed by subsequent weekly 250 mg/m^2). Overall, 39% of patients had received systemic cytotoxic chemotherapy at least 6 months prior to randomization as part of their initial definitive management. Chemotherapy was given for a maximum of six cycles, although cetuximab could be continued as maintenance therapy until disease progression or toxicity. Crossover to cetuximab was not allowed for patients on the control arm of the trial. Chemotherapy plus cetuximab significantly prolonged OS compared with chemotherapy alone with median survival of 10.1 months versus 7.4 months. Significant improvements were also seen in the PFS and objective response rates (median progression-free survival of 5.6 versus 3.3 months and 36% versus 20% objective response rates). The overall incidence of severe adverse events with the cetuximab regimen was not significantly increased compared with chemotherapy alone (82% vs 76%). This trial established a standard of care that has served as the control arm for numerous subsequent studies of first-line therapy.

Guigay et al. [39]

In this open-label phase III trial, 539 patients with recurrent or metastatic disease were randomly assigned to six cycles of cisplatin and fluorouracil plus cetuximab versus four cycles of cisplatin and docetaxel plus cetuximab. Both regimens were followed by maintenance cetuximab. At a median follow-up of 30 months, relative to the fluorouracil-based regimen, the docetaxel-based regimen demonstrated similar OS (median 14.5 vs 13.4 months, 2-year overall survival, 29% vs 21%), decreased toxicity (grade 3 or higher adverse events 34% vs 50%), fewer delays in administration (10% vs 27%), less substitution of carboplatin for cisplatin (9% vs 34%), and a higher percentage of patients initiating maintenance cetuximab (73% vs 53%).

KN-048 [40]

This more recent study was informed by past trials both in projection of outcomes for the control arm treated with the EXTREME regimen and in use of biomarker-enrichment for prospectively planned analyses. As an international trial, it accrued rapidly, and the sample size was adequate for evaluation of multiple endpoints. In this open-label phase 3 trial, 882 patients with recurrent or metastatic HNSCC not curable by surgery and without prior systemic therapy in the metastatic setting were randomly assigned on a 1:1:1 basis to receive single-agent pembrolizumab versus pembrolizumab plus a platinum and fluorouracil combination versus cetuximab plus a platinum and fluorouracil combination. PD-L1 status was assessed using CPS (combined PD-L1 score), which stains lymphocytes and other cells in addition to tumor cells. Results were evaluated for tumors with CPS>=20, tumors with CPS>=1, and the entire study population regardless of CPS. The pembrolizumab-containing combination therapy arm, in comparison with the cetuximab-containing combination therapy, demonstrated the following: improved OS in patients with CPS>= 20 with median survival 14.7 months versus 11 months and 3-year overall survival 33% versus 8%, improved OS in patients with CPS>=1 with median survival 13.6 months versus 10.4 months and 3-year OS of 26% versus 7%, and improved OS in the total study population with median survival 13 months versus 10.7 months and 3-year OS 23% versus 10%. The PFS was similar between the pembrolizumab combination and cetuximab combination arms at CPS>=20 (median 5.8 vs 5.2 months) or CPS>=1 (median 5 months each). Similar response rates were also seen (43% vs 38% for CPS>=20, 36% each for CPS>= 1) but longer duration of response in patients with any positive CPS (7.1 vs 4.2 months for CPS>=20, 6.7 vs 4.3 months for CPS>=1). There were similar rates of grades 3–5 adverse events between the treatment groups (85% vs 83%). Single agent pembrolizumab, in comparison with cetuximab-containing combination therapy, demonstrated the following: improved OS in patients with CPS>=20 (median survival 14.8 months vs 10.7 months and 3-year OS, 29% vs 9%) and in those with CPS>=1 (median 12.3 vs 10.3 months with 3-year OS, 22% vs 8%). The pembrolizumab monotherapy and standard of care arms had similar OS in the total study population (median 11.5 vs 10.7 months). There was lower PFS in the pembrolizumab arm in the total population (median 2.3 vs 5.2 months). There were also lower response rates (17% vs 36%), but longer duration of response (23 vs 5 months). There were fewer grades 3–5 adverse events (55% vs 83%). This study has been practice changing and has changed the standard of care for first-line therapy in recurrent/metastatic HNSCC.

Other ongoing trials

Several other randomized phase III trials comparing combination first-line immunotherapy with chemotherapy are ongoing or recently completed. For instance, in the KESTREL trial durvalumab with or without tremelimumab is being compared with EXTREME in recurrent/metastatic head and neck cancer (NCT02551159). There is also the Checkmate-651 trial, a randomized open-label phase III study comparing nivolumab combined with ipilimumab versus EXTREME in the first-line setting in R/M HNSCC, and a companion study called Checkmate-714 comparing ipilimumab and nivolumab to nivolumab alone in the first-line setting for R/M HNSCC in a double-blind, randomized two-arm phase II study [41,42].

Immunotherapy's potential role in definitive treatment

There are ongoing trials investigating the potential role of incorporating immunotherapy into definitive treatment for patients with locally advanced HNSCC.

JAVELIN (HN100, NCT 02952586)

This is a randomized double-blind phase 3 study of avelumab in combination with standard of care chemoradiotherapy (cisplatin plus definitive radiation therapy) versus standard of care chemoradiotherapy in the frontline treatment of patients with locally advanced HNSCC. The primary outcome is progression-free survival, and the estimated enrollment is 640 patients. A press release in March 2020, stated that the trial had been halted for futility at interim analysis.

KN-412

This study is comparing standard of care concurrent chemoradiation with cisplatin to pembrolizumab plus concurrent chemoradiation with cisplatin followed by pembrolizumab maintenance for about a year in patients with locally advanced head and neck cancer. This study has completed accrual and is awaiting final results.

Conclusion

There has been increased appreciation of the need for adequately powered studies in head and neck cancer as poor accrual to clinical trials for the many reasons outlined in the introduction has been one of the major barriers and limitations that have halted progress in the field. The task of adequately powering studies becomes even more challenging as we subset patients further with various risk stratification tools such as HPV status of their tumors, smoking history, and other tools that are still to be discovered. Over the past decade, there has been an increased recognition for the need for more international collaboration within the Head and Neck Oncology community to overcome these barriers. And consequently, we are seeing more large-scale international cooperative group trials than ever before, many of which have been discussed in this chapter. With these cooperative group trials, we have seen that conducting adequately powered studies of even subset populations of head and neck cancer have become feasible. This is best illustrated with the ECOG 1308 and RTOG 1016 trials, which enrolled enough HPV-associated HNSCC patients to adequately answer the questions these studies were designed to address. We have also seen the rise of biomarker driven studies in head and neck cancer, best illustrated by KN 048 where PD-L1 status of the tumor and immune cells in the tumor microenvironment played an integral role in the design of the study. Increased availability of contemporary normative data will further our understanding of how to even better categorize and risk stratify head and neck cancer, which will help strengthen study design moving forward.

References

[1] Global Burden of Disease Cancer Collaboration, Fitzmaurice C, Allen C, Barber RM, et al. Global, regional, and national cancer incidence, mortality, years of life lost, years lived with disability, and disability-adjusted life-years for 32 cancer groups, 1990 to 2015: a systematic analysis for the global burden of disease study. JAMA Oncol 2017;3(4):524.

[2] Sacco AG, Worden FP. Molecularly targeted therapy for the treatment of head and neck cancer: a review of the ErbB family inhibitors. Onco Targets Ther 2016;9:1927–43.

[3] Ferlay J, Soerjomataram I, Dikshit R, Eser S, Mathers C, Rebelo M, et al. Cancer incidence and mortality worldwide: sources, methods and major patterns in GLOBOCAN 2012. Int J Cancer 2015;136:E359–86.

[4] NCI. Funding for research areas, https://www.cancer.gov/about-nci/budget/fact-book/data/research-funding; 2017.

[5] American Cancer Society. Cancer facts and figures 2017. Atlanta: American Cancer Society; 2017.

[6] Haddad R, Chan A, Vermorken J. Barriers to clinical trial recruitment in head and neck cancer. Oral Oncol March 2015;51(3):203–11.

[7] ClinicalTrials.gov. Docetaxel based chemotherapy plus or minus induction chemotherapy to decrease events in head and neck cancer (DeCIDE), https://clinicaltrials.gov/ct2/show/NCT00117572; 2005. (Accessed 16 June 2014).

[8] Cohen EEW, Karrison R, Kocherginsky M, et al. DeCIDE: a phase III randomized trial of docetaxel (D), cisplatin (P), 5-fluorouracil (F) (TPF) induction chemotherapy (IC) in patients with N2/N3 locally advanced squamous cell carcinoma of the head and neck (SCCHN). ASCO Meeting Abstracts 2012;30:5500.

[9] Ang KK, Harris J, Wheeler R, et al. Human papillomavirus and survival of patients with oropharyngeal cancer. N Engl J Med 2010;363:24–35.

[10] Wysong P. Recruitment, evaluation, regionalization among novel challenges to the specialty. ENT Today; 2009. http://www.enttoday.org/details/article/496975/Recruitment_Evaluation.

[11] Boje CR. Impact of comorbidity on treatment outcome in head and neck squamous cell carcinoma – a systematic review. Radiother Oncol 2014;110:81–90.

[12] Morris LG, Sikora AG, Patel SG, et al. Second primary cancers after an index head and neck cancer: subsite-specific trends in the era of human papillomavirus-associated oropharyngeal cancer. J Clin Oncol 2011;29:739–46.

[13] Beitler J, et al. Final results of local-regional control and late toxicity of RTOG 90-03; a randomized trial of altered fractionation radiation for locally advanced head and neck cancer. Int J Radiat Oncol Biol Phys 2014;89(1):13–20.

[14] The Department of Veterans Affairs Laryngeal Cancer Study Group. Induction chemotherapy plus radiation compared with surgery plus radiation in patients with advanced laryngeal cancer. N Engl J Med 1991;324:1685–90.

[15] Forastiere A, et al. Long-term results of RTOG 91-11: a comparison of three nonsurgical treatment strategies to preserve the larynx in patients with locally advanced larynx cancer. J Clin Oncol 2012;31:845–52.

[16] Bonner J, et al. Radiotherapy plus cetuximab for locoregionally advanced head and neck cancer: 5-year survival data from a phase 3 randomized trial, and relation between cetuximab-induced rash and survival. Lancet Oncol 2010;11:21–8.

[17] Ang KK, et al. Randomized phase III trial of concurrent accelerated radiation plus cisplatin with or without cetuximab for stage III to IV head and neck carcinoma: RTOG 0522. J Clin Oncol 2014;32:2940–50.

[18] Haddad R, et al. Induction chemotherapy followed by concurrent chemoradiotherapy versus concurrent chemoradiotherapy alone in locally advanced head and neck cancer (PARADIGM): a randomized phase 3 trial. Lancet Oncol 2013;14:257–64.

[19] Cohen E, et al. Phase III randomized trial of induction chemotherapy in patients with N2 or N3 locally advanced head and neck cancer. J Clin Oncol 2014;32:2735–43.

[20] Pignon JP, et al. Meta-analysis of chemotherapy in head and neck cancer (MACH-NC): an update on 93 randomized trials and 17,346 patients. Radiother Oncol 2009;92:4–14.

[21] Bernier J, et al. Defining risk levels in locally advanced head and neck cancers: a comparative analysis of concurrent postoperative radiation plus chemotherapy trials of the EORTC 22931 and RTOG 9501. Head Neck 2005;27:843–50.

[22] Gillison ML, et al. Radiotherapy plus cetuximab or cisplatin in HPV-positive oropharyngeal cancer (NRG oncology RTOG 1016): a randomised, multicentre, non-inferiority trial. Lancet Oncol 2018.

[23] Mehanna H, et al. Radiotherapy plus cisplatin or cetuximab in low-risk human papillomavirus-positive oropharyngeal cancer (De-ESCALaTE HPV): an open-label randomised controlled phase 3 trial. Lancet Oncol 2018.
[24] Marur S, Li S, Cmelak AJ, et al. ECOG 1308: Phase II trial of induction chemotherapy followed by reduced-dose radiation and weekly cetuximab in patients with HPV-associated resectable squamous cell carcinoma of the oropharynx – ECOG-ACRIN cancer research group. J Clin Oncol 2016 Dec.
[25] Chen AM, Felix C, Wang PC, et al. Reduced-dose radiotherapy for HPV-associated SCC of the oropharynx: a single-arm, phase 2 study. Lancet Oncol 2017;18(6):803–11.
[26] Cmelak AJ, Li S, Goldwasser MA, et al. Phase II trial of chemoradiation for organ preservation in resectable stage III or IV squamous cell carcinomas of the larynx of oropharynx: results of ECOG study E2399. J Clin Oncol 2007;25(25):3971–7.
[27] Seiwert TY, Foster CC, Blair EA, et al. OPTIMA: a phase II dose and volume de-escalation trial for human papillomavirus-positive oropharyngeal cancer. Ann Oncol 2019 Feb 1;30(2):297–302.
[28] Hargreaves S, Beasley M, Hurt C, et al. Deintensification of adjuvant treatment after transoral surgery in patients with HPV-positive oropharyngeal cancer: the conception of the PATHOS study and its development. Front Oncol 2019.
[29] Simon C, Caballero CA, Fortpied C, et al. BEST OF: a phase III study assessing the best of radiotherapy (intensity modulated radiotherapy, IMRT) compared to the best of surgery (trans-oral surgery, TOS) in patients with T1-T2, N0 oropharyngeal squamous cell carcinoma (OPSCC). J Clin Oncol 2018.
[30] Cohen EEW, Soulieres D, Le Tourneau C, et al. Pembrolizumab versus methotrexate, docetaxel, or cetuximab for recurrent or metastatic head-and-neck SCC (KN-040): a randomised, open-label phase 3 study. Lancet 2019;393 (10167):156.
[31] Mehra R, Seiwert TY, Gupta S, et al. Efficacy and safety of pembrolizumab in reccurent/metasttic head and neck SCC: pooled analyses after long-term follow-up in KN-012. Br J Cancer 2018;119(2):153.
[32] Baumi J, Seiwert TY, Pfister DG, et al. Pembrolizumab for platinum and cetuximab refractory head and neck cancer: results from a single-arm phase II study. J Clin Oncol 2017;35(14):1542.
[33] Ferris RL, Blumenschein Jr G, Fayette J, et al. Nivolumab for recurrent HNSCC. N Engl J Med 2016;375(19):1856.
[34] Zandberg DP, Algazi AP, Jimeo A, et al. Durvalumab for recurrent or metastatic HNSCC: results from a single-arm phase II study in patients with greater than or equal to 25% tumor cell PD-L1 expression who have progressed on platinum-based chemotherapy. Eur J Cancer 2019;107:142.
[35] Vermorken JB, Trigo J, Hit R, et al. Open-label, uncontrolled, multicenter phase II study to evaluate the efficacy and toxicity of cetuximab as a single agent in patients with recurrent and/or metastatic HNSCC who failed to respond to platinum-based therapy. J Clin Oncol 2007;25(16):2171.
[36] ASCO Abstract. Taylor MH, Rasco DW, Brose MS, et al. A phase Ib/2 trial of lenvatinib plus pembrolizumab in patients with squamous cell carcinoma of the head and neck. J Clin Oncol 2018;36:6016.
[37] Vermorken JB, Mesia R, Reviera F, et al. Platinum-based chemotherapy plus cetuximab in head and neck cancer. N Engl J Med 2008;359(11):1116.
[38] Burtness B, Goldwasser MA, Flood W, et al. Phase III randomized trial of cisplatin plus placebo compared with cisplatin plus cetuximab in metastatic/recurrent head and neck cancer: an ECOG study. J Clin Oncol 2005; 23(34):8646–54.
[39] Guigay J, Fayette J, Mesia R, et al. TPExtreme randomized trial: TPEx versus extremre regimen in 1st line recurrent/metastatic HNSCC. J Clin Oncol 2019;37S.
[40] Rischin D, Harrington KJ, Greil R, et al. Protocol-specified final analysis of the phase 3 KEYNOTE-048 trial of pembrolizumab (pembro) as first-line therapy for recurrent/metastatic head and neck squamous cell carcinoma (R/M HNSCC). J Clin Oncol 2019;37(suppl 15) abstr 6000.
[41] Argiris A, Gillison M, Ferris RL, et al. A randomized, open-label, phase 3 study of nivolumab in combination with ipilimumab vs. EXTREME regimen (cetuximab + cisplatin/carboplatin + fluorouracil) as first-line therapy in patients with recurrent or metastatic squamous cell carcinoma of the head and neck-CheckMate 651. Ann Oncol 2016;27:1016TiP. https://doi.org/10.1093/annonc/mdw376.68.
[42] Haddad R, Gillison M, Ferris RL, et al. Double-blind, two-arm, phase 2 study of nivolumab (nivo) in combination with ipilimumab (ipi) versus nivo and ipi-placebo (PBO) as first-line (1L) therapy in patients (pts) with recurrent or metastatic squamous cell carcinoma of the head and neck. Ann Oncol 2016;27(suppl 6) 1017TiP.

CHAPTER

12

Surviving and thriving: Survivorship in the 21st century: Developing a multidisciplinary survivorship care program

Dinesh K. Chhetri[a], *Andrew Erman*[b], *Lindsay Lanciault*[b], *Alisa Zhukhovitskaya*[a], *Ryan Stephenson*[a], *David A. Rapkin*[a]

[a]Department of Head and Neck Surgery, David Geffen School of Medicine at University of California Los Angeles, Los Angeles, CA, United States [b]Department of Audiology & Speech, University of California Los Angeles, Los Angeles, CA, United States

Abstract

Many complications of head and neck cancer treatment arise long after completion of treatment. Long-term issues that adversely affect the quality of life include dysphagia, dysphonia, neck fibrosis, trismus, xerostomia, lymphedema, osteoradionecrosis, and depression. While cancer care models have traditionally focused on treatment and short-term surveillance, many of these long-term issues remain untreated or undertreated. There is now increased recognition that cancer treatment paradigms should include life-long plans for cancer survivorship. Therefore the goal of the 21st century survivorship programs is to develop a multidisciplinary group or providers dedicated to monitoring, preventing, and treating survivorship issues with the ultimate goal to maintain an optimal quality of living throughout life.

Abbreviations

SCC	squamous cell carcinoma
HNSCC	head and neck squamous cell carcinoma
OPSCC	oropharyngeal squamous cell carcinoma
HPV	human papilloma virus
CRT	chemoradiation therapy
HNCP	head and neck cancer program
QOL	quality of life

https://doi.org/10.1016/B978-0-12-820679-9.00012-8

UES	upper esophageal sphincter
ROM	range of motion
FEES	flexible endoscopic evaluation of swallowing
UCLA	University of California Los Angeles
TNE	trans-nasal esophagoscopy
MBSS	modified barium swallow studies
SLP	speech language pathologist
MBT	mind-body team
DS	distress screening
GERD	gastroesophageal reflux disease
ORN	osteoradionecrosis

Conflict of interest

The authors have no potential conflicts of interest.

Introduction

The number of head and neck cancer survivors is growing. The most common tumor type in the head and neck is squamous cell carcinoma (SCC), which has an incidence of approximately 11.2 cases per 100,000 people per year [1]. The incidence varies by primary site, with incidence of oropharyngeal primaries 3.4, laryngeal 3.3, oral cavity 3.2, and less than 1 per 100,000 people per year for other sites [1]. The risk is higher in men than in women, with an incidence of 17.9 and 5.5 per 100,000, respectively. The risk of oropharyngeal cancer is particularly more common in men, who are four times more likely to have an oropharyngeal primary and three times more likely to have a nonoropharyngeal primary than women [1]. According to SEER statistics (https://seer.cancer.gov/statfacts/html/oralcav.html), the 2019 incidence of oropharyngeal cancer was about 53,000, while laryngeal cancer was about 12,500. Oropharyngeal cancer constituted 3% of all new cancers. The 5-year survivals for these two cancer sites are 65% and 60%, respectively, with over 450,000 survivors of these cancers in the United States today.

While the incidence of nonoropharyngeal head and neck squamous cell carcinoma (HNSCC) is declining due to reduction in smoking, the incidence of human papilloma virus (HPV)-related oropharyngeal HNSCC is increasing, especially in men in developed countries [1,2]. Oropharyngeal HNSCC now accounts for approximately 40% of HNSCC in the United States. The majority of oropharyngeal HPV infections are sexually acquired. The increasing incidence of HPV-related oropharyngeal squamous cell carcinoma (OPSCC), particularly in men in developed countries, is attributed to changes in sexual behavior, with particular correlation with increasing lifetime partner number—a behavior that varies geographically. HPV-associated OPSCC has a better prognosis than non-HPV related OPSCC and SCC of multiple other head and neck sites [3,4]. Because of the younger age of diagnosis and better prognosis in the treatment of oropharyngeal cancer, more patients will be living longer after treatment, leading to an increasing population of patients with the adverse quality-of-life sequela of head and neck cancer treatment.

Treatments for head and neck cancer include surgery, radiation therapy, chemotherapy, and targeted therapies including immunotherapy, usually given in combination, leading to both acute and chronic long-term and late adverse effects. All treatments result in quality-of-life issues posttreatment. Among the most common long-term effects germane to head and neck cancer survivors are swallowing difficulties, radiation-induced neck fibrosis, lymphedema, chronic pain, fatigue, cognitive difficulties, and hearing loss. These negative effects lead to significant decrease in physical function and quality of life. In addition, there is the immediate and sustained impact of these treatments on body image and social functioning (e.g., eating and speaking) that occur based on the location of these tumors, with notable associated depression and psychosocial distress.

Most complications and side effects of head and neck cancer treatment arise long *after* completion of cancer treatment. There is now an urgent need to develop risk-based and biologically informed early interventions to mitigate the long-term and late effects of head and neck cancer treatment. Imagine a typical case of a male in his 60s with advanced HPV-positive oropharyngeal SCC who undergoes chemoradiation therapy (CRT). He develops moderate radiation mucositis and dysphagia during CRT but avoids a gastric tube (G-tube). Three months after completion of treatment, he is feeling improved, his weight loss has stabilized, and his posttreatment CT-PET scan is clear. Is his encounter with head and neck cancer now over? Unfortunately, it is just the beginning as many complications of head and neck cancer treatment arise *long after* completion of treatment. Tissue healing and scarring continues nonstop after radiation and chemotherapy, and many patients continue to suffer from voice and swallowing problems, jaw and neck stiffness, depression, and fatigue, just to name a few issues affecting quality of life. Current cancer care paradigms primarily focus on assessment, treatment, and posttreatment cancer surveillance. However, little attention is given toward proactively monitoring patients' functional status and QOL and preventing long-term organ dysfunction. This monitoring would allow us to identify and treat new and evolving treatment complications that could negatively affect QOL. Now, imagine a survivorship treatment paradigm focused on each individual's head and neck cancer and the consequences of the treatment received, where the continuing challenges after cancer treatment and the quality-of-life issues arising from continuing side effects are monitored and treated. In this new paradigm the patient may live a life of cancer survivorship enjoying life, eating, socializing, and feeling taken care of by a multidisciplinary group of experts that monitor and follow his/her physical and mental well-being. Head and neck cancer survivorship in the 21st century might follow that model.

Choice of treatment affects survivorship

Many factors must be considered when a patient and treatment team make a decision regarding choice of treatment (e.g., open vs endoscopic surgery, surgery vs radiation therapy, and chemoradiation therapy). The patient should be informed of the expectation of quality of life after therapy in addition to likelihood of disease eradication. If there are multiple therapies available, the patient will likely wish to consider the impact on quality of life for each available therapy. Foresight and forethought regarding the possible impacts on quality of life from various therapies can inform a proactive plan to optimize quality of life. One model that we and others have observed to be helpful is a multidisciplinary cancer treatment team including head and neck surgeons, radiation oncologists, medical oncologists, radiologists, pathologists, speech and language pathologists, psychologists, social workers, nutritionists, palliative medicine specialists, maxillofacial prosthodontics, and others that meet as part of a head and neck cancer program (HNCP) to discuss considerations for therapy and posttreatment quality of life. Patients may also elect to be seen in clinic after their HNCP presentation and subsequently discuss treatment options with multiple team members at the same visit and have their questions answered and have the expertise of these respective fields readily available to them. The optimal treatment is one that provides the best chance of disease eradication while preserving the best quality of life. It will be influenced by the tumor biology, cancer stage, anatomic location and involvement of vital structures, patient age, patient comorbidities, and presence of frailty. However, it should also be influenced by the patient's goals and desires regarding balancing cure at all costs versus preservation of quality of life. Certain functionality may be a higher priority for some patients than others, and this should always be discussed at the time of initial treatment recommendation.

Pathophysiology of QOL issues in survivorship

Radiation therapy particularly has the potential for negative effects on the long-term quality of life of HNSCC survivors. Ionizing radiation affects not only rapidly proliferating tumor cells but also normal tissues. In addition to direct DNA damage, external beam radiation generates reactive oxygen species leading to inflammation. Radiation therapy treats cancer by producing DNA damage from which tumor cells are not able to recover while normal cells are expected to survive. However, more rapidly dividing normal cells do become stunned and edematous. In the long term, this inflammation leads to fibrosis and scarring of tissues [5]. The mechanism of radiation induced fibrosis is similar to chronic wound healing, where acute injury leads to inflammation, recruitment of fibroblasts (and trans-differentiation into myofibroblasts), and eventually deposition of extracellular matrix [5].

Fibrosis decreases tissue compliance and results in functional and cosmetic deficits. Radiation-induced fibrosis usually manifests at least 4 months after completion of radiation therapy and then progresses. Late radiation effects are generally a function of dose, fractionation size, and volume treated [5]. Chronic toxicities continue through two major mechanisms. The first, tissue fibrosis, is thought to be related to dysregulated wound healing, oxidative stress secondary to effects of CRT, overproduction of transforming growth factor beta-1 (regulator of fibrotic process), and small vessel injury. Neuropathy is the second mechanism. The etiology for neurotoxicity is not entirely clear and may be related to chemotoxicity, brainstem neurotoxicity, peripheral devascularization, or compressive injury from fibrosis.

The side effects of tissue fibrosis and neuropathy continue unabated long term, and the adverse effects of treatment continue throughout life. In the head and neck, radiation-induced fibrosis manifests with skin induration, trismus, xerostomia via effect on salivary glands, muscle alteration and atrophy, mucosal fibrosis, lymphedema, osteoradionecrosis, decreased hyolaryngeal elevation, alteration of the upper esophageal sphincter, pharyngoesophageal strictures or stenosis, dysphonia, dysphagia, and aspiration [5–9]. Xerostomia after radiation is usually permanent and leads to long-term quality-of-life issues related to dysphagia via decreased bolus lubrication, reduced swallow tract lubrication and increased bolus transit time, dental hygiene issues, altered taste sensation, and dental decay [8–10]. Depending on the area being irradiated, radiation therapy can also result in hypothyroidism, hypopituitarism, chronic otitis media and hearing impairment, and cranial nerve injury [10].

Dysphagia is the number one QOL deficit after chemoradiation therapy for head and neck cancer. Anatomical consequences of fibrosis and scarring of the vital swallowing structures include reduced lateral channels, narrowed pharyngeal diameter, pharyngeal constrictor edema, pharyngeal stricture, and epiglottis shortening and thickening. Scarring also leads to restricted movement of the tongue base, pharyngeal constrictors, and larynx, thus affecting food bolus movement through the pharynx. Functional deficits leading to dysphagia include delayed swallow initiation, dyscoordinated bolus propulsion, reduced epiglottic inversion, dysfunctional opening of the cricopharyngeus muscle and reduced laryngeal airway protection, reduced tongue base retraction, and reduced tongue base contact with the posterior pharyngeal wall [8]. The altered epiglottis may obstruct bolus flow, and the upper esophageal sphincter (UES) may not open well. Strength of the pharyngeal squeeze to propel a bolus through the UES may be reduced. Cervical esophageal stricture may also occur. The patient may require a G-tube, have aspiration pneumonia, or demonstrate aspiration on MBSS, or patient may report subjective dysphagia [7]. Aspiration may not elicit a laryngeal cough reflex, increasing the risk of life-threatening pneumonia [8]. Severe long-term dysphagia may be seen in up to 38.5%. The incidence of dysphagia increases with patients of older age and the use of concurrent chemotherapy [7]. Patients with laryngeal, hypopharyngeal, base of tongue, and pharyngeal wall primary sites of tumor are more likely to have dysphagia compared with patients with cancer of the tonsil, oral cavity, and nasal cavity [7].

Functional consequences of neuropathy and denervation, which reduce sensory and motor function, lead to impaired sensation for pharyngeal residue, reduced laryngeal sensation, decreased or absent cough reflex, delayed triggering of functional response, and silent aspiration. Muscle atrophy leads to long-term deterioration of swallowing function. Dysgeusia may lead to decreased pleasure in meals and possibly reduced PO intake with associated weight loss and malnutrition. A patient may be more likely to aspirate because of reduced range of motion (ROM) of the laryngeal airway protective mechanisms and impaired sensation. Overall, pleasure with oral intake in the short and long term can be negatively impacted when a person has pharyngeal residue and has to swallow more times per bolus and when meals take longer.

Surgery for head and neck cancer also brings its own set of QOL issues. Essential structures such as the tongue and jaw may be resected with the surgical defect repaired with flap reconstruction. Resection with primary closure may lead to tethering of swallow structures. Tracheostomy and laryngectomy alter physiological processes required for speech and swallowing. Resection of nerves leads to sensory loss and long-term atrophy of denervated muscle tissue (with associated impact to function). Disfigurement will lead to both physiological and psychological consequences. The vast majority of surgical patents also get radiation therapy and, therefore, suffer consequences of both treatment modalities.

Quality of life after surgery depends on the type, size of surgical resections, and the function of the head and neck site that is removed. Removing a structure or part of the structure may result in dysfunction of the subsite. For example, patients with oral cavity resections may have increased bolus transit time during swallow or increased oral cavity residue after swallow [6]. Tongue base resection may result in aspiration and inefficient pharyngeal swallow. Resection of supraglottic structures involved in the protective mechanisms of the airway may result in aspiration. Resections involving the nasal cavity and sinuses may result in nasal obstruction or hyposmia. Resection of temporal bone cancers may result in hearing loss and facial nerve injury. Issues with cosmesis may result from cancers at many different subsites, while a laryngectomy results in aphonia and potential swallowing problems. How a resection site is reconstructed also has an effect on postoperative quality of life.

Surveillance and posttreatment quality-of-life monitoring

The cancer care team should carry out disease surveillance. The goals of surveillance include monitoring the patient for primary tumor or locoregional progression or recurrence and monitoring for second primary tumors. Close surveillance and repeated examinations for recurrences should be carried out after definitive therapy for head and neck cancer for up to 5 years. At UCLA, posttreatment survivorship appointments are scheduled at 3, 6, 12 months, and yearly thereafter. Patients should be seen for survivorship evaluations until end of life. Patients fall out of the survivorship clinic should cancer reoccur and through the duration of cancer treatment.

Developing a multidisciplinary survivorship care program

Understanding and anticipating the impact on quality of life after cancer treatment is leading to increased opportunity to proactively treat quality-of-life concerns in these patients [11–24]. Survivorship guidelines have been published by various groups including the American Cancer Society (ACS), the American Society of Clinical Oncology (2017), National Academy of Sciences, and American College of Surgeons (ACS) [25–28]. The ACS guidelines for head and neck cancer survivorship identified five survivorship focus areas: (1) surveillance for HNC recurrence, (2) screening for second primary cancers, (3) assessment and management of physical and psychological long-term and late effects of cancer treatment, (4) health promotion, and (5) care coordination and practice implications. See Table 1 for the long-term late side effects to be monitored identified in the ACS guidelines. However, the ACS recommendation that primary care clinicians should monitor, assess, counsel, and make appropriate referrals of these complex and time-consuming medical issues is not realistic. Developing a multidisciplinary head and neck cancer survivorship program is therefore germane and essential to a successful head and neck cancer treatment program.

Implementing and sustaining a multidisciplinary clinical pathway for head and neck cancer survivorship requires significant time, investment of money and resources (such as dedicated staff and space), a "buy-in" and participation by all specialties involved, and perseverance by a select group of health care professionals dedicated to survivorship care. Advocates for these endeavors are needed, typically those who are most likely to encounter and tend to H&N cancer survivors. Otolaryngologists and speech language pathologists are natural advocates as these specialties treat many of the complications of head and neck cancer treatment.

Many survivorship models exist, including some of the following: *Oncology specialty care programs* provide ongoing follow-up that is provided by the treating oncologist; *multidisciplinary cancer survivorship programs* offer ongoing care by a specialized team; *disease-specific survivor clinics* are programs developed for a single body area (e.g., head and neck cancer) or treatment modality; *consultative survivorship clinics* are programs in which the patient may

TABLE 1 Considerations for assessments in HNC survivorship care.

Quality-of-life considerations for HNC survivorship care
Accessory nerve palsy
Cervical dystonia
Shoulder dysfunction
Trismus
Dysphagia
Gastroesophageal reflux disease (GERD)
Lymphedema
Fatigue
Dysgeusia
Auditory and vestibular dysfunction
Sleep disturbance or sleep apnea
Dysphonia
Hypothyroidism
Dental disease including caries and periodontitis
Xerostomia
Osteoradionecrosis (ORN)
Oral infections or candidiasis
Psychosocial disorders – Negative impact on body self-image – Depression – Anxiety

only be seen once (this model can be ideal for providing a treatment summary and care plan); *integrated survivorship clinics* are embedded within an oncology setting, and in this model the provider is a survivorship specialist; and the *shared-care survivor model*, in which care is coordinated between the oncologist and primary care provider (in this model, survivorship follow-up is eventually transferred to the primary care provider). Developing a survivorship program requires interest from various groups—the more support the better. At the University of California, Los Angeles (UCLA), regular meetings were held to forge a common pathway and understand what role each specialty would play in support of survivorship care. Literature searches were performed to understand relevant assessments and the timing of survivorship evaluation and management. The UCLA Head and Neck Survivorship Program aligns most closely with the disease-specific survivor model. The program is directed by a laryngologist whose practice is largely devoted to treating dysphagia and dysphonia, with many of these patients having a history of HNC. The UCLA Speech Pathology Clinic closely supports this Survivorship Program. Therefore the UCLA Head and Neck Cancer Survivorship Clinic is staffed by professionals who have a focus on the function and quality of life. We promote regular visits to the survivorship clinic, much like periodic dental appointments, to identify and treat the deleterious consequences of the cancer and treatment that may occur over time. We recommend survivorship assessments at 3 and 6 months after completion of treatment and annually thereafter.

Developing cancer survivorship program requires multiple steps. Most likely, those providers committed to survivorship will develop and participate in the program. The skill set and focus of those providers and institutional resources will have an impact on the survivorship model it selects. The providers must consider the scope of the assessment: For example, will the patients be screened for the multitude of posttreatment sequelae described in the American Society of Clinical Oncology Head and Neck Cancer Survivorship Guideline, or will the evaluation be narrowed? Questionnaires may be utilized to collect patient information relative to overall health and quality of life. The European Organization for Research and Treatment of Cancer Quality of Life Questionnaire—Head and Neck (EORTC QLQ-H&N35) is a questionnaire targeted to survey health-related quality-of-life issues for HNC patients [29]. The MD Anderson Symptom Inventory Head and Neck Cancer (MDASI-HN) measures symptom distress experienced by HNC patients, related to their disease and/or treatment [30]. The Patient Health Questionaire-2 (PHQ-2) is a two-item screening survey for depression [31]. The MD Anderson Dysphagia Inventory (MDADI), Swallowing Quality of Life (SWAL-QOL), and Eating Assessment Tool (EAT-10) questionnaires relate specifically to swallow function [32–34]. The Voice Handicap Index (VHI) surveys patients for dysphonia. The program will need documentation templates for electronic medical record specific to survivorship that can be shared [35]. Education tools should be developed, including an after-visit summary that provides referrals and the reason for the referral. Web-based videos may be useful to present the concept of survivorship, what symptoms and signs to monitor over time, and the areas assessed. Patient education is critical to achieve patient buy-in, both in attending the survivorship clinic and adopting the recommendations. Finally, it may be wise to consider which outcomes should be captured to understand the effectiveness of the survivorship program and to continue tailoring the program's effectiveness toward achieving its goals.

Evaluation and treatment of survivorship QOL issues

A HNC survivorship clinic will proactively assess multiple anatomic and physiologic areas of potential concern. As such, this endeavor may be time intensive. At UCLA, we have elected to share this human resource burden between the Head and Neck and Speech Pathology Clinics. The patient is initially seen by a speech pathologist, who screens the patient's swallow and voice function, trismus, lymphedema, head and neck and shoulder range of motion, and weight. An example of a mock speech pathology survivorship assessment is found in the appendix to this chapter (see Appendix). This information is available to the laryngologist for his portion of the evaluation, which includes a physical exam, endoscopic exam of the pharynx and larynx, and functional exams of voice and swallow as needed such as videostroboscopy, flexible endoscopic evaluation of swallowing (FEES), and transnasal esophagoscopy (TNE). Therefore these assessment tools are recommended. A TNE system is highly useful to grossly assess UES function and to evaluate if a cervical esophageal stricture is present. Other assessments that may be considered depending on an individual patient's needs include modified barium swallow studies (MBSS), high-resolution manometry (HRM), and thyroid function blood test panels. We have a goal to develop electronic medical record capability to allow comparison of a patient's survivorship data across visits. This could allow identification of subtly worsening physiologic processes or determination that impaired function is stable and assessment of treatment efficacy over time.

The otolaryngologist focused on survivorship will regularly monitor the patient's head and neck anatomy and function and provide surgical interventions related to voice, swallowing, and airway function. Generally, many options are available for treatment of scarring and narrowing of the aerodigestive anatomy, including pharyngoplasty, epiglottidectomy, cricopharyngeal myotomy, and esophageal dilation [36,37]. Options are less available for motor and sensory deficits. Over time, anatomic and functional changes are closely monitored, and appropriate referrals are made in a timely manner to other specialists such as speech pathology, nutrition, dentists, physical therapy, and reconstructive surgery.

Speech language pathologists (SLP) working with survivors should be experienced and knowledgeable about the voice, speech, and swallowing problems that may occur with HNC treatment. An instrumental assessment is required to assess swallowing dysfunction in this population given the posttreatment anatomical, motor, and sensory changes that may be anticipated. FEES has the advantage of providing information about pharyngeal secretion management. Anatomical abnormalities contributing to dysphagia may be observed more clearly during FEES than MBSS. However, a modified barium swallow study (MBSS) is optimal for observing oral swallow behaviors and assessing pharyngeal contractile behavior and upper esophageal function. At UCLA the laryngologist screens patients with FEES and refers as necessary for MBSS. The laryngologist and speech pathologist meet to review MBSS and endoscopic results to determine optimal treatment plans. The SLP uses instrumental assessments to determine if compensatory swallow behaviors are beneficial and/or if the patient is a candidate for swallow therapy. Compensation and therapy both have the overall goals of protecting the lower airway and/or improving bolus clearance. SLPs may recommend diet modifications (e.g., pureed food instead of chewable food) to reduce aspiration risk or improve swallow efficiency. SLPs or physical therapy may also treat trismus. SLPs may provide education about behavioral compensations to ameliorate the problems caused by xerostomia.

If dysphonia is identified, depending on the source, surgical options aimed at improving laryngeal closure or restoring vibratory capacity could be considered. Working with a speech pathologist who is skilled in voice therapy has been considered the leading treatment option for dysphonia secondary to radiation changes [38]. Dysarthria or speech impairments occurring as a result of postradiation are often best managed by a speech pathologist. Voice and speech therapy could include exercises to improve vocal fold mobility or training compensatory strategies to achieve a more functional voice or improve one's overall speech intelligibility and naturalness.

The survivorship clinic team members will be equipped to deal with some of the problems that may be identified. Challenges beyond the scope of these team members will need to be addressed by other professionals. Therefore, it is necessary to have an extensive list of referrals to give patients. This may include but is not limited to oncologic dentistry, nutrition, psychology, gastroenterology, pulmonology, maxillofacial prosthodontists, audiology, physical therapy, and lymphedema therapy.

Compliance related to survivorship care

A fundamental aspect of a survivorship program's success is patient compliance. Adherence to the program can be broken down into three levels. First, patients must attend survivorship clinic appointments. Next, patients must follow-up with referrals (e.g., lymphedema therapy, and physical therapy) resulting from the survivorship clinic visit. Finally, patients need to adhere to ongoing recommendations (e.g., performing daily swallow exercises and the use of dental fluoride trays). Compliance, by definition, can have a negative connotation since it is defined as the act of conforming or yielding to another. With respect to the medical definition, compliance is defined as following the advice of a health care professional. These definitions do not obviously include patient involvement in the care plan. Patients who do not follow recommendations may be labeled noncompliant, which may negatively and unintentionally impact the provider-patient relationship. When patients do not follow recommendations, extra face-to-face time is required to understand obstacles interfering with patient implementation of the care plan.

Unfortunately, there is insufficient research on patient compliance within the survivorship program context. When looking solely at attendance to posttreatment surveillance appointments, travel time to treatment site was found to be a significant component [39]. Telehealth may be useful to provide survivorship services for patients who live far away from the treatment site. Although relatively sparse, there has been some published data assessing compliance with respect to routine dental care after radiation. Regrettably, findings suggested subpar compliance [40]. This problem of noncompliance has been identified in physician-directed management across a variety of chronic diseases including diabetes and posttransplant care [41,42]. When attempting to understand why a patient is unable to comply with medical recommendations, one could postulate a variety intricate factors that might be at play. This could include lack of medical literacy, limited comprehension of risk secondary to noncompliance, or absence of observable improvement. Additionally, socioeconomic factors, such as being able to afford out-of-pocket expenses or transportation costs, and psychosocial stress (e.g., depression or anxiety) may be barriers to care.

Motivational interviewing is an evidence-based counseling method that helps a patient find their own specific internal motivation to change their behavior. It has been effectively used in rehabilitation to help patients follow clinicians' recommendations to achieve their specific goals [43]. Perhaps when working with patients in a survivorship program, this approach could be implemented by providers to engage the patient in their own care goals.

The mindbody connection: A biopsychosocial perspective on survivorship

From early in the modern history of treating HNC, providers have recognized the exceptionally drastic psychosocial consequences of this cluster of diseases and the morbidities and toxicities of their treatments. Almost 50 years ago, Rozen et al., writing from the perspective of maxillofacial prosthodontics, declared: [44]

> The prosthetic rehabilitation of a maxillofacial cancer patient is not complete without consideration of the patient's psychosocial problems. When a patient is told he has cancer, he is filled with apprehension and

> thoughts of impending mutilation and death. This initial shock is compounded when therapy is begun and the patient is faced with the permanent loss of a body part or, at the very least, a limitation of normal function. It is not surprising, then, that the patient who arrives at the Maxillofacial Rehabilitation Clinic is often depressed and exhausted-physically, financially, and emotionally. His problems cannot be solved by prosthetic care alone. A successful prosthesis will not necessarily return him to a useful life. Failure to keep appointments, unusual or confusing complaints, and inappropriate responses to treatment reflect the intrapsychic and interpersonal problems during rehabilitation of these patients.

Nonetheless the community of providers caring for HNC patients has been slow to act on this recognition and slower still to integrate mental health professionals into routine care [45,46]. In this context the national movement toward a survivorship-oriented model of health care delivery and mission definition is historically significant, marking a sea change in medical culture [47]. Until recently, a reductionistic approach to caring for medical patients in general and for HNC patients in particular was the norm. Providers were trained to focus on those aspects of the patient that their specialty training equipped them to address and often their clinical focus ignored the reality that a whole person was experiencing the illness, diagnostic workup, treatment, and adjustment to posttreatment life. While this approach had advantages in terms of developing deeper and more nuanced understandings of pathophysiology of disease and developing innovative treatments, its shortcomings have become increasingly salient [48]. More and more, it is recognized that treatment of most cancer patients is best delivered by a multidisciplinary team centered on and caring "for the whole patient" [49,50]. This team necessarily includes mental health professionals (e.g., clinical social workers, psychologists, and psychiatrists) who may participate in patient care from diagnosis through end of life [51,52].

As a foundation for conceptualizing and implementing the aspirations of a survivorship-oriented approach to the care of HNC patients in general, and for implementing psychosocial dimensions of care in particular, the Biopsychosocial Model (BPS) developed by Engel and Romano is a powerful and excellent fit [53]. In contrast to the more traditional biomedical model, which is reductionistic and employs a linear conceptualization of causality (pathophysiology causes disease and is best remediated with biomedical interventions), the BPS approach is an ecological or complex system-oriented perspective. The BPS approach views a person as consisting of linearly and nonlinearly interacting dimensions, which include biological, psychological, and social systems dimensions. From the BPS perspective, pathophysiology, psychopathology, and sociological dysfunction co-arise and are mutually influencing. So too are physiological, psychological, and sociological resiliency and health. Symptoms emerge from simple or complex behavior at any or all of these levels of activity. Similarly, provision of care and healing may involve interventions addressing any or all of these levels. It is worth noting that the BPS perspective does not conflict with the linear causality-based explanations of the traditional biomedical model; it incorporates the biomedical model into a framework having many more possibilities for understanding a patient's condition, more options for healing interventions, and more parameters or metrics with which to assess the safety, efficacy, and cost of those interventions. Provision of health care in general and of psychosocial care in particular from a BPS perspective is harmonious with central aspirations of a survivorship program. This includes providing care emphasizing quality of life, emphasizing the patient's experience of themselves (which views the patient as a whole person), and providing care that includes the well-being of the patient's family and significant others.

Elsewhere in this chapter, we have noted morbidities and toxicities relating to HNC and its treatments. Here, we can note some of those sequelae that can particularly insult a patient psychologically and socially. These include pain, nausea, anxiety, depression, sleep disturbance, fatigue, and stress including financial stress, cognitive problems, and sexual dysfunction. Less obvious and perhaps less frequent but highly potent sequelae can occur, such as the following:

- Traumatizing exposure to one's interior anatomy, as when a patient views their orbit following an orbital exenteration or views their interior, mid-facial anatomy following a total rhinectomy.
- Degradation of highly valued structures and functions central to a person's sense of identity or status, as when a previously physically attractive person experiences deformity, when a person with a previously beautiful and effective voice loses that capacity for self-expression and potency.
- These losses can be experienced as real subjectively and intersubjectively, that is, the patient may be the only person who perceives their changes as grave losses and/or their family and social environment may join in such perceptions, ratifying them and making them interpersonally "real."
- Deformity and loss of function can lead to disruption of valued social relationships and activities, social unacceptability and rejection, anticipatory anxiety, and social isolation, which can occur across a range of settings from highly intimate (e.g., sexuality and physical closeness), to family ("I'm afraid I will frighten my grandchildren so I don't want them to see me"), to work settings, and to public settings ("I never go to restaurants, parties, malls, or out to the movies anymore").
- Difficulty breathing, anticipated and/or experienced, which can induce panic.
- Intense grief reactions that the person may not be able to work through or resolve and so become chronic, leading to depression.
- All of these changes can induce stress, anxiety, depression, loss of self-esteem, and social withdrawal in caregivers and spouses (spousal caregivers appear to be particularly vulnerable), children, coworkers, friends, and the community.

The stressors caused by patient's HNC and subsequent treatment emerge into a "final common pathway" with an attack upon the fabric of experience upon which patients and people in their lives rely upon for the creation of meaning and identity. When the capacity to create meaning deteriorates in an individual or social system, demoralization and despair typically ensue, sometimes progressing to fatal levels [54]. Within the multidisciplinary provider team, it is the primary responsibility of the Mindbody Team (MBT) to address these pathologies.

In the UCLA multidisciplinary Head and Neck Cancer Program (HNCP), the MBT consists of a patient navigator, clinical social workers (one ambulatory and one inpatient), and clinical and health psychologists (part-time volunteer faculty). We hope to build this team as our survivorship program extends to more phases of patient care. The patient navigator is our front-line provider, interacting with all HNCP patients. When problematic psychosocial issues arise, the patient navigator or any other member of the provider team communicates with one of the social workers or psychologists. Typically the social workers receive the first-tier referrals, and the psychologists receive referrals when a patient or family issue appears to require more time or a more intense intervention. As is the case in many large institutions, our MBT is backed-up by an integrative oncologic care center and by the potential for obtaining consults from other departments such as psychiatry. All MBT members attend the HNCP's weekly case conferences, and the psychologists participate in a high percentage of the patient-family conferences with their treating specialists. Within these patient-family-provider conferences, the psychologists facilitate communication with providers and provide counseling, emotional support, brief psychotherapy, and referrals for more extended therapy, as needed.

A psychosocial distress screening (DS) program can make substantial contributions to patients' experience of survivorship and can function as the "connective tissue" for the psychosocial care dimension of a survivorship program [11,13,55]. The term "distress" in cancer patients was carefully chosen to avoid potential stigmatization associated with terms such as anxiety and depression [56].

"Psychosocial distress in cancer has been defined as an unpleasant emotional experience of a psychological (cognitive, behavioral, and emotional), social, and/or spiritual nature that may interfere with the ability to cope effectively with cancer and its treatment. Distress extends along a continuum, ranging from common normal feelings of vulnerability, sadness, and fears to problems that can become disabling, such as depression, anxiety, panic, social isolation, and spiritual crisis" [57]. Reported prevalence of psychosocial distress in cancer patients ranges from 29.6% to 43.4% [58,59]. It appears probable that HNC patients are likely to experience higher levels of distress than patients with many other cancers. In general, distress responds well to appropriate treatment, but if allowed to remain untreated, it is associated with worsened comorbidities, morbidity, and mortality [60]. DS is recommended for all cancer patients and has been required for accreditation by the American College of Surgeons Committee on Cancer since 2015 [42,58]. However, implementation of DS for all cancer patients has been slow at many institutions across the country [60].

One widely accepted model for a DS program identifies five steps: (1) screening, (2) evaluation, (3) referral, (4) follow-up, and (5) documentation and quality improvement [13]. (1) Screening means use of any of several validated DS instruments that can be administered and scored rapidly to all cancer patients when they enter a treatment system [14]. (2) Evaluation refers to more extensive interview and assessment for patients identified by the screening instrument as experiencing a significant level of stress. (3) Referral means appropriate referral of evaluated patients to resources that can address their needs. (4) Follow-up relates to follow-up by a survivorship team provider (mostly likely an MBT member) to determine if the patient successfully made contact with the resource referred to and the degree of patient satisfaction with the referral. (5) Documentation refers to charting in an official record of patient care, which should enable all of the patient's providers to be apprised of their distress level and efforts to reduce it. Quality improvement, included in step 5, refers to collection of data from the previous steps to evaluate the effectiveness and weaknesses of the existing DS program and to feed that information back into program adjustment and redesign.

Consideration of the mindbody connection includes effects on individual providers and on the functioning of the provider team as a whole. An illuminating feature of the BPS approach is that, with its emphasis on connected and interacting levels of functioning, it implies that providers will necessarily be influenced by their patients' experiences. As our HNC patients deal successfully and unsuccessfully with the challenges and changes they encounter, patients and their families inevitably communicate the range of their emotional states, including distress of all sorts, to their providers and, through those providers, to the provider team as a whole. Clearly, it is necessary for the culture of the survivorship provider team to validate the full range of provider experiences that can arise, including provider distress and the need to build in support to cope with that distress.

Conclusions and future directions

The number of HNC survivors is expected to rise with an increasing proportion of younger patients and more women. The diagnosis of HNC too often forebodes emotional, psychological, and financial strain that patients will face as they endure the physical consequences of the disease process and its treatment.

Survivorship starts at the time of cancer diagnosis. The choice of treatment has direct effects on the posttreatment quality of life. The time of diagnosis is invariably overwhelming emotionally, psychologically, and unfortunately oftentimes logistically and financially for many patients. The cancer care team should assume the responsibility of guiding the patient through this process not only in the options of cancer treatment and prognosis but also with quality-of-life considerations, risk of disability, and anticipated functional changes. Treatment side effects of chemoradiation therapy continue unabated throughout life. Surgical changes to anatomy and physiology can result in permanent functional changes. Survivorship, therefore, continues for the rest of the patient's life.

Research is needed to understand the pathophysiology and time course of development of fibrosis and organ dysfunction. The demographic, clinical, biological, and treatment-related risk factors leading to progression of posttreatment symptoms, speech and swallowing difficulties, and physical neck dysfunction need to be better understood. Tissue engineering approaches might develop methodology to counter long-term tissue fibrosis and/or return function to dysfunctional organs.

Many patients come to regional centers for treatment then return back to their homes. It might be hard to routinely follow-up at the tertiary care center, and further care may be assumed by local health care providers who may not have the specialized resources of a survivorship clinic. HNC patients are optimal candidates for telehealth interventions due to communication barriers that can arise from the disease process and its treatment. Telehealth is the use of electronic communication to exchange medical information for the benefit of a patient's health. There are a growing number of applications of telehealth, and its use will likely only continue to become more widespread. Within the context of HNC survivorship, telehealth will likely prove to be a valuable adjunct to the necessary close surveillance, but workarounds will be needed to address some patients' reduced abilities to hear and speak.

Developing a successful survivorship program for HNC patients requires patience and persistence in what may seem at times like a slow process. Often, necessary resources (e.g., staff, staff time, physical facilities, and financing) may appear to be scanty and hard to obtain. To navigate these obstacles, it is quite helpful to identify sources of support and of resistance in the larger institution. But patience and persistence do pay off (it's usually a marathon, not a sprint), and though the work can be challenging, providing coherent and sustained survivorship care for our patients is exceptionally rewarding. In the 21st century, multidisciplinary survivorship programs hopefully will flourish and become routine part of the care of the patient with HNC.

Appendix

Speech Pathology Clinic Survivorship Evaluation

PATIENT: ***
MRN: ***
DOB: ***
Date of Onset: 9/5/19 (date of G-tube placement, exacerbation of chronic condition)
Date of Evaluation: 10/2/19
TIME OF SERVICE: 1000
LOCATION: Speech Pathology Clinic
Pt swallowing complaint: "My swallowing has gotten worse. I needed a feeding tube because I kept losing weight. Since the G-tube, I've only taken water. I drink maybe a half a cup a water a day. I haven't tried anything else."
Pt communication complaint: Some mild decline in speech function (see below).
Other complaints: The patient complained of trismus and restricted head and neck ROM.
Current Diet:
G-tube and some water. Reporting drinking only a half a cup a day.
RELEVANT MEDICAL HISTORY
Age/Sex: 61 yo male
Cancer History:
Cancer location: Right BOT SCCa
Date/Type of surgery for H&N cancer: November 2012 TORS and right MRND
XRT type and dates: Adjuvant CRT (IMRT Erbitux); February 2012. Developed a recurrence in July 2012 and was treated with CRT.
Systemic therapy type and dates: Hyperbaric treatment: October–November 2018 (20 treatments).
Pulmonary History:
H/o pneumonia in 2014 and a recent episode last month
H/o pulmonary problems denied
Other history related to communication, cognitive and/or swallowing problems:

- July 2018: Developed symptoms of osteoradionecrosis of right mandible as manifested by worsening trismus and right jaw pain. Started hyperbaric treatment (see above for details of treatment).
- 9/5/19: G-tube placement due to malnutrition (persistent weight loss).

Prior Swallow Therapy:
Participated in swallow preservation during XRT: No
Other swallow therapy—dates and reason: Therapy on these dates: January–April 2015. He is continuing to perform his exercises daily.
Last MBSS date and impressions, and recommendations:
4/24/18 MBSS.
Aspiration risk: The patient is at risk to aspirate small volumes of material after the swallow from pharyngeal residue that spills into his lower airway after the swallow. This risk exists for all food and liquid consistencies. Aspiration risk can be minimized if the patient hawks and expectorates remaining residue with each bolus; he performs this strategy independently.
Efficiency of the oropharyngeal swallow: The patient has difficulty forming a cohesive bolus and transporting it toward his pharynx. Pharyngeal contraction, hyolaryngeal elevation/excursion, and opening of the UES are all

poor. Collectively this leads to diffuse pharyngeal residue. Majority of liquid (thin and nectar thick) residue can be cleared through the pharynx with multiple swallows. The patient independently swallows multiple times per bolus to clear residue from his pharynx. He will be more efficient with liquids vs *the more viscous food consistencies. Sensation: No laryngeal cough reflex. This suggests a sensory deficit.*

Recommendations:

Blenderized liquid diet.

Continue to hock and expectorate after the swallow.

Continue with thorough oral care.

Seek immediate medical attention if s/s of aspiration pneumonia occur- patient was given written information.

The patient is currently scheduled to see Dr. Chhetri for the Survivorship Clinic on 12/27/19

RESULTS

Weight:

Pt reported weight is stable.

Current weight: See impressions section for this value.

Trismus:

Pt reports reduced oral opening that does interferes with dental care.

Maximal inter-incisal opening as of today's service: 21 mm.

Dental Assessment:

Pt uses fluoride trays inconsistently.

Dental visit report: goes to the dentist office every year.

Last visited the dental office: Last visit about one year ago.

Dental observations:

Molars adequate for mastication: yes

Incisors adequate for cutting food: yes

Oral Peripheral Examination (Remarkable Findings):

Velar elevation restricted on left

Reduced lingual mass on the right side

Xerostomia:

Pt reports dry mouth sometimes interferes with swallowing and/or speaking.

Mucosa observation

Oral mucosa has a dry appearance.

Swallowing Assessment:

Consistencies and amounts given:

Thin liquid: serial sips x2 trials.

Clinical Swallow Results:

Pt reported the following swallow problems: The patient described progressive difficulty swallowing with associated weight loss. This resulted in recent G-tube placement. The patient has only been drinking small amounts of water and he stated, "I can only drink water now and even that is difficult. I use the G-tube."

Swallow trials showed:
Pt took serial sips of water. A total of ~4–5 oz across two trials. Pt performed an immediate cough and hawk to expectorate after each sequence (likely performing a previously learned supraglottic swallow maneuver + hawk).

Speech/Voice Assessment:
Pt reported speech problems of "I'm having a hard time pronouncing certain words. It only really impacts me when I'm speaking on the phone. My speech seems to be declining."
Perception of Pt's Intelligibility:
Mildly impaired
Perception of Pt's Speech Naturalness:
Mild-Moderately impaired
Speech errors noted:
Hypernasality and slight imprecision of lingual phonemes
Speech improvement with trial therapy:
Technique(s) attempted and result: Alternate word choice, effective. Increasing volume, effective.

Vocal quality:
Pt denies voice changes
Voice perception:
Rough (mild)

Lymphedema:
Pt reports variable edema of the right side of his neck
Lymphedema observations:
Low hanging submental tissue that patient said occurred with or after CA Tx

Reduced Head and Neck ROM:
Pt reports restricted H&N ROM
Head and Neck ROM observations:
Restricted on the right side
Restricted on the left side

Shoulder Mobility ROM:
Pt denied restricted shoulder ROM
Shoulder ROM observations:
Pt can raise arms overhead

Impressions:
Swallow Function:

Signs of dysphagia noted during today's clinical swallow evaluation
Pt appears safe and efficient to continue to take water by mouth as tolerated
MBSS is recommended to assess swallow safety and/or swallow efficiency

Speech Function:

Pt demonstrates speech errors that have a small impact on communication function
Speech therapy is recommended (short course)

Voice Function:

Voice is unremarkable.

Weight:

Current weight: 170bs

Pt reports stable weight

Oral Opening/Trismus:

Oral opening is restricted and interferes with eating and/or dental care

Xerostomia:

Oral mucosa appears dry

Pt reports dry mouth condition can interfere with swallowing and/or speaking

Dental Care:

Pt by report is NOT receiving regular dental care

Pt has fluoride trays but is not using them as recommended

Lymphedema:

Variable lymphedema reported by the patient

Head and neck ROM:

Head and neck mobility appears restricted

Shoulder ROM:

No obvious ROM issues

Treatment Diagnosis:

Swallowing Function:

Swallowing NOMS: Level 2: Individual is not able to swallow safely by mouth for nutrition and hydration, but may take some consistency with consistent maximal cues in therapy only. Alternative method of feeding required.

Speech/Voice Function:

Motor Speech NOMS: Level 6: The individual is successfully able to communicate intelligibly in most activities, but some limitations in intelligibility are still apparent in vocational, avocational, and social activities. The individual rarely requires minimal cueing to produce complex sentences/messages intelligibly. The individual usually uses compensatory strategies when encountering difficulty.

Recommendations/Plan:

Diet recommendation:

Continued non-oral means of receiving nutrition

Discussed with the patient that he should consume more water by mouth daily. This would be for therapeutic swallowing purposes.

Dental recommendation:

Pt should receive dental care every three months (dental cleaning 4 × a year; exam by dentist 2 × a year)

Referrals for Dr. Chhetri to consider:

- Dental referral
- PT referral for: Lymphedema, Reduced head & neck ROM

Speech therapy referral for:

MBSS

Speech therapy to improve naturalness and/or intelligibility (short course)

Trismus

Patient Education:
Yes. Topic: Evaluation results and Evaluation recommendations
Person Taught: Patient
Barrier: None
Needs: Knows well
Teaching Method: Verbal instruction
Outcome: Able to repeat and/or demonstrate taught information

References

[1] Fakhry C, Krapcho M, Eisele DW, D'Souza G. Head and neck squamous cell cancers in the United States are rare and the risk now is higher among white individuals compared with black individuals. Cancer 2018; 124(10):2125–33.

[2] Gillison ML, Chaturvedi AK, Anderson WF, Fakhry C. Epidemiology of human papillomavirus-positive head and neck squamous cell carcinoma. J Clin Oncol 2015;33(29):3235–42.

[3] Masterson L, Moualed D, Liu ZW, et al. De-escalation treatment protocols for human papillomavirus-associated oropharyngeal squamous cell carcinoma: a systematic review and meta-analysis of current clinical trials. Eur J Cancer 2014;50(15):2636–48.

[4] O'Rorke MA, Ellison MV, Murray LJ, Moran M, James J, Anderson LA. Human papillomavirus related head and neck cancer survival: a systematic review and meta-analysis. Oral Oncol 2012;48(12):1191–201.

[5] Straub JM, New J, Hamilton CD, Lominska C, Shnayder Y, Thomas SM. Radiation-induced fibrosis: mechanisms and implications for therapy. J Cancer Res Clin Oncol 2015;141(11):1985–94.

[6] Mittal BB, Pauloski BR, Haraf DJ, et al. Swallowing dysfunction – preventative and rehabilitation strategies in patients with head-and-neck cancers treated with surgery, radiotherapy, and chemotherapy: a critical review. Int J Radiat Oncol Biol Phys 2003;57(5):1219–30.

[7] Caudell JJ, Schaner PE, Meredith RF, Locher JL, Nabell LM, Carroll WR, Magnuson JS, Spencer SA, Bonner JA. Factors associated with long-term dysphagia after definitive radiotherapy for locally advanced head-and-neck cancer. Int J Radiat Oncol Biol Phys 2009;73(2):410–5.

[8] Eisbruch A, Lyden T, Bradford CR, Dawson LA, Haxer MJ, Miller AE, Teknos TN, Chepeha DB, Hogikyan ND, Terrell JE, Wolf GT. Objective assessment of swallowing dysfunction and aspiration after radiation concurrent with chemotherapy for head-and-neck cancer. Int J Radiat Oncol Biol Phys 2002;53(1):23–8.

[9] Sroussi HY, Epstein JB, Bensadoun RJ, et al. Common oral complications of head and neck cancer radiation therapy: mucositis, infections, saliva change, fibrosis, sensory dysfunctions, dental caries, periodontal disease, and osteoradionecrosis. Cancer Med 2017;6(12):2918–31.

[10] Yeh S-A. Radiotherapy for head and neck cancer. Semin Plast Surg 2010;24(02):127–36.

[11] Williams C. Psychosocial distress and distress screening in multidisciplinary head and neck cancer treatment. Otolaryngol Clin North Am 2017;50:807–23.

[12] Clark PG, Bolte S, Buzaglo J, Golant M, Daratsos L, Loscalzo M. From distress guidelines to developing models of psychosocial care: current best practices. J Psychosoc Oncol 2012;30:694–714.

[13] Lazenby M, Tan H, Pasacreta N, Ercolano E, McCorkle R. The five steps of comprehensive psychosocial distress screening. Curr Oncol Rep 2015;17.

[14] Pirl WF, Fann JR, Greer JA, Braun I, Deshields T, Fulcher C, et al. Recommendations for the implementation of distress screening programs in cancer centers: report from the American Psychosocial Oncology Society (APOS), Association of Oncology Social Work (AOSW), and Oncology Nursing Society (ONS) joint task force. Cancer 2014;120:2946–54.

[15] Holland JC, Jacobsen PB, Riba MB. Distress management. Cancer Control 2001;8:88–93.

[16] Posluszny DM, Dougall AL, Johnson JT, Argiris A, Ferris RL, Baum A, et al. Posttraumatic stress disorder symptoms in newly diagnosed patients with head and neck cancer and their partners. Head Neck 2015;37:1282–9.

[17] Carmignani I, Locatello LG, Desideri I, Bonomo P, Olmetto E, Livi L, et al. Analysis of dysphagia in advanced-stage head-and-neck cancer patients: impact on quality of life and development of a preventive swallowing treatment. Eur Arch Otorhinolaryngol 2018;275:2159–67.

[18] Landeen KC, Spanos WC, Gromer L. Topical superoxide dismutase in posttreatment fibrosis in patients with head and neck cancer. Head Neck 2018;40:1400–5.
[19] Nilsen ML, Lyu L, Belsky MA, Mady LJ, Zandberg DP, Clump 2nd DA, Skinner HD, Peddada SD, George S, Johnson JT. Impact of neck disability on health-related quality of life among head and neck cancer survivors. Otolaryngol Head Neck Surg 2019;162(1):64–72.
[20] Deng J, Murphy BA. Lymphedema self-care in patients with head and neck cancer: a qualitative study. Support Care Cancer 2016;24:4961–70.
[21] Nixon J, Purcell A, Fleming J, McCann A, Porceddu S. Pilot study of an assessment tool for measuring head and neck lymphoedema. Br J Community Nurs 2014;19(Suppl S6):S8–S11.
[22] Pigott A, Nixon J, Fleming J, Porceddu S. Head and neck lymphedema management: evaluation of a therapy program. Head Neck 2018;40:1131–7.
[23] Purcell A, Nixon J, Fleming J, McCann A, Porceddu S. Measuring head and neck lymphedema: the "ALOHA" trial. Head Neck 2016;38:79–84.
[24] Messing BP, Ward EC, Lazarus C, Ryniak K, Kim M, Silinonte J, et al. Establishing a multidisciplinary head and neck clinical pathway: an implementation evaluation and audit of dysphagia-related services and outcomes. Dysphagia 2019;34:89–104.
[25] Cohen EEW, LaMonte SJ, Erb NL, Beckman KL, Sadeghi N, Hutcheson KA, et al. American cancer society head and neck cancer survivorship care guideline. CA Cancer J Clin 2016;66:203–39.
[26] Nekhlyudov L, Lacchetti C, Davis NB, Garvey TQ, Goldstein DP, Nunnink JC, et al. Head and neck cancer survivorship care guideline: American society of clinical oncology clinical practice guideline endorsement of the American cancer society guideline. J Clin Oncol 2017;35:1606–21.
[27] National Academies of Sciences, Engineering, and Medicine. Long-term survivorship care after cancer treatment. In: Proceedings of a workshop, Washington, DC; 2018.
[28] Commission on Cancer, American College of Surgeons. Cancer program standards 2016: ensuring patient-centered care. Chicago, IL: American College of Surgeons; 2016.
[29] Bjordal K, Hammerlid E, Ahlner-Elmqvist M, de Graeff A, Boysen M, Evensen JF, Biorklund A, de Leeuw JR, Fayers PM, Jannert M, et al. Quality of life in head and neck cancer patients: validation of the European Organization for Research and Treatment of Cancer quality of life questionnaire-H&N35. J Clin Oncol 1999;17:1008–19.
[30] Rosenthal DI, Mendoza TR, Chambers MS, Burkett VS, Garden AS, Hessell AC, Lewin JS, Ang KK, Kies MS, Gning I, Wang XS, Cleeland CS. The M. D. Anderson symptom inventory-head and neck module, a patient-reported outcome instrument, accurately predicts the severity of radiation-induced mucositis. Int J Radiat Oncol Biol Phys 2008;72(5):1355–61.
[31] Kroenke K, Spitzer RL, Williams JB. The patient health questionnaire-2: validity of a two-item depression screener. Med Care 2003;41(11):1284–92.
[32] Chen AY, Frankowski R, Bishop-Leone J, Hebert T, Leyk S, Lewin J, Goepfert H. The development and validation of a dysphagia-specific quality-of-life questionnaire for patients with head and neck cancer: the M. D. Anderson dysphagia inventory. Arch Otolaryngol Head Neck Surg 2001;127(7):870–6.
[33] McHorney CA, Robbins J, Lomax K, Rosenbek JC, Chignell K, Kramer AE, Bricker DE. The SWAL-QOL and SWAL-CARE outcomes tool for oropharyngeal dysphagia in adults: III. Documentation of reliability and validity. Dysphagia 2002;17(2):97–114.
[34] Belafsky PC, Mouadeb DA, Rees CJ, Pryor JC, Postma GN, Allen J, Leonard RJ. Validity and reliability of the Eating Assessment Tool (EAT-10). Ann Otol Rhinol Laryngol 2008;117(12):919–24.
[35] Rosen CA, Lee AS, Osborne J, Zullo T, Murry T. Development and validation of the voice handicap index-10. Laryngoscope 2004;114(9):1549–56.
[36] Jamal N, Erman A, Chhetri DK. Trans-oral partial epiglottidectomy to treat dysphagia in post-treatment head and neck cancer patients: a preliminary report. Laryngoscope 2014;124(3):665–71.
[37] Bergeron JL, Chhetri DK. Indications and outcomes of endoscopic CO_2 laser cricopharyngeal myotomy. Laryngoscope 2014;124(4):950–4.
[38] Villari CR, Courey MS. Management of dysphonia after radiation therapy. Oncol Clin North Am 2015;(48):601–9.
[39] Deutschmann MW, Sykes KJ, Harbison J, Cabrera-Muffly C, Shnayder Y. The impact of compliance in posttreatment surveillance in head and neck squamous cell carcinoma. JAMA Otolaryngol Head Neck Surg 2015;141(6):519–25.

[40] Toljanic JA, Heshmati RH, Bedard JF. Dental follow-up compliance in a population of irradiated head and neck cancer patients. Oral Surg Oral Med Oral Pathol Oral Radiol Endod 2002;93(1):35–8.
[41] Anderson R, Funnell M. Compliance and adherence are dysfunctional concepts in diabetes care. Diabetes Educ 2000;26(4):597–604.
[42] Dew MA, DiMartini AF, DeVito DA, Myaskovsky L, Steel J, Unruh M, Switzer GE, Zomak R, Kormos RL, Greenhouse JB. Rates and risk factors for nonadherence to the medical regimen after adult solid organ transplantation. Transplantation 2007;83(7):858–73.
[43] Barberan-Garcia A, et al. Personalised rehabilitation in high-risk patients undergoing elective major abdominal surgery. Ann Surg 2018;267(1):50–6.
[44] Rozen RD, Ordway DE, Curtis TA, Cantor R. Psychosocial aspects of maxillofacial rehabilitation. Part 1. The effect of primary cancer treatment. J Prosthet Dent 1972;28(4):423–8.
[45] Rowland JH, Hewitt M, Ganz PA. Cancer survivorship: a new challenge in delivering quality cancer care. J Clin Oncol 2006;24(32):5101–4.
[46] Levit L, Balogh E, Nass S, Ganz PA, Institute of Medicine (IOM), editors. Delivering high-quality cancer care: charting a new course for a system in crisis. Washington, D.C.: The National Academies Press; 2013.
[47] Hewitt M, Greenfield S, Stovall E, editors. From cancer patient to cancer survivor: Lost in transition/committee on cancer survivorship: Improving care and quality of life, National Cancer Policy Board, Institute of Medicine and National Research Council. Washington, DC: The National Academies Press; 2005.
[48] Bodenheimer T. Coordinating care—a perilous journey through the health care system. N Engl J Med 2008; 358(10):1064–71.
[49] St. John MA. Multidisciplinary approach to head and neck cancer. Otolaryngol Clin North Am 2017;50(4): xvii–xviii.
[50] Epstein RM, Street RL. The values and value of patient-centered care. Ann Fam Med 2011;9(2):100–3.
[51] Stanton AL. Psychosocial concerns and interventions for cancer survivors. J Clin Oncol 2006;24(32):5132–7.
[52] Heineman T, St. John MA, Wein RO, Weber RS. It takes a village: the importance of multidisciplinary care. Otolaryngol Clin North Am 2017;50(4):679–87.
[53] Engel GL. The clinical application of the biopsychosocial model. Am J Psychiatry 1980;137(5):535–44.
[54] Frankl VE. Man's search for meaning. New York: Washington Square Press; 1984.
[55] Bultz BD, Carlson LE. Emotional distress: the sixth vital sign in cancer care. J Clin Oncol 2005;23(26):6440–1.
[56] Holland JC, Gooen-Piels J. Guidelines for recognition of psychosocial distress. In: Kufe D, Pollock R, Weichselbaum R, et al., editors. Holland-frei cancer medicine. 6th ed. Hamilton, ON: BC Decker; 2003.
[57] National Comprehensive Cancer Network. NCCN clinical practice guidelines in oncology: distress management. version 1.2019. Plymouth Meeting, PA: National Comprehensive Cancer Network; 2019.
[58] Fashoyin-Aje LA, Martinez KA, Dy SM. New patient-centered care standards from the commission on cancer: opportunities and challenges. J Support Oncol 2012;10(3):107–11.
[59] Ercolano E, Hoffman E, Tan H, Pasacreta N, Jensen MP, McCorkle R. Managing psychosocial distress: lessons learned in optimizing screening program implementation. Oncology 2018;32(10):488–90. 492–483.
[60] Commission on Cancer ACoS. Cancer program standards 2012: ensuring patient-centered care, version 1.2.1. Chicago, IL: American College of Surgeons; 2012. p.140.

Index

Note: Page numbers followed by *f* indicate figures and *t* indicate tables.

D

I

FSC
www.fsc.org
MIX
Papier | Fördert
gute Waldnutzung
FSC® C083411

Kolibri 360 GmbH
Ferdinand-Jühlke-Straße 7
99095 Erfurt, Deutschland
produktsicherheit@kolibri360.de